Integrating Evolutionary Biology into Medical Education

Integrating Evolutionary Biology into Medical Education

for maternal and child healthcare students, clinicians, and scientists

EDITED BY

Jay Schulkin and Michael L. Power

OXFORD
UNIVERSITY PRESS

OXFORD
UNIVERSITY PRESS

Great Clarendon Street, Oxford, OX2 6DP,
United Kingdom

Oxford University Press is a department of the University of Oxford.
It furthers the University's objective of excellence in research, scholarship,
and education by publishing worldwide. Oxford is a registered trade mark of
Oxford University Press in the UK and in certain other countries

First Edition published in 2020
Impression: 1

Published in the United States of America by Oxford University Press
198 Madison Avenue, New York, NY 10016, United States of America

British Library Cataloguing in Publication Data
Data available

Library of Congress Control Number: 2019947645

ISBN 978–0–19–881415–3

DOI: 10.1093/oso/9780198814153.003.0001

Printed and bound by
CPI Group (UK) Ltd, Croydon, CR0 4YY

To our many colleagues who we have collaborated with in our research into pregnancy, maternal health and human well being. And to Rosalind Schulkin, mother, rational agent, medical realist, wonderful medical decision maker.

FOREWORD

Integrating Evolution into Medical Education for Maternal and Child Healthcare Providers

Sadly, too few medical students or residents study evolution. Yet, in an era of powerful new genomic technologies an evolutionary approach to medicine would direct physicians to ask and answer critical questions about why we get sick. In this book, editors Jay Schulkin, PhD, and Michael L. Power, PhD, build a compelling case for integrating evolutionary biology into undergraduate and postgraduate medical education, as well as its intrinsic value to medicine. Chapter by chapter the authors—experts in anthropology, biology, ecology, physiology, public health, and various disciplines of medicine—lay out the rationale for clinically-relevant evolutionary thinking. They do this within the broader context of medicine but through the focused lens of maternal and child health, with an emphasis on female reproduction and the early-life biochemical, immunological, and microbial responses influenced by evolution.

Reproduction is the engine that drives evolution, and much of human evolution reflects a patchwork quilt of "best practices" needed to accommodate the large fetal brains and associated long gestations of Homo sapiens. The book's authors provide powerful examples of how humans have been shaped by natural selection to maximize reproductive fitness, even when such adaptation may compromise the health of a given individual, either early in development or later in life. They detail how "mismatches" between genes forged in our distant evolutionary past and our modern environment promote certain diseases including obesity and its associated metabolic syndrome. The tightly woven, highly readable narrative illustrates how a medical education that considers evolved traits can deepen our understanding of the complexities of the human body, variability in health, susceptibility to disease, and ultimately help guide treatment, prevention, and public health policy.

As a medical school dean, I recognize the barriers to integrating evolutionary biology into an already tightly packed medical curriculum with its many competing priorities. However, as a maternal-fetal medicine researcher and obstetric practitioner committed to finding solutions for preterm delivery, recurrent pregnancy loss, and reducing maternal morbidity and mortality, I believe that evolutionary medicine can help us provide better health for humankind and better outcomes for mothers and babies.

Evolutionary changes to species happen over millennia, so our genotypes have not kept pace with modernity's vastly different diets, sedentary lifestyles, narrowed social support structures, and unique mental stresses. Chapter 1 describes how two different pathways—evolutionary mismatches and developmental mismatches—lead to an inability to physiologically cope with external challenges and increase disease risk in humans. Examples, including substitution of bottle feeding with cow's milk formula for breast feeding and uteroplacental vascular insufficiency, provide a greater understanding of maternal metabolism and fetal growth regulation.

The public health perspective on evolution is explored in Chapter 2 that focuses on human adaptation to new environments and stressors within the lifespan and between generations (life history theory) as well as the energy trade-offs that favor genetic fitness, affect metabolic capacity differently and shape disease vulnerability. This public health framework presents a new opportunity to address the social determinants of health and improve prevention.

Today in many developed countries menstruation begins at younger ages, women have far fewer children and breastfeed less than their ancestors. Chapter 3 poses the question of how an evolutionary medicine perspective may help mitigate some of the reproductive health challenges that women face today? The authors argue that a greater understanding of the cultural conditions under which women's reproductive biology evolved would benefit treatment decisions, and also promote non-clinical options to improve the health of both mothers and their babies, such as more social support to start and maintain lactation.

Chapter 4 introduces the concept of the "4th trimester," the early postpartum period critical for both child and maternal health. It discusses how health care providers could more effectively meet the needs of new mothers and their families if they better understood the mismatch between the contemporary demands of child rearing—including infant feeding practices, physical recovery from childbirth and return to work to name a few—and our evolutionary past where a village supported mothers and enabled infants to thrive.

Human reproduction involves a "tug of war" between mother and fetus with the goal of optimally transmitting genetic information. The mother must allocate enough resources to maintain her own health while providing adequate oxygen and nutrients to the fetus. Chapter 5 discusses how preeclampsia, gestational diabetes, and preterm birth may be maladaptive consequences of the struggle to achieve this delicate balance.

The effects of reproduction on health and longevity are complex. Chapter 6 uses the physiological adjustments of pregnancy, including the maternal allocation of metabolic energy to fetal growth and the biological trade-offs this entails, to illustrate an evolutionary view of homeostasis.

Small non-coding RNA molecules (miRNAs), important in the regulation of gene expression and continuously released in the maternal circulation throughout pregnancy, have emerged as a hot topic in women's health. Chapter 7 focuses on the role miRNA may play in various cancers of the female reproductive system, including breast and ovarian tumors, in the complications of pregnancy and the possible functions of miRNA contained in milk. The authors strongly contend that biomedical research investigating therapies to suppress or replace miRNAs would benefit from viewing these molecules from an evolutionary perspective.

The role of evolutionary medicine in understanding the global obesity epidemic is the focus of Chapter 8. Epigenetics and the coevolution of our microbiome are highlighted as forces propelling the rapid intergenerational increases in weight at the population level; the authors propose that case studies of obesity in pregnancy could help teach clinicians about evolutionary principles, concepts of population health, and the potential for pregnancy care to affect outcomes.

Chapter 9 distinguishes the movement known as "medical Darwinism" that emerged in the late nineteenth century from current Darwinian medicine as a shift from disease to *vulnerability* to disease. In the former, disease was seen as a hereditary defect in a previously perfect human machine, whereas the contemporary approach to evolutionary medicine views the body as a product of natural selection with trade-offs and vulnerabilities that often lead to disease.

Animals share human vulnerabilities to many common and deadly diseases including atherosclerosis, cancer, autoimmune disorders and biobehavioral conditions. Chapter 10 describes the Timbergean approach to clinical medicine. Nikolaas Timbergen, a Nobel Laureate animal behaviorist, advocated for including species-spanning phylogenic information in theories of causation. Recognizing that human pathology evolved from hundreds of millions of years of animal health and disease, the Tibergean approach encourages physicians and medical students to broadly consider questions about human disease in non-human animals to improve the quality of hypotheses and investigation.

Chapter 11 continues delving into four categories central to Timbergen's perspective on studying the cause of any biologic phenomenon: how animal behavior develops over a lifetime (ontogeny), how its physiological machinery works (mechanism), why a species evolves a trait to solve a reproductive or survival problem (adaptive value) and the evolutionary history of selective pressures across many generations (phylogeny). Using infectious disease and chronic inflammatory diseases as examples, the authors conclude that reproductive fitness and evolutionary history should be considered to more completely understand the pathophysiology of disease.

Future physicians will be challenged by increasing amounts of genomic information to incorporate into their clinical decision making. Chapter 12 contends that evolutionary medicine provides the foundation required to integrate genomic advances into research and practice. Medical students would benefit from greater genetic literacy, including an understanding of genetic variation as a driver of evolution and the benefits and potential harms of increasingly precise genetic technologies such as the DNA editing tool CRISPR.

Finally, Chapter 13 introduces the concept of "ecological rationality" in judgment and decision making, its practical relevance to patient care and medical education, and its compatibility to with an evolutionary approach to medicine. The authors suggest using evolutionary medicine to build a framework based on ecological rationality to develop educational tools that can enhance the health decisions of both physicians and patients.

In short, this excellent narrative makes a strong case for incorporating evolutionary thinking early in medical training to help guide the types of critical questions physicians ask—or should be asking.

Charles J. Lockwood, MD, MHCM

Preface

This book is expanded from a small conference we held when we were the Research department at the American college of Obsterticians and Gynecologist. The conference was held at the College in Washington DC. Some of the people in this book were with us for that one delightful day.

As for us, we have worked together for now 20 years, merging worlds, and foraging new ones for ourselves. The lifeblood of the mind is in the inquiry and the fleeting satisfaction that emerges in explanation, discovery and the more lasting in the continued nurtured colleagueship. And most importantly to be part of scientific and medical universe that promotes women's health.

We thank our many colleagues, those that contributed to this book, and have collaborated with us on matters important for human health.

Our work is supported by the Maternal and Child Health Bureau of HRSA through cooperative agreement UA6MC31609. Many thanks for their support.

Contents

List of Contributors

Heide Aungst, Division of Human Genetics, Center for Prevention of Preterm Birth, Cincinnati Children's Hospital Medical Center and Department of Pediatrics, University of Cincinnati College of Medicine, Cincinnati, OH, USA.

Heather Brockway, Division of Human Genetics, Cincinnati Children's Hospital Medical Center, CCHMC, USA.

Peter D. Gluckman, Liggins Institute, University of Auckland, Auckland, New Zealand.

Mark A. Hanson, Institute of Developmental Sciences, University of Southampton, Southampton, UK.

Barbara. N. Horowitz, Visiting Professor, Harvard Department of Human Evolutionary Biology, Professor of Medicine, UCLA Division of Cardiology, USA.

Charles J. Lockwood, Senior Vice President, USF Health, Dean, Morsani College of Medicine, Professor, Obstetrics & Gynecology, and Public Health University of South Florida, Tampa, FLA, USA.

Felicia M. Low, Institute for Clinical Sciences, Singapore.

Sam Mesiano, Department of Reproductive Biology, Case Western Reserve University, Cleveland, OH, USA.

Shabnam Mousavi, Max Planck Institute for Human Development, Berlin, Germany.

Louis J. Muglia, Director, Department of Obstetrics and Gynecology, University of Cincinnati College of Medicine, Cincinnati, OH, USA.

Robert Perlman, The University of Chicago, IL, USA.

Michael L. Power, Smithsonian Conservation Biology Institute, Smithsonian Institution, Washington DC, USA.

Caroline W. Quaglieri, Medical College of Wisconsin, Milwaukee, WI, USA.

Eda G. Reed, Conservation Ecology Center, Smithsonian Conservation Biology Institute, Washington DC, USA.

Karen R. Rosenberg, Department of Anthropology, University of Delaware, Newark, DE, 19716, USA.

Robert Rossi, Department of Obstetrics and Gynecology, University of Cincinnati College of Medicine, Cincinnati, OH, USA.

Carsten Schradin, Université de Strasbourg, CNRS, IPHC UMR 7178, F-67000 Strasbourg, France; School of Animal, Plant and Environmental Sciences, University of the Witwatersrand, Johannesburg, South Africa.

Jay Schulkin, Research professor, Department of Obgyn, University of Washington, and department of neuroscience, Georgetown University, USA.

Rainer H. Straub, Dept. of Internal Medicine I, Lab of Experimental Rheumatol. & Neuroendocrine Immunology, University Hospital, BIOPARK 1, Am Biopark 9, 93053 Regensburg, Germany.

Alison M. Stuebe, Department of Obstetrics and Gynecology, University of North Carolina at Chapel Hill UNC, USA.

Wenda R. Trevathan, Department of Anthropology, New Mexico State University, Las Cruces, NM 88003-8001, Mexico.

Kristin P. Tully, Carolina Global Breastfeeding Institute, University of North Carolina at Chapel Hill, UNC, USA.

Jonathan C. K. Wells, Childhood Nutrition Research Centre, UCL Great Ormond Street Institute of Child Health, 30 Guilford Street, London, UK.

Louise Wilkins-Haug, Professor, Harvard Medical School, Chief, Maternal Fetal Medicine Division, Department of Obstetrics, Gynecology and Reproductive Medicine, Brigham and Women's Hospital, Boston MA 02115, USA.

Fabio Zampieri, Department of Cardiac, Thoracic and Vascular Sciences, University of Padua Medical School, Via Aristide Gabelli 86, 35,121 Padua, Italy.

Chloe Zera, Assistant Professor, Harvard Medical School, Maternal Fetal Medicine Division, Department of Obstetrics and Gynecology, Beth Israel Hospital, Boston MA 02115, USA.

"No biological problem is solved until both the proximate and the evolutionary causation has been elucidated. Furthermore, the study of evolutionary causes is as legitimate a part of biology as is the study of the usually physicochemical proximate causes." **E. Mayr, 1982**

"Nothing in biology makes sense except in the light of evolution."
Theodosius Dobzhansky, 1973

Introduction

Michael L. Power and Jay Schulkin

When you go to the doctor, either for a standard check-up or because of an acute or chronic medical condition, you aren't likely to wonder about how well versed your physician is in evolutionary biology. For the most part evolutionary biology will have little direct bearing on your doctor visit, either on any diagnoses or any clinical remedies your physician may prescribe. Clinical medicine is an inherently practical field, and in many of its manifestations requires little underlying theoretical knowledge. A surgeon needs to know how the human body is put together, but does not necessarily need to know why that particular conformation evolved. However, evolutionary thought does influence and pervade medical science, or at least it should, according to leading figures in medicine and science. An evolutionary perspective provides a more complete conception of human disease, health, and well-being.

Medical science is inherently a biological science. A firm understanding of biology is a necessity to engage in medical science, medical research, and, to a lesser but still significant extent, practice clinical medicine. Evolution is recognized as a fundamental organizing principle of modern biology, as illustrated in the epigraph of this introduction. Over the past century, the field of evolutionary biology became an established and productive perspective in the biological science community, and yet it can be argued that evolution is the area of biology that is the least well-integrated into medicine and medical education (Gluckman et al., 2009; Perlman, 2013).

Because evolutionary biology is the central tenet of biology, and the science of biology is a fundamental aspect of medicine, it would seem essential to have evolutionary biology as an integral component of medical education (Nesse & Williams, 1998; Nesse et al., 2010; Gluckman et al., 2005; Gluckman et al., 2011). However, integrating evolutionary biology into medical education faces several roadblocks. The standard medical education is already replete with complex subjects and a long period of training (Pennisi, 2011). The addition of the evolutionary perspective to this curriculum would certainly seem daunting, and many medical educators express concern over potential controversy if evolution is introduced into the curriculum of their schools (Hidaka et al., 2015). The perception of the value of an evolutionary perspective in medical training appears to vary widely among physician-scientists and medical educators. Few would deny it has any utility; but for many the benefits may not outweigh the costs to overcome the barriers and the potential for friction among staff and students. Medical education needs strategies and teaching aids to lower the barriers to incorporating evolution into medical training.

I.1 History of medicine

The science of medicine is a relatively young field. The art of medicine is much older. Medicine as a

Power, M. L. and Schulkin, J., *Introduction* In: *Integrating Evolutionary Biology into Medical Education—for maternal and child healthcare students, clinicians, and scientists*. Edited by: Jay Schulkin and Michael L. Power, Oxford University Press (2020).
 DOI: 10.1093/oso/9780198814153.003.0001

profession dates back to at least 3000 BC. Surgery was commonly performed in ancient Egypt. Traditional Chinese medicine still uses the herbs, acupuncture, and physical techniques that first originated thousands of years ago. The ancient Egyptian *Kahun Gynaecological Papyrus,* which covers a variety of female health issues including fertility, pregnancy, and contraception, is the oldest surviving medical text (ca 1800 BC). Also from Egypt, the *Edwin Smith Papyrus* (ca 1600 BC) is a surgical treatise dealing with trauma, and may be a copy of a much older document possibly dating back to 3000 BC. One of the founding texts of traditional Indian medicine (*The Compendium of Suśruta*) is thought to be more than two thousand years old. The original *Hippocratic Oath* mentions only the art of medicine, while the modern revised version explicitly mentions both science and art (Box I.1).

Aristotle (384–322 BC), called by some the first scientist and the first empiricist, held that knowledge

Box I.1 Hippocratic oath: Past and present

Version of the Hippocratic Oath from circa 275 AD

I swear by Apollo the Healer, by Asclepius, by Hygieia, by Panacea, and by all the gods and goddesses, making them my witnesses, that I will carry out, according to my ability and judgment, this oath and this indenture.

To hold my teacher in this art equal to my own parents; to make him partner in my livelihood; when he is in need of money to share mine with him; to consider his family as my own brothers, and to teach them this art, if they want to learn it, without fee or indenture; to impart precept, oral instruction, and all other instruction to my own sons, the sons of my teacher, and to indentured pupils who have taken the physician's oath, but to nobody else.

I will use treatment to help the sick according to my ability and judgment, but never with a view to injury and wrong-doing. Neither will I administer a poison to anybody when asked to do so, nor will I suggest such a course. Similarly I will not give to a woman a pessary to cause abortion. But I will keep pure and holy both my life and my art. I will not use the knife, not even, verily, on sufferers from stone, but I will give place to such as are craftsmen therein.

Into whatsoever houses I enter, I will enter to help the sick, and I will abstain from all intentional wrong-doing and harm, especially from abusing the bodies of man or woman, bond or free. And whatsoever I shall see or hear in the course of my profession, as well as outside my profession in my intercourse with men, if it be what should not be published abroad, I will never divulge, holding such things to be holy secrets.

Now if I carry out this oath, and break it not, may I gain for ever reputation among all men for my life and for my art; but if I break it and forswear myself, may the opposite befall me. Translation by James Loeb.

Revision in 1964 by Louis Lasagna, Academic Dean of the School of Medicine at Tufts University

I swear to fulfill, to the best of my ability and judgment, this covenant:

I will respect the hard-won scientific gains of those physicians in whose steps I walk, and gladly share such knowledge as is mine with those who are to follow.

I will apply, for the benefit of the sick, all measures [that] are required, avoiding those twin traps of overtreatment and therapeutic nihilism.

I will remember that there is art to medicine as well as science, and that warmth, sympathy, and understanding may outweigh the surgeon's knife or the chemist's drug.

I will not be ashamed to say "I know not," nor will I fail to call in my colleagues when the skills of another are needed for a patient's recovery.

I will respect the privacy of my patients, for their problems are not disclosed to me that the world may know. Most especially must I tread with care in matters of life and death. If it is given me to save a life, all thanks. But it may also be within my power to take a life; this awesome responsibility must be faced with great humbleness and awareness of my own frailty. Above all, I must not play at God.

I will remember that I do not treat a fever chart, a cancerous growth, but a sick human being, whose illness may affect the person's family and economic stability. My responsibility includes these related problems, if I am to care adequately for the sick.

I will prevent disease whenever I can, for prevention is preferable to cure.

I will remember that I remain a member of society, with special obligations to all my fellow human beings, those sound of mind and body as well as the infirm.

If I do not violate this oath, may I enjoy life and art, respected while I live and remembered with affection thereafter. May I always act so as to preserve the finest traditions of my calling and may I long experience the joy of healing those who seek my help.

is derived from sensory experience. Aristotle's father, Nicomachos, was the physician for the king of Macedonia (Alexander the Great's grandfather), and Aristotle likely received training in early life in medicine and biology from his father. Aristotle wrote prolifically on many subjects, and in his extensive works on biology he advocated direct observation and comparison. His works and thought regarding empiricism in biology influenced the ancient Greek physician, Galen (ca 130 AD), who dissected animals in order to understand human anatomy (human dissection was forbidden). Pedanius Dioscorides (ca. 30–90 AD) served as a physician and botanist in the Roman army during the time of Nero, which allowed him to travel extensively. He published a five-volume book of his travels titled *On Medical Material*, which described the medical properties and uses of about 600 plants as well as the medicinal properties of milk, honey, and various chemicals such as mercury and arsenic. The book was in essence the first pharmacopeia that was widely used for 1600 years, and stressed the Aristotelian concept of reliance on personal experience, investigation and experimentation.

It was during the 1500s that the ancient writings of the classical Greeks (Hippocrates, Aristotle, and Galen, among others) were carefully reexamined. Scholars and scientists performed dissections of animals and cadavers and the results were compared with the words of these great thinkers of the past. Amidst much turmoil and attack by the Galenists, who valued deference to the authority of the writings of the ancient Greeks, the new scientific way of thinking which relied on observation and experiment gained sway and began to correct the mistakes from the past. In many ways this was, in truth, a return to the philosophy and empiricism of Aristotle. The art of medicine began to be supplemented by the science of medicine.

Throughout history many doctors have also been scientists. Charles Darwin's grandfather, Erasmus Darwin (1731–1802), was a physician who wrote extensively on biology, and introduced concepts that anticipated the evolutionary perspective. Charles Darwin himself started to follow his grandfather and father into medicine, but did not finish medical school, instead going on to become the most influential natural historian of the modern era. Initially the field of anthropology was heavily involved in the biological study of humans, especially anatomy, and many physicians also considered themselves anthropologists. The French surgeon/neurologist Paul Broca (1824–1880), for whom Broca's area of the frontal cortex is named, formed the Anthropological Society of Paris in 1859. Similarly, Rudolf Ludwig Carl Virchow (1821–1902), during his day referred to by some as the "Pope of Medicine," co-founded the German Anthropological Society in 1869 and the Society for Anthropology, Ethnography and Prehistory in 1870. Medicine and what today we call biological anthropology were understandably intertwined, as both involve the biological investigation of human beings.

In the 1800s, the science of physiology began to be integrated into medicine. Claude Bernard (1813–1878), one of the most influential physiologists of that time, formulated the concept of the internal milieu, which was later expanded by Starling (1923) and then Walter Cannon (1932) into the concept of homeostasis. One of Bernard's stated goals during his life was to establish the use of the scientific method in medicine. In 1865, he produced *An Introduction to the Study of Experimental Medicine*, which urged scientists to integrate research and knowledge-seeking into medical practice, rather than relying on natural inclination (Bernard, 1957). Bernard's work in experimental medicine set a standard for research, as did the work of other scientists of this era such as Robert Koch and Louis Pasteur. However, Bernard's plea to integrate science into medicine remained relatively unheeded until the publication of the Flexner Report in 1910.

I.2 The Flexner Report

The Flexner Report was the culmination of a concerted effort by a number of leading physician-scientists to change medical education in the United States (US) starting in the late 1800s (Flexner, 1910). The death bequest of Baltimore merchant and banker Johns Hopkins in 1873 left the bulk of his fortune for the purpose of building both a medical school and hospital that would be integrated as one institution. This was a revolutionary concept, combining the practice of medicine with medical education and research (Chesney, 1963). In 1889, after completion of the construction of Johns Hopkins Hospital, the Trustees of the Hopkins estate secured the services of four outstanding members of the medical community as the founding physicians: William Osler (1849–1919) the first chief of the Department of Medicine at Johns Hopkins Hospital, William Halstead (1852–1922) the first chief of surgery, Howard Kelly (1858–1943) the physician who established gynecology as a true specialty, and the pathologist William Welch (1850–1934). William Osler created the first residency program, in which medical school graduates receive advanced training (Osler, 1922). A new concept, he brought medical students out of the lecture halls and into the wards to witness and assist in bedside medicine. Third-year medical students were required to take patient histories, perform physical examinations, and do laboratory tests on patient samples. Learning by doing was a vital tenet of his educational philosophy, and his wish for an epitaph on his tombstone was "He brought medical students into the wards for bedside teaching."

Abraham Flexner (1866–1959) was a graduate of Johns Hopkins University with a Bachelor of Arts in Classics. After graduation he returned to his hometown of Louisville, KY and became a high school teacher. He was disillusioned by the standard educational practice of lectures and the rigid structures imposed on students. After four years of teaching he founded his own experimental school that emphasized a hands-on approach to education. In 1908 he published a book (*The American College*, Flexner, 1908), which was highly critical of the educational structures of the day, especially the emphasis on lectures. The book brought him to the attention of Henry Pritchett, president of the Carnegie Foundation for the Advancement of Teaching. In 1908, the Council on Medical Education of the American Medical Association asked the Carnegie Foundation to survey American medical education. Pritchett selected Flexner to conduct the survey. In 1910, *Carnegie Foundation Bulletin Number Four*, more commonly known as the Flexner Report, was published. In his report, Flexner explicitly held up Johns Hopkins School of Medicine as the ideal for US medical education, stressing the integration of clinical practice, medical education and research within a university, and the hands-on approach to education led by Osler (see Bliss, 1999, 2011).

The Flexner Report produced a change in US medical culture, advocating for medicine to transform itself from an authority-driven art to become an integrative scientific discipline that relies on evidence and the scientific method (Perlman, 2011; Schulkin & Matteson, 2011). However, the Flexner report was a product of its times, and along with valuable recommendations that led to a better integration of science and research into medicine there were overtly racist and sexist opinions and recommendations. Flexner believed African American physicians should be subordinate to white physicians, but that African American communities would be better served by well-educated African American physicians rather than by poorly trained white ones. He also saw little need for institutions dedicated to the training of women as physicians.

In the decades after the Flexner Report a large number of US medical schools closed, dramatically decreasing the number of slots available for medical students. The number of white men enrolled in medical schools decreased, but the decline in enrollment was disproportionately felt among women and African Americans, a predictable outcome for those times. Many of the surviving institutions intentionally decreased their enrollment of women, favoring the training of men. Additionally, of the seven historically black medical schools operating at the time, five closed. The decrease in medical schools and hence reduction in the number of medical students led to an increase in the quality of medical education and its graduates, but also to an increase in inequality of opportunity for women and African Americans.

I.3 Evolution and medicine

The Flexner Report resulted in an increased emphasis on integrating science and experimental paradigms into medical training, but unfortunately did not lead to an incorporation of the relatively new science of evolution. In 1926, Professor Morton of the Yale Medical School would ask "what is the relationship of evolution to medicine?" and would answer "The problem of evolution permeates every fundamental branch of medical education" (Morton, 1926). However, evolution has had a mixed history of acceptance by the medical profession. The role of religion explains some of the medical and scientific communities' rejection of Darwinism from its inception; but not all. For example, Rudolf Virchow, the "Pope of Medicine" and an early founder of the discipline of anthropology was both an ardent anti-Darwinist and an avowed agnostic. His objections to Darwinism were couched in the scientific method and his belief that evolution was an unproven hypothesis that lacked empirical support. Of course, Virchow also strongly disbelieved the germ theory of disease, for much the same reasons.

Virchow was strongly utilitarian in his approach to science. His attitude towards research was if it doesn't cure patients, what good is it? This attitude illustrates an intrinsic tension between the practice of medicine and the broader science of medicine, which can affect the perceived value of an evolutionary perspective in medical education. For much of the clinical practice of medicine an evolutionary perspective is not particularly relevant or helpful. Medicine often deals mostly with proximate mechanisms, not ultimate, evolutionary ones. Yet many scholars argue that an evolutionary perspective is as crucial to a basic medical science education as anatomy and microbiology (Nesse et al., 2010; Nesse & Stearns, 2008).

According to the American Association for the Advancement of Science, 98% of scientists believe that all living things have evolved over time; however, there are few data on the level of physicians' beliefs (Pew Research Center, 2015). Anecdotal evidence suggests that the percentage of physician believers is not as high as scientists. Evolutionary biology is not often taught in medical schools, due to lack of time and faculty expertise (Nesse & Schiffman, 2003). A survey of deans of medical schools found an appreciation for the potential importance of evolution to the education of medical students, but also found strong structural and perceptual barriers to incorporating evolutionary perspective courses and training into the curriculum. Very few faculties at medical schools have advanced training in evolutionary biology, and the number of hours of classroom time devoted to evolution is generally small (Hidaka et al., 2015; Figure I.1).

Those who advocate integrating evolutionary biology into medical education could be accused of adding one more piece of science to be stuffed into the minds of young physicians by medical educators who are not fully qualified to teach the subject. However, including the evolutionary perspective in medical education arms physicians with an understanding of the ultimate (evolutionary) answers to clinical questions, rather than simply proximate (basic mechanistic) explanations, providing physicians with a more complete understanding of the etiology of disease (Tinbergen, 1969; Mayr, 1982; Nesse & Williams, 1994; Perlman, 2013).

An evolutionary perspective provides a lens on the "why" in addition to the "how" of health and disease. As a result, physicians will become more scientifically literate, beyond the basic textbook and memorization methods of medical school, and gain novel insight not available in the typical curriculum (Jenkins & Antolin, 2011; MacCallum, 2007). The integration of evolution into pre- and post-graduate medical education may help direct new research pursuits (Gluckman & Bergstrom, 2011; Antolin et al., 2012). Studying the evolutionary answers to clinical questions may actually give insight into possible proximate mechanisms, aiding the basic

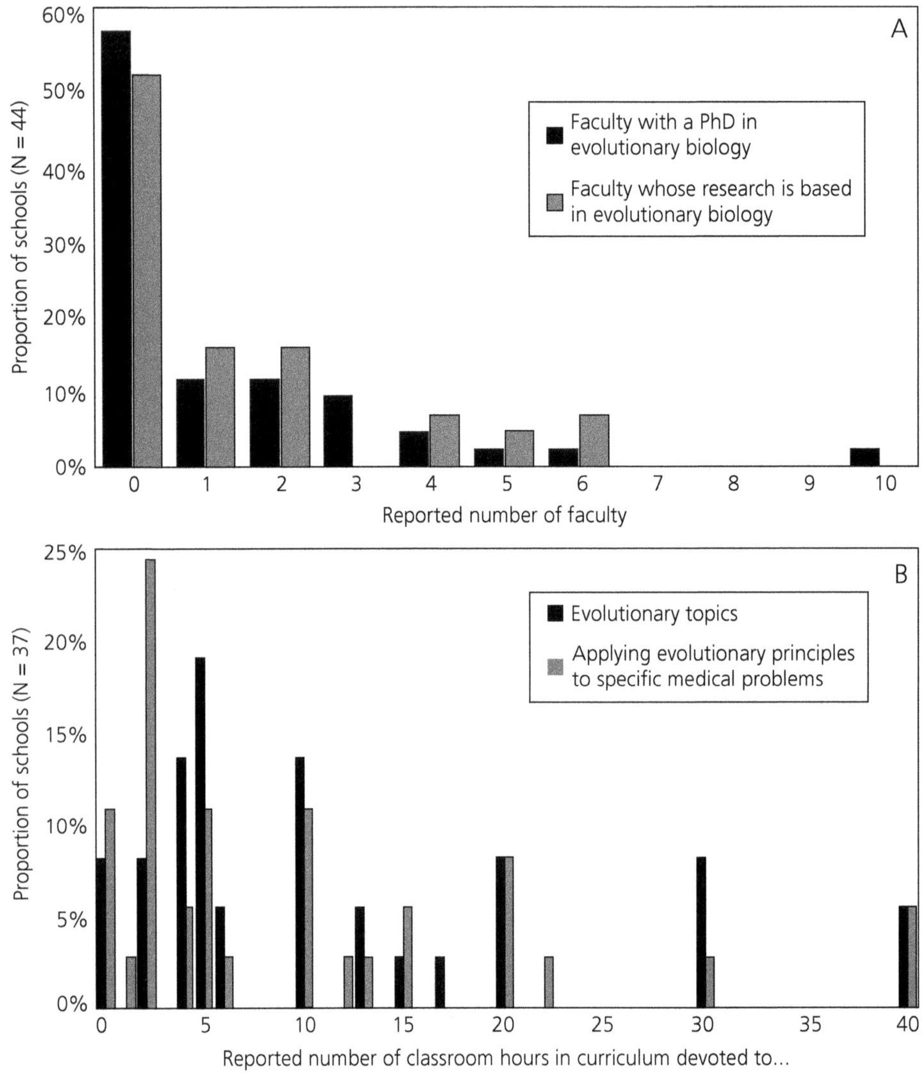

Figure I.1 Faculty at US medical schools with advanced training in evolutionary biology, and the number of hours of classroom time devoted to evolution in medical schools
Excerpted from Hidaka et al., 2015

and clinical research process (Nesse & Williams, 1998).

Evolutionary processes are obviously important to modern medicine; witness the spread of antibiotic resistant bacteria. That phenomenon is understandable, indeed, predictable from an evolutionary perspective. The medical and public health response to slow the increase in antibiotic resistant pathogens requires an evolutionary perspective to be effective (Wells et al., 2017). The importance of our coevolved microbiomes, the microbial communities that live in different compartments of our body, is another example of evolutionary thought and modern medicine coming together. The health of our microbiomes contributes to our own health and disruptions of our microbiomes can trigger disease.

The relation of the gut microbiome to our health and well-being is an important new frontier in biology and medicine, one in which an evolutionary perspective brings important insights (Power & Schulkin, 2009; Rook et al., 2017).

I.4 Mismatch paradigm

Evolved adaptations are responses to challenges faced in the species' past. Selection doesn't predict; it is the result of the interactions of individual phenotypes with the environment. Given sufficient consistency of the environment over a period of time selection will favor traits that, on average, produce the most descendants within that environment, although even then the adaptive traits may represent compromises between different selective pressures. Evolution doesn't produce perfection; it produces successful solutions within a context of a multitude of interacting selective pressures. However, environments can sometimes change much faster than the time scales over which evolution acts. The incongruity between an evolved phenotype and the current environment forms the basis of "mismatch" theory or paradigm (Hood & Jenkins, 2008; Gluckman & Hanson, 2006; Hanson & Gluckman, 2008), a core principle of what has been termed evolutionary medicine (Nesse, 2001). Evolution may produce a successful phenotype that may not be suited for a future environment (Gluckman et al., 2011; Perlman, 2013), resulting in pathology.

Scenarios like this may result in a "mismatch" of trait and environment, resulting in pathology or decreased reproductive success. Many human conditions have been attributed to this "mismatch theory" of pathology (Stearns & Medzhitov, 2015). The strength of the evidence supporting diseases of "mismatch" varies depending on the conceptual evolutionary framework, as well as the investigative research conducted to bolster the theoretical pathway of selection for vulnerable traits of disease. Examples of potential mismatch pathology and challenges relating to pregnancy and women's health are examined in the book (Table I.1).

The obesity epidemic, with its associated metabolic diseases, has been suggested to be an example of the mismatch paradigm and human pathology

Table I.1 Select examples of mismatch examined in this volume and their associated chapters.

Example of mismatch	Chapter in this book
Evolutionary vs developmental mismatch in pregnancy and fetal and infant development	Gluckman
Bipedalism, atricial birth, birth weight	Gluckman, Wells, Stuebe, and Tully
Obesity and nutritional impacts	Gluckman, Zera, and Wilkins-Haug
Childbirth complications, biological and social	Trevathan, Rosenberg, Aungst et al.
Biological vs sociocultural optimal times for childbearing	Trevathan and Rosenberg
Postpartum maternal-infant needs, breastfeeding trends, sleep	Stuebe and Tully

(Power & Schulkin, 2009; Power, 2012). Many aspects of our evolved biology appear ill suited to the modern life style of increased sedentary behavior and plentiful, energy-rich food. Another example is the increase in the prevalence of allergic diseases, such as asthma, irritable bowel syndrome, type 1 diabetes, and multiple sclerosis. The emerging consensus implicates reduced exposure to commensal and other non-harmful microbes and to alterations in microbiomes at an early age for this increase (Bloomfield et al., 2016). Many aspects of modern life contribute to reduced exposure to non-harmful and often co-evolved microorganisms that are a critical component to the education and maturation of the infant's immune system. For example, the increase in cesarean sections and decrease in breastfeeding reduce the transmission of maternal microbes to her offspring. Note that this newer consensus is a significant modification to the older, popular "hygiene hypothesis" which suggested that exposure to infectious/pathogenic microorganisms reduced the risk of later allergic disease. It is now thought that exposure to non-harmful microbes, especially to the coevolved microorganisms that comprise our various microbiomes that reduces risk.

A final example is something the reader may be doing this very minute: sitting in a chair. Sitting in a chair usually results in a right-angle posture between the thighs and the torso, which exerts strain on the pelvis and lumbar spine. Sitting in a chair can exert 30% more pressure on the intervertebral discs than standing, and strains the spinal column, back muscles, lower back nerves, and diaphragm. If you consider the association between the number of hours spent sitting and a host of metabolic diseases (including cardiovascular disease), then it becomes understandable that some people have suggested excessive sitting in a chair is as dangerous to health as smoking (Cranz, 2017).

I.5 Evolutionary medicine

Evolutionary medicine is an integrative discipline, combining the fields of evolutionary biology, experimental research, and clinical medicine. Contributors to the field of evolutionary medicine today come from many schools of study. Anthropologists, biologists, physiologists, and ecologists work alongside historians of science and physician scientists. Desire for a full integration of evolutionary thought into the education of medical students and seasoned physicians has been voiced by most of the field's contributors (Nesse & Stearns, 2008; Nesse et al., 2010; Stearns, 2012; Perlman, 2013).

Historians of evolutionary medicine suggest that work on integrating evolutionary biology into medicine dates back to over a century ago, but sank into oblivion for social and political reasons (i.e., Nazi eugenics movement, outlawing of teaching evolution in the classroom, etc.). First re-explored fully in the early 1990s, the integration of these subjects brought about a new perspective on the practice of medicine and on how to educate medical students (Nesse & Williams, 1994; Williams & Nesse, 1991). Evolutionary medicine's resurgence in the 1990s can be attributed to the development of genetic research and the influx of infectious diseases that required a new way of studying evolution within a medical context (Ewald, 1994; Zampieri, 2009).

Incorporating an evolutionary perspective into medicine will most likely not change the daily activity of physicians, but it will help guide the thinking and types of questions a physician-scientist should ask, by adding an evolutionary dimension (Nesse & Stearns, 2008; Brune & Hochberg, 2013; Mitteroecker et al., 2016). Our hope is that in the future, the approach to medicine that evolutionary thinking provides will be a seamless component of medical practice, rather than be considered a separate school of thought (Stearns, 2012). We agree with Stearns that the best outcome would be for evolution to be so well integrated into medical thought that the term evolutionary medicine is considered redundant.

I.6 Why this volume?

There are a number of books focusing on incorporating the evolutionary perspective into medical science. Some are organized as traditional text books (e.g., *Principles of Evolutionary Medicine* (Gluckman et al., 2017) and *Evolutionary Medicine* (Stearns & Medzhitov, 2015)), others as edited compendiums (e.g., *Evolutionary Medicine* (Trevathan et al., 1999) and *Evolution in Health and Disease* (Stearns & Koella, 2008)), and some more resemble a monograph reflecting the personal experience and journey of a scientist integrating evolutionary thought into their biomedical profession (e.g., *Evolution of Infectious Disease* (Ewald, 1994) and *Evolution & Medicine* (Perlman, 2013)). All have their place and function in the effort to increase evolutionary thought in biomedicine and medical education. This edited volume was produced to be an additional tool that medical educators can use to enhance the integration of evolutionary biology into the medical curriculum. Our aim was to create a readable as well as educational narrative outlining the value of evolutionary thought to existing medical science and to the future of biomedicine.

Although some chapters are broadly about the importance and function of an evolutionary perspective in medicine, many focus on aspects of female reproduction. All aspects of our biology have been affected by evolution, but it can be argued that reproductive biology may be the most directly affected. Mothers transmit biochemical, immunological and microbial information to their offspring from implantation through weaning. The health of an individual is significantly affected by early life experience, and the physiological and metabolic responses to these early life experiences are part of our evolved biology. Our understanding of the proximate side of biology and human medicine continues to outstrip our understanding of the ultimate, evolutionary processes under which our biology arose. To fully understand the significance of how the modern world has changed our experiences and medical practice (including exposures to novel challenges due to technological advance, and the lack of exposure to "old friends" such as coevolved microorganisms) we argue that an evolutionary perspective is a necessary tool.

I.7 Structure of the book: There are three sections of the book

The first section is "Life History: Biological and Cultural Continuity" and has four chapters. The first two are foundational for medical decision making and the second two display the biological and cultural continuity of evolutionary medicine and considerations of human health, viability and wellbeing.

The second is "Biological Regulation" and has four chapters. The first two are broadly biological views about regulation and the second two are specific instances for where evolutionary medicine ought to be a consideration; one is narrow and the other broad in scope.

The final and third section is "Perspectives: Past, Present, and Future" in which there are five chapters. The first is a glimpse of the past, and the second two are demonstrably under the influence of the great ethologist Nikolaas Tinbergen. The penultimate chapter provides a perspective of the genetic revolutions continuing impact on medical decision making and medical practice, and the last chapter links medical decision making and the influence of ecological and evolutionary considerations.

References

Antolin, M. F., et al. (2012). Evolution and medical education in undergraduate education: A prescription for all biology students. *Evolution*, 66(6): 1991–2006.

Bernard, C. (1865, 1957). *An Introduction to the Study of Experimental Medicine*. New York: Dover Press.

Bliss, M. (1999). *William Osler: A Life in Medicine*. Oxford: Oxford University Press.

Bliss, M. (2011). *The Making of Modern Medicine*. Chicago: University of Chicago Press.

Bloomfield S. F., Rook G. A. W., Scott E. A., Shanahan F., Stanwell-Smith R., & Turner P. (2016). Time to abandon the hygiene hypothesis: new perspectives on allergic disease, the human microbiome, infectious disease prevention, and the role of targeted hygiene. *Perspectives Public Health*, 136: 213–24.

Brüne, M. & Hochberg, Z. (2013). Evolutionary medicine—the quest for a better understanding of health, disease and prevention. *BMC Medicine*, 11: 116.

Cannon, W. B. (1932, 1967). *The Wisdom of the Body*. New York: Norton Press.

Chesney, A. M. (1963). *The Johns Hopkins Hospital and the School of Medicine*. Volumes 1, 2, 3. Baltimore: Johns Hopkins University Press.

Cranz, G. (2017). Rethinking the chair and sitting. In: *Sedentary Behavior and Health: Concepts, Assessments, and Interventions*, W. Zhu and N. Owen (eds.), pp. 45–53. Champaign: Human Kinetics Publishers.

Ewald, P. W. (1994). *Evolution of Infectious Disease*. New York: Oxford University Press.

Flexner. A. (1910). *Medical Education in the United States and Canada*. New York: Carnegie Foundation.

Flexner, A. (1908). *The American College: A Criticism*. New York: The Century Co.

Gluckman, P., Beedle A., & Hanson, M. (2017). *Principles of Evolutionary Medicine*. New York: Oxford University Press.

Gluckman, P. D., Hanson, M. A., & Spencer, H. G. (2005). Predictive adaptive responses and human evolution. *TRENDS in Ecology and Evolution*, 20: 527–33.

Gluckman, P. D. & Hanson, M. (2006). *Mismatch: Why our world no longer fits our bodies*. New York: Oxford University Press.

Gluckman, P. D., et al. (2011). How evolutionary principles improve the understanding of human health and disease. *Evolutionary Applications*, 4(2): 249–63.

Hanson, M. & Gluckman, P. D. (2008). Developmental origins of health and disease: new insights. *Basic Clinical Pharmacology Toxicology*, 102(2): 90–3.

Hidaka, B., et al. (2015). The status of evolutionary medicine education in North American medical schools. *BMC Medical Education*, 15(38): 1–9.

Hood, E. & Jenkins, K. P. (2008). Evolutionary medicine: a powerful tool for improving human health. *Evolution: Education and Outreach*, 1(2): 114–20.

Jenkins, K. P. & Antolin, M. F. (2011). Evolution and Medicine. *Evo Edu Outreach*, 4: 556–8.

MacCallum, C. J. (2007). "Does medicine without evolution make sense?" *PLoS Biol*, 5(4): e112.

Mayr, E. (1982). *The Growth of Biological Thought: Diversity, Evolution, and Inheritance*. Cambridge: Harvard University Press.

Mitteroecker, P. et al. (2016). Cliff-edge model of obstetric selection in humans. *PNAS*, 113: 14680–85.

Morton, D. J. (1926). The relation of evolution to medicine. *Science*, 64(1660): 394–6. Magazine article: https://science.sciencemag.org/content/64/1660/394.long

Nesse, R. & Williams, G. C. (1994). *Why We Get Sick: The New Science of Darwinian Medicine.* New York: Vintage Books.

Nesse, R. M. & Williams, G. C. (1998). Evolution and the origins of disease. *Scientific American*, November. article. http://www-personal.umich.edu/~nesse/Articles/Nesse-EvolMed-SciAmer-1998.pdf

Nesse, R. M. (2001). How is Darwinian medicine useful? *Western Journal of Medicine*, 174(5): 358–60.

Nesse, R. M. & Schiffman, J. D. (2003). Evolutionary Biology in the Medical School Curriculum. *BioScience*, 53(6): 585–7.

Nesse, R. M., & Stearns, S. C. (2008). The great opportunity: Evolutionary applications to medicine and public health. *Evolutionary Applications*, 1: 28–48.

Nesse, R. M., et al. (2010). Making evolutionary biology a basic science for medicine. *Proceedings of the National Academy of Sciences*, 107: 1800–807.

Osler, W. (1922). *The Evolution of Modern Medicine*. New Haven: Yale University Press.

Perlman, R. L. (2011). Evolutionary biology: A basic science for medicine in the 21st century. *Perpsectives in Biology and Medicine*, 54: 75–88.

Perlman, R. L. (2013, 2014). *Evolution and Medicine*. Oxford: Oxford University Press.

Pennisi, E. (2011). Darwinian medicine's drawn-out dawn. *Science*, 334(6602): 1486–7.

Power M. L. (2012). The human obesity epidemic, the mismatch paradigm, and our modern "captive" environment. *Am J Hum Biol*, 24(2): 116–22.

Power, M. L. & Schulkin, J. (2009). *The Evolution of Obesity*. Baltimore; Johns Hopkins University Press.

Rook, G. et al. (2017). Evolution, human microbiome interactions and life history plasticity. *Lancet*, 390: 521–30.

Schulkin, J. & Matteson, K. A. (2011). The Flexner report and evidence based medicine over the last 100 years. In: *Changing Landscape of Academic Women's Heatlh Care in the United States*. W. F. Rayburn and J. Schulkin (eds.). New York: Springer.

Stearns, S. & Koella, J. (2008). *Evolution in Health and Disease*. Oxford: Oxford University Press.

Stearns, S. & Medzhitov, R. (2015). *Evolutionary Medicine.* Sunderland: Sinauer Associates.

Stearns, S. C. (2012). Evolutionary medicine: Its scope, interest and potential. *Proceedings of the Royal Society of London B: Biological Sciences*, 279(1746): 4305–21.

Tinbergen, T. (1969). *The Study of Instinct*. Oxford: Oxford University Press.

Trevathan, W. R., Smith, E. O., & McKenna, J. J. (eds.) (1999). *Evolutionary Medicine.* New York: Oxford University Press.

Wells, J. C. K et al. (2017). Evolutionary public health. *Lancet*, 390: 500–09.

Williams, G. C. & Nesse, R. M. (1991). The dawn of Darwinian medicine. *Quarterly Review of Biology*, 66: 1–22.

Zampieri, F. (2009) Origins and history of Darwinian medicine. *Humana Mente*, 9: 13–38.

SECTION I

LIFE HISTORY: BIOLOGICAL AND CULTURAL CONTINUITY

CHAPTER 1

Evolutionary Medicine, Pregnancy, and the Mismatch Pathways to Increased Disease Risk

Peter D. Gluckman, Felicia M. Low, and Mark A. Hanson

1.1 Evolutionary medicine

While medical thought has traditionally focused on the proximate mechanisms underlying how disease develops and progresses to inform diagnosis and therapy, the emergent field of evolutionary medicine aims to disentangle more fundamental questions addressing the ultimate reasons *why* our bodies are vulnerable to disease. It uses an evolutionary framework to improve understanding of how humans have evolved, how they develop, and how their ecological context leads to impacts on human health and disease (Williams & Nesse, 1991; Stearns, 2012). This perspective can inspire myriad questions about disease susceptibility: why are autoimmune diseases more common now? Why do we get back pain? Why do we get myopia, or need dental braces? Why is there a global epidemic of obesity and type 2 diabetes? Some of these questions are addressed in this chapter. More broadly, insights gained from probing ultimate causation of disease have informative value for health care professionals and their patients, for public health, for medical research, and even for policy making (Carroll et al., 2014; Wells et al., 2017).

A 2009 report from the American Association of Medical Colleges recognized the importance of foundational competency in evolutionary biology (AAMC-HHMI Scientific Foundation for Future Physicians Committee, 2009), and since then modules on evolutionary medicine have increasingly been introduced into core medical school curricula worldwide. However, despite an appreciation of the importance of evolutionary principles by medical school faculties, overall curricular coverage especially in North America and much of Europe remains relatively low (Hidaka et al., 2015). This is not for lack of compelling teaching resources—there are many examples of key evolutionary concepts being embedded within a clinical context in ways that are engaging and relevant (Antolin et al., 2012; Buklijas et al., 2011). Textbooks have been available since 2009 (Stearns & Medzhitov, 2015; Gluckman et al., 2009; Gluckman et al., 2016b), as are other books aimed at students (e.g., Perlman, 2013; Stearns & Koella, 2008), and continued integration of evolution and medical education is likely to have benefits for critical thinking and learning (Graves et al., 2016; Hidaka et al., 2015; Antolin et al., 2012; Buklijas et al., 2011).

Gluckman, P. D., Low, F. M., and Hanson, M. A., *Evolutionary Medicine, Pregnancy, and the Mismatch Pathways to Increased Disease Risk* In: *Integrating Evolutionary Biology into Medical Education—for maternal and child healthcare students, clinicians, and scientists*. Edited by: Jay Schulkin and Michael L. Power, Oxford University Press (2020). © Oxford University Press.
DOI: 10.1093/oso/9780198814153.003.0002

Table 1.1 Several principles of evolutionary medicine.

- Selective processes (i.e., natural selection) operate to maximize reproductive success and hence Darwinian fitness, with any impact on the organism's health or longevity being a secondary effect.
- Both our evolutionary history as a species, and our individual developmental history, influence our susceptibility to disease.
- Our lifestyles and environments are now immensely different from those in which the modern human phenotype has evolved, thus increasing the likelihood of mismatch and, consequently, greater disease risk.
- Evolutionary processes are subject to multiple constraints, and trade-offs are involved. For example, a particular trait that may be undergoing strong selection cannot continue to evolve once its benefits no longer outweigh its costs.
- The consequences of such trade-offs may not be apparent for some time. For example, antagonistic pleiotropy describes the concept that genes or traits that have been selected for their beneficial (that is, Darwinian fitness-promoting) effects in early life, may have detrimental later effects (Williams, 1957).

Table 1.2 Evolutionary processes that lead to altered disease risk in humans.

- Evolutionary novelty and evolutionary mismatch: Exposure to an environment or a challenge that is either entirely new, or is particularly extreme in nature to the extent that the individual is unable to adapt to cope.
- Developmental mismatch (factors associated with the life-history strategy): Encompasses a range of concepts including trade-offs and antagonistic pleiotropy, developmental plasticity, and anticipatory/predictive responses.
- Excessive or uncontrolled defense mechanisms.
- Human-microbe interactions and co-evolution.
- Evolutionary constraints.
- Balancing selection leading to persistence of a harmful allele.
- Sexual selection and competition.
- Cladal, demographic, and migratory histories.

Evolutionary medicine is underpinned by several fundamental principles (Table 1.1; Gluckman et al., 2016b). Evolutionary processes can affect disease risk via multiple, non-mutually exclusive pathways (Table 1.2; Gluckman et al., 2016b, Gluckman et al., 2011a).

This chapter focuses on the first two pathways—evolutionary mismatch, and developmental mismatch underpinned by life-history associated factors. Both pathways involve the organism being exposed to external challenges that can eventually pose major public health challenges, but they differ on the basis of why the environmental context poses a challenge. Evolutionary mismatch arises when an organism encounters a new environment with which its lineage has not evolved to cope. Developmental mismatch stems from an organism's ability to adaptively adjust its developmental trajectory, through the process of developmental plasticity, to produce a fitness enhancing phenotype in a particular environment early in life; later encounter of a different environment with which its physiology cannot cope results in developmental mismatch.

These two pathways underpin the Developmental Origins of Health and Disease (DOHaD) paradigm, for which a considerable body of evidence has amassed from experimental studies in animals, and clinical and epidemiological studies in humans (Gluckman & Hanson, 2006b; Barouki et al., 2012; Gluckman et al., 2008). The examples used also provide a greater understanding of the significance of metabolic regulation in pregnancy and of fetal growth regulation. They provide an insight into how evolutionary arguments can change the proximate understandings of physiology and pathophysiology. This chapter discusses how the two mismatch pathways have deleterious consequences for disease risk, and contextualizes the DOHaD phenomenon within an evolutionary framework, discussing the adaptive advantage of anticipatory responses and the evidence for this in humans.

An additional important aspect relating to evolutionary processes has arisen from recent insights in the field of epigenetics, in particular transgenerational epigenetic mechanisms (Uller, 2014; Skinner, 2015). It is increasingly apparent that genetic inheritance is not the only form of information that can be passed across generations, with emerging evidence of epigenetic mechanisms allowing for both direct and indirect transfer of information across generations in humans (Vaiserman et al., 2017; Hur et al., 2017). Beyond that, environmental factors can be common across generations, and cultural evolution—as evidenced in societal transformation and technological innovation—can create rapid environmental change, accelerating the risks of both evolutionary and developmental mismatch (as discussed later).

1.2 Evolutionary mismatch

Some animals adjust their activities and behaviours to maintain a constant environment. This phenomenon is known as niche construction (Odling-Smee et al., 2003). For example, some termite species build large mounds with two flat North-South-orientated sides to modulate the mound's internal temperature, while beavers dam a river for feeding and safety purposes. In contrast, contemporary humans engage in niche modification, continually modifying their environment beyond adjusting it simply to maintain an equilibrium. This unparalleled and cumulatively innovative capacity in humans is facilitated by our skills in cognition, language, learning, and manual dexterity, while our social structures enable knowledge dissemination with speed and ease. The agricultural and industrial revolutions over the past few centuries, accompanied by an accelerated pace of technological advancement and human population growth, is a relatively recent phenomenon in the context of our evolutionary history (Fogel, 2004). The speed of the changes in our social, nutritional, and built environments is in stark contrast to the slow pace of evolutionary change, which generally operates on the timescale of millennia. As a consequence, contemporary humans are now increasingly exposed to environments that are completely novel in evolutionary terms (Low & Gluckman, 2016).

There are two ways by which evolutionary mismatch can arise. The first is exposure to a wholly novel environment ("evolutionary novelty"). Because our lineage has never before encountered such environments, there has never been selection pressure to adapt to them and so we would have no evolved adaptive responses. In other words, the genetic substrate is mismatched to the environment. For example, some aspects of the modern day environment such as being indoors, artificial light, reading, underuse of peripheral vision, and increased education, are associated with the development of myopia in children. Although there appears to be some genetic component to myopia susceptibility, it is thought that the additional exposure to environments with multiple risk factors reveals this trait (Morgan & Rose, 2005). Another example is malocclusion (misaligned teeth), which is exceedingly rare in hunter-gatherers who consume harder, uncooked foods but has emerged following the advent of agriculture and softer cooked/processed foods (Rose & Roblee, 2009; Pinhasi et al., 2015). Reduced chewing has led to reduced jaw size, resulting in dental crowding. A third example is the development of scurvy due to a diet deficient in vitamin C. Humans are unable to synthesize this vitamin, having inherited a non-functional gene involved in the vitamin C biosynthetic pathway from the common ancestor of monkeys, apes, and humans about 61 million years ago (Drouin et al., 2011), but ease of access to fruits and vegetables had rendered this mutation neutral. However, individuals exposed to novel environments without access to dietary vitamin C, such as eighteenth-century sailors who had embarked on long sea voyages, would then experience the unfortunate consequences of the mismatch (Buklijas et al., 2011).

The second way by which evolutionary mismatch may occur is exposure to a particular aspect of the environment that is significantly more extreme than previously experienced throughout the organism's evolutionary history. This is exemplified by the transition from hunter-gatherer diets characterized by high variety and richness in fiber and micronutrients, to a hypercaloric diet comprising energy-dense, highly processed foods. This dietary shift is prevalent in urbanized populations especially in Western societies, but increasingly so in low–middle income countries undergoing a nutritional transition to the Western diet (Popkin et al., 2012). Some types of food such as trans fatty acids and high fructose corn syrup are themselves evolutionary novelties borne from technological advances in food manufacture. It is therefore not surprising that the changes in both quantity and quality of our nutrition have exceeded the adaptive capacity of our evolved metabolic physiologies, contributing to the pervasiveness of obesity, type 2 diabetes (T2D), and other co-morbidities (Gluckman & Hanson, 2006a; Gluckman et al., 2016a).

An example involving the neonate and infant is substitution of breast milk with cow's milk-based formula, which has a different energy content and composition of macronutrients, micronutrients, and bioactive compounds such as hormones and immunological factors (Martin et al., 2016). Furthermore, bottle feeding is itself a novel feeding method that can affect the appropriate development of satiety as it is the feeder rather than the infant who is controlling intake (Li et al., 2012; Azad et al., 2018). This evolutionary novelty is experienced during an especially plastic stage of the life course, and has been associated with greater weight gain, altered immune function, altered eating behaviors, and potentially altered neurodevelopment and behavioral development (Le Huërou-Luron et al., 2010; Owen et al., 2005; Lechner and Vohr, 2017).

Nutrition related evolutionary mismatch also may be encountered *in utero*, for example, in the form of exposure to high levels of glucose via gestational diabetes mellitus (GDM). GDM is likely to be an evolutionary novelty given the different diets during the bulk of our evolutionary past and the probable rarity of maternal obesity. In addition, the severe consequences of untreated GDM in the absence of modern healthcare, namely fetal macrosomia and its attendant high risks of both perinatal and maternal morbidity and mortality, would have been strong selective pressure against any inheritable components of GDM.

Pregnancy is associated with numerous hormonally-mediated physiological and metabolic changes (Zeng et al., 2017). As a normal pregnancy advances, maternal insulin resistance increases to facilitate glucose transfer to the fetus; meanwhile, maternal lipogenesis that occurs during early–mid gestation is followed by lipolysis in late gestation to allow fetal fat accumulation (Sonagra et al., 2014; Barbour et al., 2007). Insulin resistance and hyperlipidemia are amplified by GDM and obesity, resulting in increased nutrient supply to the fetus. Elevated maternal blood glucose leads to increased passage of glucose to the fetus, because there is no evolved mechanism by which the placenta can limit its transfer given the lack of past selective pressure (Ma et al., 2013). Sustained fetal hyperglycemia then leads to hyperinsulinemia and adipogenesis, with lifetime consequences through to adulthood. GDM is a risk factor for post-pregnancy T2D in the mother, and adiposity and impaired glucose metabolism in her offspring (Metzger, 2007; Bellamy et al., 2009). Nutritional transition in many regions and countries has been linked to increased prevalence of gestational diabetes (Leng et al., 2015; Ferrara, 2007).

Humans are social creatures for whom social structures and environments play an important role; it is thought that we have evolved to live in groups comprising 100–150 people (Dunbar, 2008). Even when urbanization began to become more widespread, social groupings and institutional organizations had largely remained around this group size. However, this started to change with the development of telephonic communication and motorized transport, and more complex societal organization and breakdown of strong religious and social identities. Then, since the end of the twentieth century, our social environments have undergone particularly dramatic changes such as the ubiquity of electronic communication and online social networks, with consequent impacts on the numbers and complexities of social interactions, on societal hierarchies, and on the very nature of autonomy. A central concept of the development of human social structures is living in a single group with clear rules, boundaries, or identities; however this is increasingly no longer the case, and it is now common to concurrently live in multiple groups (Wilson, 2002). There are suggestions that the human neocortex has not evolved to adequately cope with these fundamental societal shifts, and the mismatch is being reflected in increased mental health and behavioral disorders (Gluckman & Allen, 2018; Seabrook et al., 2016; Elhai et al., 2017).

1.3 Environmentally induced responses in development

A full understanding of the concepts of evolutionary mismatch and developmental mismatch requires an appreciation for the operation and utility of evolved adaptive cues. Over the life course of an organism, the period of early development is perhaps the most plastic in terms of responsiveness to external cues. Exposure to a particular environmental cue may elicit a range of responses depending on the nature of the cue. Those that are especially severe may result in irreversible developmental disruption and teratogenesis, such as the rubella virus which, in humans, causes cataracts, neuronal and cardiac problems, or the drug thalidomide which causes defects of the extremities.

More commonly, environmental cues may instead elicit more subtle responses that involve potentially adaptive modifications to the developmental program. The ability of the organism to interpret signals from the environment in early life and adapt its development accordingly is known as developmental plasticity (Bateson et al., 2004; Gluckman et al., 2005b). The timing of the cue also matters, as they generally have to be present during a critical or sensitive period of development to exert an effect. Teratogenic challenges operate within relatively narrow and specific windows (Gilbert-Barness, 2010), while in mammals developmental plasticity tends to be greatest from gestation through to early weaning.

The outcomes of the processes of developmental plasticity can be observed at a number of phenotypic levels, from that of DNA (e.g., a regulatory change potentially associated with epigenetic alterations), to the cell or organ (e.g., reduced number of nephrons), through to behavioral aspects (e.g., impact of learning on neurodevelopment and behavior). Developmental plasticity is observed across numerous taxa, suggesting that it is a normative phenomenon that has evolved because it is adaptive, which in evolutionary semantics refers to being fitness promoting. Developmental plasticity may be non-directional in organisms with high reproductive output, especially asexual reproducers such as bacteria. Bacteria can produce many phenotypically variable offspring randomly or in response to environmental stressors (Philippi & Serger, 1989), relying on the probability that such "bet-hedging" will result in at least some offspring that are well suited to survive. In contrast, developmental plasticity in sexually reproducing species and particularly those with lower fecundity such as humans is generally directional. Developmentally plastic responses may lead to an immediate survival or fitness advantage, or may involve a degree of prediction or anticipation that only becomes apparent later. Depending on the cue, both immediate and delayed responses may be made (Low et al., 2012). These concepts are discussed in detail later.

1.3.1 Key features of developmental plasticity

An important feature of developmental plasticity is that the nature of the responses evolved is directly dependent on—and thus limited by—the range of experiences encountered during the species' lineage. Another feature is that developmental plasticity decreases with age (Godfrey et al., 2010). This is because maintaining maximal plasticity through the entire life course not only entails extremely high energetic costs, but also is inherently incompatible with the increasing physiological and anatomical constraints that arise as development—itself inherently irreversible—proceeds.

It is also crucial to note that compared to species from other taxa and other mammals, humans present a unique case because the period of reproductive competence, at least in females, is much shorter than the life span. Indeed, modeling of the selection forces operating through the human life cycle has demonstrated that survival to reproductive age is a key determinant of fitness (Jones, 2009). Two particular aspects of our life history traits and behaviors are relevant here: the first is that humans are born in an altricial state and have a relatively long period of juvenile dependency, yet the period of reproductive competence in females is relatively short. The second is that human fecundity is low and parental investment in offspring is high. Thus, there appears to be little adaptive advantage of anticipatory responses having value beyond the first few decades of life. Selective processes operate without consideration of health or longevity, and the manifestation of non-communicable diseases past reproductive age in humans is unlikely to have evolutionary consequences.

1.4 Developmental mismatch

The concepts of evolutionary and developmental mismatch should not be conflated with each other. Although drawing on similar evolutionary principles, developmental mismatch is distinct from evolutionary mismatch in that it involves the developing organism responding to cues in ways that evolved to be adaptive.

1.4.1 Immediate adaptive responses

Cues experienced during development may elicit an immediate adaptive response (IAR) that may incur a trade-off where disadvantage is experienced in later life (Low et al., 2012). An ecological example is amphibian tadpoles living in a desiccating pond. These tadpoles experience crowding and undergo early metamorphosis, which enables survival to live on land as an adult, but with the trade-off of being a smaller size that is more prone to predation (Denver et al., 1998). In mammalian fetuses, cues such as hypoxia or poor maternal nutrition may result in restricted fetal growth. This IAR conserves resources to promote survival at least until birth; when affected in late gestation, the fetus reduces somatic growth and diverts resources towards the brain and heart (Gluckman & Hanson, 2005). As this is the developmental stage at which the fetus would normally be laying down fat but is not doing so, the infant is often born with insulin sensitivity (Mericq et al., 2005). This is thought to promote rapid postnatal adiposity with the availability of fats from breast milk, and serve as a preventive measure to buffer the highly energy expending brain from future resource depleting challenges, such as gastrointestinal disease, particularly during weaning (Kuzawa, 2010). Notably, as will be discussed later, fetal growth restriction often occurs concomitantly with a predictive adaptive response (PAR) that leads to insulin resistance in later childhood (Mericq et al., 2005; Ibanez et al., 2006). As an IAR, fetal growth restriction incurs postnatal trade-offs of increased risks of neonatal morbidity and mortality. Another example of an IAR is premature birth in response to maternal infection, maternal stress, or poor maternal nutrition. In these instances, the potential benefits of being born into a relatively safer postnatal environment are judged to outweigh the potential risks of premature delivery.

1.4.2 Predictive adaptive responses

The class of responses made by a developing organism that alter its physiology for adaptive value, not at the time of the cue but in a predicted future environment, is known as a PAR (Gluckman et al., 2005a; Bateson et al., 2014) or anticipatory maternal effects (Marshall & Uller, 2007). A classic demonstration is provided by the meadow vole, which develops a thicker coat of fur during gestation in anticipation of colder weather when born in autumn, and develops a thinner coat that is better suited to a warmer environment when it is born in spring. The cue is changes in daylight hours, and this is transduced to the offspring via alterations in maternal melatonin levels (Lee & Zucker, 1988). Here, the coat thickness promotes later survival and thus fitness even though it has no immediate benefit at birth.

Anticipatory responses may also be induced by variably occurring cues such as predator presence. For example, pregnant snowshoe hares exposed to predator induced stress produce offspring with altered stress levels and attention behavior—traits that confer a clear postnatal advantage for survival (Sheriff et al., 2015). Juvenile water fleas exposed to predator-specific chemical cues called kairomones develop long spines on their heads and tails, and a thicker carapace, as protective defense mechanisms (Dzialowski et al., 2003; Laforsch et al., 2004). The desert locust is especially sensitive to chemical and other signals of population density during the post-hatching and early larval instar stages. When population density is not anticipated to be high, the offspring adopt a solitarious phase that favors behavioral independence and low mobility. However when signals of crowding are received, the offspring develop into a gregarious morph that is smaller, highly mobile, interactive with other locusts, and voracious in appetite (Pener & Simpson, 2009; Simpson et al., 1999). Such polyphenisms resulting from developmental influences are well documented in many other insects (Simpson et al., 2011).

1.4.3 Malprediction

Although PARs have evolved for adaptive purposes, if the organism finds itself in an environment that is different from that predicted during development, then a situation of developmental mismatch arises, leading to increased risk of a maladaptive outcome. It is important to note from the outset that PARs do not cause disease. Instead, in the context of an unpredicted later environment they lead to an apparent greater sensitivity to that environment due to being less physiologically adapted to it (Vickers et al., 2000; Vickers et al., 2003). For example, developmental mismatch may stem from behavioural factors, such as later-life migration from a poor nutritional environment that was experienced *in utero*, to an energy replete environment. Socioeconomic factors including urbanisation and rapid nutritional transition may result in marked changes to nutritional environments within a life course (Popkin, 2001).

In the ecological examples of PARs described earlier, the predictions made are likely to be relatively accurate. In humans, the situation may be somewhat more complex. PARs bestow an adaptive advantage by promoting childhood survival and adult reproduction by matching the organism's physiology to the predicted environment. The fetus or infant receives information on its later environment via metabolic or hormonal cues from its mother, which enables the individual to continually make predictions of the future environment that it will live in at least until reproductive age. The relationship between the maternal state and fetal state is likely modulated by placental mechanisms that allow for inertia in the system, so as to avoid transient fluctuations in maternal state that do not reflect the longer-term environment (Kuzawa, 2005). In sheep, which also have a relatively long gestational period of 21 weeks, fetuses subjected only to prolonged but not short-term maternal undernutrition showed a failure of catch-up growth (Mellor & Murray, 1982).

In mammals, predictions may be inaccurate and result in developmental mismatch. An example is uteroplacental vascular insufficiency, where diminished maternal blood flow through the placenta reduces oxygen and nutrient supply to the fetus. This typically induces fetal growth restriction (Krishna & Bhalerao, 2011), but also leads to erroneous predictions of a low nutrition postnatal environment. Being born small is associated with increased risk of T2D (Whincup et al., 2008), with impaired glucose tolerance apparent even at age four (Milovanovic et al., 2014). This example also illustrates that PARs do not necessarily occur in isolation, but are often associated with an IAR. Indeed it seems likely that all IARs would be associated with a PAR, although PARs may occur without an obvious IAR. It also illustrates the compounding effect of PARs, where adults with lower birth weight are at higher relative risk of T2D than their heavier counterparts with every additional unhealthy lifestyle behavior adopted (Li et al., 2015). Less severe cues where birth weight is not affected, such as a moderately unbalanced maternal diet, have been linked to childhood adiposity (Godfrey et al., 2011). Investigations of the underlying molecular pathways have revealed the involvement of epigenetic change in biologically plausible genes.

Simulation modelling has demonstrated that PARs can still evolve even if fidelity of the predictions is not extremely high (Jablonka et al., 1995). There may also be an asymmetry in terms of the outcomes of incorrect predictions, in a way that leads to biases towards prediction of a certain environment. For example, incorrectly predicting a nutritionally rich environment but ending up with poor access to nutrition will have a greater fitness impact that the converse scenario of incorrectly predicting a nutritionally constrained environment but ending up in a food-rich environment. Thus, and as further discussed next, PARs in humans may have evolved with a bias of favoring constrained environments. Another important feature to note is that, unless developmental mismatch occurs, PARs may not have any overt phenotypic effects (Bateson et al., 2014).

1.4.4 PARs in humans

1.4.4.1 Specific considerations

There have been debates on the validity of the PAR framework (Wells, 2007; Wells, 2012) and they highlight the need to expand on several aspects of the associated conceptual framework. In particular, consideration of PARs in humans requires taking into account several unique life history traits. Maternal constraint refers to the set of mechanisms that prevent fetal size from reaching the full genetic potential (Ounsted et al., 1986; Gluckman & Hanson, 2004). This is an evolved mechanism to better match the size at birth to maternal physical characteristics; it enables passage of the large fetal head through a narrow pelvic canal, reducing the risk of labour dystocia and hence neonatal and maternal mortality (Gluckman & Hanson, 2004). These anatomical features are specific to humans and have themselves evolved as outcomes of the evolution of bipedalism and a large brain.

The phenomenon of maternal constraint is near universal among humans, with population-wide clinical data showing that lowest perinatal mortality rates are in effect at the 80–84th birth-weight centiles rather than at the median, indicating a level of constraint on birth size (Vasak et al., 2014). Maternal constraint promotes fetal bias towards the induction of PARs that predict lower postnatal nutritional environments, and has implications for amplifying the risks posed by the modern day nutritionally replete environment. Here, evolutionary mismatch is in effect being amplified by developmental mismatch.

Early nutrition and maternal stress are likely the most important PAR-inducing cues in humans, and both can be seen as reflections of survival related cues in the evolutionary past. Predictions of a postnatal environment with inadequate food access can, if prolonged, prompt reduced somatic growth in the form of stunting, and a metabolic and behavioral profile that tends towards energy conservation and strategies to lay down fat. If the prediction is accurate and the infant grows in a nutritionally poor environment, the individual is then relatively matched to their environment, with fewer consequences in later life than if the infant had predicted a high nutrition environment. However, if the prediction is faulty and the later environment turns out to be nutritionally abundant, the developmental mismatch is greater because of the presence of PARs. The individual then becomes at greater risk of obesity, T2D, and other co-morbidities (Gluckman & Hanson, 2006a). The effects of poor early life nutrition have been extensively investigated in animal models and humans; in the presence of a nutritionally rich postnatal environment, the organism is sensitized to obesity, hyperphagia, sedentary behavior, and insulin resistance in childhood (Sloboda et al., 2007; Vickers et al., 2000; Vickers et al., 2003; Hofman et al., 2004). It should be noted that relative insulin resistance in childhood, which reflects a PAR, may in fact be preceded by insulin sensitivity in infancy as a consequence of fetal growth restriction, a concomitant IAR (Mericq et al., 2005).

PARs can also alter life history traits such as rate of sexual maturation (Sloboda et al., 2007). An accelerated life history strategy means greater likelihood of a mismatch with age of psychosocial maturity, and the implications for adolescent morbidity and mortality have been discussed elsewhere (Gluckman & Hanson, 2006c; Gluckman et al., 2011b).

Humans have low fecundity and exhibit a high level of maternal investment in each pregnancy, making the evolution of PARs especially important for survival of the offspring to reproductive age. Compared to other highly fecund, polytocous species in which maternal-offspring fitness trade-offs often favor the mother, where fertility is blocked or severely reduced by undernutrition, there is greater preservation of offspring fitness in humans. For example, humans are still able to maintain fertility (albeit reduced) under conditions of severe undernutrition—fertility rates fell by 45% during the Chinese famine of 1959–1961, but a rate of 3.1 births per woman was still achieved (Peng, 1987).

1.4.4.2 Empirical evidence

While evolutionary medicine provides a substantive base from which to generate hypotheses, the inherently historical nature of evolution and the decades-long generation time in humans have posed particular challenges in the transition to hypothesis testing. A number of considerations can aid in formulating a robust approach to testing evolutionary hypotheses (Nesse, 2011). These include clarifying both the object of explanation and how the explanation should be defined, considering the categories of explanation in which the hypothesis may be placed, and recognising the value of modelling studies as well as comparative and experimental methods. Given the challenges of designing evolutionary experiments in humans, other approaches such as studies of natural experiments, and demographic or historical patterns have to be relied on.

A cohort of survivors of severe acute malnutrition in infancy has provided a valuable opportunity to investigate the operation of PARs in humans. As a result of the nutritional stress encountered, some of these individuals had developed marasmus, while others had developed the oedematous condition kwashiorkor. Both syndromes can be observed within a population undergoing the same nutritional challenge, but their metabolic profile and other phenotypic outcomes are distinct (Shetty, 2006). For example, while marasmic infants undergo severe wasting (rapid loss of body weight resulting in a low weight-for-height), those who have developed kwashiorkor are characterized by other clinical symptoms such as oedema and skin lesions. Fatality rates are higher in this latter group (Mengesha et al., 2016; Ogbeide & Osuhor, 1984). Kwashiorkor-afflicted infants also tend to have lower protein and lipid turnover compared to their marasmic counterparts, which results in a large unused energy store, with hepatic steatosis being found at post-mortem (Badaloo et al., 2006; Jahoor et al., 2008; Di Giovanni et al., 2016). This suggests that under conditions of extreme nutritional deficit, marasmic infants tend to be metabolically "thrifty," but kwashiorkor infants—being unable to mobilize their nutritional stores—are instead metabolically "profligate."

Post-rehabilitative studies of a cohort of Jamaican children previously diagnosed with severe acute malnutrition showed that, as adults, those who had developed marasmus tended to be shorter and to have impaired glucose metabolism, as evidenced by poorer insulin sensitivity (Francis-Emmanuel et al., 2014). Molecular investigations revealed altered epigenetic states as measured by methylation at multiple CpG loci in muscle tissue (Sheppard et al., 2017), including genes involved in glucose metabolism, body size, and body composition. This shows that the metabolic differences have persisted, and that these children have fundamentally different physiologies. To determine if these metabolic differences had arisen before severe malnutrition was experienced, a retrospective analysis was conducted (Forrester et al., 2012). On average, children with kwashiorkor had significantly higher birth weights than those who developed marasmus (333 g, 95% confidence interval 217, 449, $p< 0.001$).

Thus, the totality of the evidence suggests that the marasmic children, who had made an IAR of restricting fetal growth, had also predicted a more nutritionally deplete postnatal environment and accordingly made PARs for a thrifty metabolic phenotype. The sequential impact could be seen at first as assisted survival of the malnutrition episode beyond childhood. Then, exposure in adulthood to the nutritional transition environment in Jamaica led to emergence of developmental mismatch, which would have been exaggerated by the evolutionary mismatch of energy-dense, poor quality foods. The consequences of the inaccurate PARs were then seen as greater glucose intolerance and, in turn, greater risk of T2D (Boyne et al., 2017; Forrester et al., 2012). In contrast, children who developed kwashiorkor did not predict a future of low nutrition, and were ill equipped to exploit their nutritional stores upon exposure to malnutrition, eventually leading to increased mortality. However among kwashiorkor children that did survive, metabolic parameters in adulthood were more favorable than in marasmic survivors despite living in the same adult nutritional environment. It is important to note that because of the availability of a nutritional rehabilitation unit and medical care, the mortality in kwashiorkor group had been minimal, thus avoiding a selection bias.

Taken together, this set of studies constitutes a natural experiment that demonstrates a survival difference—that is, a clear fitness effect—that has evidence of biological plausibility at the molecular level, and is in agreement with other experimental data. It provides strong support for the hypothesis that differential developmental trajectories *in utero*, which manifest as discordances in birth weights, is associated with differential metabolic profiles with implications for different coping strategies during nutritional insults.

1.5 Conclusions

Perspectives from evolutionary medicine can greatly assist in our understanding of human physiology, health, and disease, at the levels of both the individual and the population. Importantly, this can also inform public health interventional policies (Hanson & Gluckman, 2015; Low et al., 2018). The examples discussed have demonstrated how evolution of bipedalism and altricial birth in humans because of brain growth can explain changing patterns of disease. The examples also provide evolutionary explanations for the smaller growth of the first-born child and for the rapid rise in rates of GDM being seen worldwide. While perhaps not directly affecting clinical management, the evolutionary perspective can do much to change the research agenda, and enhance patient and clinical understandings.

Two of the pathways to increased disease risk that are based on evolutionary principles involve mismatch, one based on the species' evolutionary history, and the other based on the individual's developmental history. Evolutionary mismatch is underpinned by the niche modifying activities of humans that can place us in evolutionarily novel situations. Human life-history traits confer relative significance to offspring fitness compared to other species, and has led to an evolved ability to make an IAR and/or PAR in response to maternal stressors. PARs by themselves remain covert until developmental mismatch occurs.

Evolutionary and developmental mismatch clearly play a major role in the global epidemic of obesity and its complications. Greater attention needs to be paid to the impacts of evolutionarily novel nutritional, social, and technological environments on human health. Minimizing both types of mismatch may require adapting certain behaviors within a changed environment. Examples include public health efforts to improve pre-conceptional and prenatal nutrition, promote breastfeeding, and reduce the rates of GDM. Experimental evidence that developmentally primed phenotypes may be averted through early life interventions suggests that PARs can be reversed (Vickers & Sloboda, 2012). A context-dependent, unified approach that accounts for nutritional, socioeconomic, and policy dimensions over the life course will likely be most effective (Low et al., 2018).

References

AAMC-HHMI Scientific Foundation for Future Physicians Committee (2009). *Scientific Foundations for Future Physicians*. Washington, DC: American Association of Medical Colleges and Howard Hughes Medical Institute.

Antolin, M. F., K. P. Jenkins, C. T. Bergstrom, B. J. Crespi, S. De, A. Hancock, K. A. Hanley, T. R. Meagher, A. Moreno-Estrada, R. M. Nesse, G. S. Omenn, & S. C. Stearns (2012). Evolution and medicine in undergraduate education: A prescription for all biology students. *Evolution*, 66(6): 1991–2006.

Azad, M. B., L. Vehling, D. Chan, A. Klopp, N. C. Nickel, J. M. McGavock, A. B. Becker, P. J. Mandhane, S. E. Turvey, T. J. Moraes, M. S. Taylor, D. L. Lefebvre, M. R. Sears, & P. Subbarao (2018). Infant feeding and weight gain: Separating breast milk from breastfeeding and formula from food. *Pediatrics*, 142(4): e20181092.

Badaloo, A. V., T. Forrester, M. Reid, & F. Jahoor (2006). Lipid kinetic differences between children with kwashiorkor and those with marasmus. *Am J Clin Nutr*, 83(6): 1283–8.

Barbour, L. A., C. E. McCurdy, T. L. Hernandez, J. P. Kirwan, P. M. Catalano, & J. E. Friedman (2007). Cellular mechanisms for insulin resistance in normal pregnancy and gestational diabetes. *Diabetes Care*, 30(Supplement 2): S112–S119.

Barouki, R., P. D. Gluckman, P. Grandjean, M. Hanson, & J. J. Heindel (2012). Developmental origins of non-communicable disease: Implications for research and public health. *Environmental Health*, 11: 42.

Bateson, P., D. Barker, T. Clutton-Brock, D. Deb, B. D'Udine, R. A. Foley, P. Gluckman, K. Godfrey, T. Kirkwood, M. Mirazon Lahr, J. McNamara, N. B. Metcalfe, P. Monaghan, H. G. Spencer, & S. E. Sultan (2004). Developmental plasticity and human health. *Nature*, 430(6998): 419–21.

Bateson, P., P. Gluckman, & M. Hanson (2014). The biology of developmental plasticity and the Predictive Adaptive Response hypothesis. *The Journal of Physiology*, 592(11): 2357–68.

Bellamy, L., J.-P. Casas, A. D. Hingorani, & D. Williams (2009). Type 2 diabetes mellitus after gestational diabetes: A systematic review and meta-analysis. *The Lancet*, 373(9677): 1773–9.

Boyne, M. S., P. Francis-Emmanuel, I. A. Tennant, D. S. Thompson, & T. E. Forrester. (2017). Cardiometabolic risk in marasmus and kwashiorkor survivors. In *Handbook of Famine, Starvation, and Nutrient Deprivation: From Biology to Policy*, V. Preedy & V. B. Patel (eds.) (pp. 1–23). Cham: Springer International Publishing.

Buklijas, T., F. M. Low, A. S. Beedle, & P. D. Gluckman (2011). Developing a curriculum for evolutionary medicine: Case studies of scurvy and female reproductive tract cancers. *Evolution: Education and Outreach*, 4(4): 595–602.

Carroll, S. P., P. S. Jørgensen, M. T. Kinnison, C. T. Bergstrom, R. F. Denison, P. Gluckman, T. B. Smith, S. Y. Strauss, & B. E. Tabashnik (2014). Applying evolutionary biology to address global challenges. *Science*, 346(6207): 1245993.

Denver, R., N. Mirhadi, & M. Phillips (1998). Adaptive plasticity in amphibian metamorphosis: Response of *Scaphiopus hammondii* tadpoles to habitat desiccation. *Ecology*, 79: 1859–72.

Di Giovanni, V., C. Bourdon, D. X. Wang, S. Seshadri, E. Senga, C. J. Versloot, W. Voskuijl, R. D. Semba, I. Trehan, R. Moaddel, M. I. Ordiz, L. Zhang, J. Parkinson, M. J. Manary, & R. H. J. Bandsma (2016). Metabolomic changes in serum of children with different clinical diagnoses of malnutrition. *Journal of Nutrition*, 146(12): 2436–44.

Dunbar, R. (2008). Mind the gap; or why humans are not just great apes. *Proceedings of the British Academy*, 154: 403–23.

Drouin, G., J.-R. Godin, & B. Pagé (2011). The genetics of vitamin C loss in vertebrates. *Current Genomics*, 12(5): 371–78.

Dzialowski, A. R., J. T. Lennon, W. J. O'Brien, & V. H. Smith (2003). Predator-induced phenotypic plasticity in the exotic cladoceran *Daphnia lumholtzi*. *Freshwater Biology*, 48(9): 1593–602.

Elhai, J. D., R. D. Dvorak, J. C. Levine, & B. J. Hall (2017). Problematic smartphone use: A conceptual overview and systematic review of relations with anxiety and depression psychopathology. *Journal of Affective Disorders*, 207: 251–9.

Ferrara, A. (2007). Increasing prevalence of gestational diabetes mellitus: A public health perspective. *Diabetes Care*, 30(Supplement 2): S141–6.

Fogel, R. W. (2004). *The Escape from Hunger and Premature Death, 1700–2100: Europe, America and the Third World*. Cambridge: Cambridge University Press.

Forrester, T. E., A. V. Badaloo, M. S. Boyne, C. Osmond, D. Thompson, C. Green, C. Taylor-Bryan, A. Barnett, S. Soares-Wynter, M. A. Hanson, A. S. Beedle, & P. D. Gluckman (2012). Prenatal factors contribute to emergence of kwashiorkor or marasmus in response to severe undernutrition: Evidence for the predictive adaptation model. *PLoS One*, 7(4): e35907.

Francis-Emmanuel, P. M., D. S. Thompson, A. T. Barnett, C. Osmond, C. D. Byrne, M. A. Hanson, P. D. Gluckman, T. E. Forrester, & M. S. Boyne (2014). Glucose metabolism in adult survivors of severe acute malnutrition. *The Journal of Clinical Endocrinology & Metabolism*, 99(6): 2233–40.

Gilbert-Barness, E. (2010). Teratogenic causes of malformations. *Annals of Clinical & Laboratory Science*, 40(2): 99–114.

Gluckman, P. & K. Allen (2018). Understanding wellbeing in the context of rapid digital and associated transformations: Implications for research, policy and measurement. A discussion paper. Auckland: International Network for Government Science Advice.

Gluckman, P., A. Beedle, T. Buklijas, F. Low, & M. Hanson (2016a). Nutritional and metabolic adaptation. In *Principles of Evolutionary Medicine* (pp. 205–36). Oxford: Oxford University Press.

Gluckman, P., A. Beedle, T. Buklijas, F. Low, & M. Hanson (2016b). *Principles of Evolutionary Medicine*. Oxford: Oxford University Press.

Gluckman, P. & M. Hanson (2005). *The Fetal Matrix: Evolution, Development, and Disease*. Cambridge: Cambridge University Press.

Gluckman, P. & M. Hanson (2006a). *Mismatch: Why Our World No Longer Fits Our Bodies*. Oxford: Oxford University Press.

Gluckman, P. D., A. S. Beedle, & M. A. Hanson (2009). *Principles of Evolutionary Medicine*. Oxford: Oxford University Press.

Gluckman, P. D. & M. A. Hanson (2004). Maternal constraint of fetal growth and its consequences. *Seminars in Fetal & Neonatal Medicine*, 9(5): 419–25.

Gluckman, P. D. & M. A. Hanson (eds.) (2006b). *Developmental Origins of Health and Disease*. Cambridge: Cambridge University Press.

Gluckman, P. D. & M. A. Hanson (2006c). Evolution, development and timing of puberty. *Trends in Endocrinology and Metabolism*, 17(1): 7–12.

Gluckman, P. D., M. A. Hanson, C. Cooper, & K. L. Thornburg (2008). Effect of in utero and early-life conditions on adult health and disease. *New England Journal of Medicine*, 359(1): 61–73.

Gluckman, P. D., M. A. Hanson, & H. G. Spencer (2005a). Predictive adaptive responses and human evolution. *Trends in Ecology & Evolution*, 20(10): 527–33.

Gluckman, P. D., M. A. Hanson, H. G. Spencer, & P. Bateson (2005b). Environmental influences during development and their later consequences for health and disease: Implications for the interpretation of empirical studies. *Proceedings of the Royal Society B*, 272(1564): 671–7.

Gluckman, P. D., F. M. Low, T. Buklijas, M. A. Hanson, & A. S. Beedle (2011a). How evolutionary principles improve the understanding of human health and disease. *Evolutionary Applications*, 4(2): 249–63.

Gluckman, P. D., F. M. Low, & K. L. Franko. (2011b). Puberty and adolescence: Transitions in the life course. In *Improving the Transition: Reducing Social and Psychological Morbidity During Adolescence*, (pp. 19–34). Auckland: Office of the Prime Minister's Science Advisory Committee.

Godfrey, K. M., P. D. Gluckman, & M. A. Hanson (2010). Developmental origins of metabolic disease: Life course and intergenerational perspectives. *Trends Endocrinol Metab*, 21(4): 199–205.

Godfrey, K. M., A. Sheppard, P. D. Gluckman, K. A. Lillycrop, G. C. Burdge, C. McLean, J. Rodford, J. L. Slater-Jefferies, E. Garratt, S. R. Crozier, B. S. Emerald, C. R. Gale, H. M. Inskip, C. Cooper, & M. A. Hanson (2011). Epigenetic gene promoter methylation at birth is associated with child's later adiposity. *Diabetes*, 60(5): 1528–34.

Graves, J. J. L., C. Reiber, A. Thanukos, M. Hurtado, & T. Wolpaw (2016). Evolutionary science as a method to facilitate higher level thinking and reasoning in medical training. *Evolution, Medicine, and Public Health*, 2016(1): 358–68.

Hanson, M. A. & P. D. Gluckman (2015). Developmental origins of health and disease—Global public health implications. *Best Practice & Research Clinical Obstetrics & Gynaecology*, 29(1): 24–31.

Hidaka, B. H., A. Asghar, C. A. Aktipis, R. M. Nesse, T. M. Wolpaw, N. K. Skursky, K. J. Bennett, M. W. Beyrouty, & M. D. Schwartz (2015). The status of evolutionary medicine education in North American medical schools. *BMC Medical Education*, 15: 38.

Hofman, P. L., F. Regan, W. E. Jackson, C. Jefferies, D. B. Knight, E. M. Robinson, & W. S. Cutfield (2004). Premature birth and later insulin resistance. *New England Journal of Medicine*, 351: 2179–86.

Hur, S. S. J., J. E. Cropley, & C. M. Suter (2017). Paternal epigenetic programming: Evolving metabolic disease risk. *Journal of Molecular Endocrinology*, 58(3): R159–R168.

Ibanez, L., K. Ong, D. B. Dunger, & F. de Zegher (2006). Early development of adiposity and insulin resistance after catch-up weight gain in small-for-gestational-age children. *Journal of Clinical Endocrinology & Metabolism*, 91(6): 2153–8.

Jablonka, E., B. Oborny, I. Molnar, E. Kisdi, J. Hofbauer, & T. Czaran (1995). The adaptive advantage of phenotypic memory in changing environments. *Philosophical Transactions of the Royal Society of London—Series B: Biological Sciences*, 350(1332):133–41.

Jahoor, F., A. Badaloo, M. Reid, & T. Forrester (2008). Protein metabolism in severe childhood malnutrition. *Annals of Tropical Paediatrics*, 28(2): 87–101.

Jones, J. H. (2009). The force of selection on the human life cycle. *Evolution and Human Behavior*, 30(5): 305–14.

Krishna, U. & S. Bhalerao (2011). Placental insufficiency and fetal growth restriction. *Journal of Obstetrics and Gynaecology of India*, 61(5): 505–11.

Kuzawa, C. W. (2005). "Fetal origins of developmental plasticity: Are fetal cues reliable predictors of future nutritional environments?. *American Journal of Human Biology*, 17(1): 5–21.

Kuzawa, C. W. (2010). Beyond feast–famine: Brain evolution, human life history, and the metabolic syndrome. In *Human Evolutionary Biology*, M. P. Muehlenbein (ed.), (pp. 518–27). Cambridge: Cambridge University Press.

Laforsch, C., W. Ngwa, W. Grill, & R. Tollrian (2004). An acoustic microscopy technique reveals hidden morphological defenses in *Daphnia*. *Proceedings of the National Academy of Sciences of the United States of America*, 101(45): 15911–4.

Le Huërou-Luron, I., S. Blat, & G. Boudry (2010). Breast- v. formula-feeding: Impacts on the digestive tract and immediate and long-term health effects. *Nutrition Research Reviews*, 23(1): 23–36.

Lechner, B. E. & B. R. Vohr (2017). Neurodevelopmental outcomes of preterm infants fed human milk. *Clinics in Perinatology*, 44(1): 69–83.

Lee, T. M. & I. Zucker (1988). Vole infant development is influenced perinatally by maternal photoperiodic history. *American Journal of Physiology*, 255(5 Part 2): R831–R838.

Leng, J., P. Shao, C. Zhang, H. Tian, F. Zhang, S. Zhang, L. Dong, L. Li, Z. Yu, J. C. N. Chan, G. Hu, & X. Yang (2015). Prevalence of gestational diabetes mellitus and its risk factors in Chinese pregnant women: A prospective population-based study in Tianjin, China. *PLOS ONE*, 10(3): e0121029.

Li, R., J. Magadia, S. B. Fein, & L. M. Grummer-Strawn (2012). Risk of bottle-feeding for rapid weight gain during the first year of life. *Archives of Pediatrics & Adolescent Medicine*, 166(5): 431–6.

Li, Y., S. H. Ley, D. K. Tobias, S. E. Chiuve, T. J. VanderWeele, J. W. Rich-Edwards, G. C. Curhan, W. C. Willett, J. E. Manson, F. B. Hu, & L. Qi (2015). Birth weight and later life adherence to unhealthy lifestyles in predicting type 2 diabetes: Prospective cohort study. *BMJ: British Medical Journal*, 351: h3672.

Low, F. M. & P. D. Gluckman. (2016). Evolutionary Medicine: Mismatch. In *The Encyclopedia of Evolutionary Biology*, R. M. Kliman (eds). Oxford: Academic Press.

Low, F. M., P. D. Gluckman, & M. A. Hanson (2012). Developmental plasticity, epigenetics and human health. *Evolutionary Biology*, 39(4): 650–65.

Low, F. M., P. D. Gluckman, & M. A. Hanson. (2018). A life course approach to public health: Why early life matters. In *Oxford Textbook of Nature and Public Health: The Role of Nature in Improving the Health of a Population*, M. van den Bosch and W. Bird (eds.), (pp. 11–25). Oxford: Oxford University Press.

Ma, R. C. W., J. C. N. Chan, W. H. Tam, M. A. Hanson, & P. D. Gluckman (2013). Gestational diabetes, maternal obesity and the NCD burden. *Clinical Obstetrics and Gynecology*, 56(3): 633–41.

Marshall, D. J. & T. Uller (2007). When is a maternal effect adaptive?. *Oikos*, 116(12): 1957–63.

Martin, C. R., P.-R. Ling, & G. L. Blackburn (2016). Review of infant feeding: Key features of breast milk and infant formula. *Nutrients*, 8(5): 279.

Mellor, D. J. & L. Murray (1982). Effects on the rate of increase in fetal girth of refeeding ewes after short periods of severe undernutrition during late pregnancy. *Research in Veterinary Science*, 32(3): 377–82.

Mengesha, M. M., N. Deyessa, B. S. Tegegne, & Y. Dessie (2016). Treatment outcome and factors affecting time to recovery in children with severe acute malnutrition treated at outpatient therapeutic care program. *Global Health Action*, 9: 30704.

Mericq, V., K. K. Ong, R. A. Bazaes, V. Pena, A. Avila, T. Salazar, N. Soto, G. Iniguez, & D. B. Dunger (2005). Longitudinal changes in insulin sensitivity and secretion from birth to age three years in small- and appropriate-for-gestational-age children. *Diabetologia*, 48(12): 2609–14.

Metzger, B. E. (2007). Long-term outcomes in mothers diagnosed with gestational diabetes mellitus and their offspring. *Clinical Obstetrics and Gynecology*, 50(4): 972–9.

Milovanovic, I., F. Njuieyon, S. Deghmoun, D. Chevenne, C. Levy-Marchal, & J. Beltrand (2014). SGA children with moderate catch-up growth are showing the impaired insulin secretion at the age of 4. *PLOS ONE*, 9(9): e100337.

Morgan, I. & K. Rose (2005). How genetic is school myopia?. *Progress in Retinal and Eye Research*, 24(1): 1–38.

Nesse, R. M. (2011). Ten questions for evolutionary studies of disease vulnerability. *Evolutionary Applications*, 4(2): 264–77.

Odling-Smee, F. J., K. N. Laland, & M. W. Feldman (2003). *Niche Construction: The Neglected Process of Evolution*. Princeton, NJ: Princeton University Press.

Ogbeide, O. & P. C. Osuhor (1984). Morbidity and mortality patterns among malnourished children in Benin City, Nigeria. *Tropical Doctor*, 14(4):178–80.

Ounsted, M., A. Scott, & C. Ounsted (1986). Transmission through the female line of a mechanism constraining human fetal growth. *Annals of Human Biology*, 13(2): 143–51.

Owen, C. G., R. M. Martin, P. H. Whincup, G. D. Smith, & D. G. Cook (2005). Effect of infant feeding on the risk of obesity across the life course: A quantitative review of published evidence. *Pediatrics*, 115(5): 1367–77.

Pener, M. P. & S. J. Simpson. (2009). Locust phase polyphenism: An update. In *Advances in Insect Physiology*, S. J. Simpson & M. P. Pener (eds.) pp. 1–272. London: Academic Press.

Peng, X. (1987). Demographic consequences of the Great Leap Forward in China's provinces. *Population and Development Review*, 13(4): 639–70.

Perlman, R. L. (2013). *Evolution & Medicine*. Oxford, UK: Oxford University Press.

Philippi, T. & J. Serger (1989). Hedging one's evolutionary bets, revisited. *Trends in Ecology and Evolution*, 4(2): 41–4.

Pinhasi, R., V. Eshed, & N. von Cramon-Taubadel (2015). Incongruity between affinity patterns based on mandibular and lower dental dimensions following the transition to agriculture in the near East, Anatolia and Europe. *PLOS ONE*, 10(2): e0117301.

Popkin, B. M. (2001). The nutrition transition and obesity in the developing world. *The Journal of Nutrition*, 131(3): 871S–873S.

Popkin, B. M., L. S. Adair, & S. W. Ng (2012). Global nutrition transition and the pandemic of obesity in developing countries. *Nutrition Reviews*, 70(1): 3–21.

Rose, J. C. & R. D. Roblee (2009). Origins of dental crowding and malocclusions: An anthropological perspective. *Compendium of Continuing Education in Dentistry*, 30(5): 292–300.

Seabrook, M. E., L. M. Kern, & S. N. Rickard (2016). Social networking sites, depression, and anxiety: A systematic review. *JMIR Ment Health*, 3(4): e50.

Sheppard, A., S. Ngo, X. Li, M. Boyne, D. Thompson, A. Pleasants, P. Gluckman, & T. Forrester (2017). Molecular evidence for differential long-term outcomes of early life severe acute malnutrition. *EBio Medicine*, 18: 274–80.

Sheriff, M. J., E. K. McMahon, C. J. Krebs, & R. Boonstra (2015). Predator-induced maternal stress and population demography in snowshoe hares: The more severe the risk, the longer the generational effect. *Journal of Zoology*, 296(4): 305–10.

Shetty, P. 2006. Malnutrition and undernutrition. *Medicine*, 34(12): 524–9.

Simpson, S. J., A. R. McCaffery, & B. F. Hägele (1999). A behavioural analysis of phase change in the desert locust. *Biological Reviews*, 74(4): 461–80.

Simpson, Stephen J., Gregory A. Sword, & N. Lo (2011). Polyphenism in insects. *Current Biology*, 21(18): R738–R749.

Skinner, M. K. (2015). Environmental epigenetics and a unified theory of the molecular aspects of evolution: A neo-Lamarckian concept that facilitates neo-Darwinian evolution. *Genome Biology and Evolution*, 7(5): 1296–302.

Sloboda, D. M., R. Hart, D. A. Doherty, C. E. Pennell, & M. Hickey (2007). Age at menarche: Influences of prenatal and postnatal growth. *Journal of Clinical Endocrinology and Metabolism*, 92(1): 46–50.

Sonagra, A. D., S. M. Biradar, D. K, & J. Murthy D.S (2014). Normal pregnancy—a state of insulin resistance. *Journal of Clinical and Diagnostic Research*, 8(11): CC01–CC03.

Stearns, S. C. (2012). Evolutionary medicine: Its scope, interest and potential. *Proceedings of the Royal Society B: Biological Sciences*, 279(1746): 4305–21.

Stearns, S. C. & J. C. Koella (eds.) (2008). *Evolution in health and disease*, Oxford: Oxford University Press.

Stearns, S. C. & R. Medzhitov (2015). *Evolutionary Medicine*. Sunderland: Sinauer Associates.

Uller, T. (2014). Evolutionary Perspectives on Transgenerational Epigenetics. In *Transgenerational Epigenetics*, T. Tollefsbol, (eds.), pp. 175–85. Oxford: Academic Press.

Vaiserman, A. M., A. K. Koliada, & R. L. Jirtle (2017). Non-genomic transmission of longevity between generations: Potential mechanisms and evidence across species. *Epigenetics & Chromatin*, 10(1): 38.

Vasak, B., S. V. Koenen, M. P. H. Koster, C. W. P. M. Hukkelhoven, A. Franx, M. A. Hanson, & G. H. A. Visser (2015). Human fetal growth is constrained below optimal for perinatal survival. *Ultrasound in Obstetrics & Gynecology*, 45(2): 162–7.

Vickers, M. H., B. H. Breier, W. S. Cutfield, P. L. Hofman, & P. D. Gluckman (2000). Fetal origins of hyperphagia, obesity, and hypertension and postnatal amplification by hypercaloric nutrition. *American Journal of Physiology*, 279(1): E83–E87.

Vickers, M. H., B. H. Breier, D. McCarthy, and P. D. Gluckman (2003). Sedentary behavior during postnatal life is determined by the prenatal environment and exacerbated by postnatal hypercaloric nutrition. *American Journal of Physiology*, 285(1): R271–R273.

Vickers, M. H. & D. M. Sloboda (2012). Strategies for reversing the effects of metabolic disorders induced as a consequence of developmental programming. *Frontiers in Physiology*, 3: 242.

Wells, J. C. (2007). Flaws in the theory of predictive adaptive responses. *Trends in Endocrinology and Metabolism*, 18(9): 331–7.

Wells, J. C. (2012). A critical appraisal of the predictive adaptive response hypothesis. *International Journal of Epidemiology*, 41(1): 229–35.

Wells, J. C. K., R. M. Nesse, R. Sear, R. A. Johnstone, & S. C. Stearns (2017). Evolutionary public health: Introducing the concept. *The Lancet*, 390(10093): 500–09.

Whincup, P. H., S. J. Kaye, C. G. Owen, et al. (2008). Birth weight and risk of type 2 diabetes: A systematic review. *JAMA*, 300(24): 2886–97.

Williams, G. C. (1957). Pleiotropy, Natural Selection, and the Evolution of Senescence. *Evolution*, 11(4): 398–411.

Williams, G. C. & R. M. Nesse (1991). The dawn of Darwinian medicine. *The Quarterly Review of Biology*, 66(1): 1–22.

Wilson, D. S. (2002). *Darwin's Cathedral: Evolution, Religion and the Nature of Society*. Chicago: University of Chicago Press.

Zeng, Z., F. Liu, & S. Li (2017). Metabolic adaptations in pregnancy: A review. *Annals of Nutrition and Metabolism*, 70(1): 59–65.

CHAPTER 2

Evolutionary Public Health: How Interventions may Benefit from Insights Generated by Life History Theory

Jonathan C. K. Wells

2.1 Introduction

Public health efforts were already being initiated thousands of years ago. In Greece in the fourth century BC, for example, the military state of Sparta issued a royal decree requiring mothers to breastfeed for 2 years to ensure a healthy fighting force a generation later, while adolescent girls were encouraged to undertake physical training to bear strong sons. Similarly, the Romans had substantial understanding of the importance of nutrition for health, and the need to combat infectious disease. The forts defending their empire were built to a common plan across Europe, and were systematically equipped with a hospital, granaries to preserve high quality food supplies, and latrines flushed by rainwater to minimise the transmission of infections (Wells, 2016).

In the medieval period, the precise source of disease typically remained elusive, but effective public health responses could still emerge. In 1666, the "Black Death" plague reached the small village of Eyam in the rural English Midlands, transported from London in a shipment of material to a tailor. As the death toll in Eyam rapidly rose, the villagers voluntarily took the decision to isolate themselves, to prevent other villages from being infected. They broke off all direct contact, and placed coins soaked in vinegar (disinfectant) at the village boundary, in return for food being left for them by their neighbors (Whittles & Didelot, 2016).

The foundations of contemporary public health were established in the nineteenth century, due to increasing awareness that the living conditions of the entire population affected the health of all. In the UK, for example, the Victorians recognized associations between squalid living conditions, poor physical growth and development, and epidemics of infectious disease. Early scientific studies revealed that infections are transmitted by various vectors exploiting unhealthy environments, and that efforts to mitigate this scenario must be targeted at the entire population. It was vital to prevent the maintenance of disease reservoirs amongst poorer groups that might transmit disease back to other groups. Successive studies also linked a range of individual pathogens with specific diseases, though these efforts took time to consolidate and at the dawn of the twentieth century miasma theory was still being invoked to explain certain epidemics

Wells, J. C. K., *Evolutionary Public Health: How Interventions may Benefit from Insights Generated by Life History Theory* In: *Integrating Evolutionary Biology into Medical Education—for maternal and child healthcare students, clinicians, and scientists*. Edited by: Jay Schulkin and Michael L. Power, Oxford University Press (2020).
DOI: 10.1093/oso/9780198814153.003.0003

such as infant diarrhea (Dwork, 1987). From the mid-nineteenth century, major efforts were undertaken to improve sanitation and hygiene, while the benefits of breastfeeding were recognized alongside the major mortality risks associated with formula feeding (Toubas, 2007).

The nineteenth century was also characterized by ground-breaking work on evolutionary theory by Darwin and Wallace (Darwin, 1859), leading to a radically new scientific framework for biological enquiry. Through the twentieth century, evolutionary theory consolidated, linking with the new discipline of genetics and stimulating the neo-Darwinian synthesis. In the second half of the century, evolutionary thinking permeated almost all avenues of biological enquiry, and now provides the theoretical backbone for multiple disciplines. Evolutionary theory is already widely used in several areas of public policy, including the management of fisheries and efforts to combat agricultural pests (Jørgensen et al., 2007).

That human medicine remained a relative exception to this "merger" between theory and practice until late in the twentieth century may be attributed to several factors. On the one hand, Darwin himself paid relatively little attention to the extent to which infectious disease might be a selective pressure impacting humans. Early studies of human evolution were more concerned with our distant ancestry than with our current health. Moreover, early attempts to apply Darwinian thinking to human society were crude and attracted a strong backlash, particularly in the aftermath of the Second World War (Kuper, 1999). On the other hand, medicine as a discipline is poorly theorized, and this is compounded by its partition into numerous sub-disciplines that have remained relatively isolated. Only very recently have more holistic social and biological perspectives and theoretical models of any kind begun to emerge. Darwinian medicine only really developed from the late 1980s, stimulated by the pioneering work of Nesse and Williams (Nesse & Williams, 1994). These authors highlighted the untapped potential of evolutionary theory to shed new light on variability in human health and susceptibility to disease.

Initially, the main aims of Darwinian medicine were to improve understanding of variability in disease risk. For many, evolutionary theory remains synonymous with the theory of natural selection, whereby different genetic variants are favored according to the reproductive success of their carriers. Undoubtedly, this perspective helps understand genetic variability in the distribution of disease risk factors, and research has provided compelling evidence that individual alleles influence disease risk (Frazer et al., 2009), prompting major interest in the potential for gene-based "personalized medicine." However, humans have not been found to demonstrate substantial genetic variability in association with local ecological conditions. Rather, humans display subtle shifts in allele frequency across large geographical ranges (Hancock et al., 2010). Moreover, the Human Genome Project identified rather fewer human genes than had initially been expected (Willyard, 2018). Of those identified, it is largely rare variants that are most strongly correlated with disease, whereas more commonly occurring variants explain a very small component of disease risk (Cohen et al., 2004).

In the field of public health, the most significant progress to date relates to several issues concerning the prevention or treatment of infections. Arguably the key theme so far has been that imperfect vaccines or drugs can paradoxically select for more virulent pathogens, which may even evolve resistance to such treatments (Gandon et al., 2001; Arias and Murray, 2009).

Infections remain a severe source of disease amongst our species, but it is also true, at the beginning of the twenty-first century, that the primary sources of morbidity and mortality are rapidly shifting to non-communicable diseases, where the key risk factors are living conditions, lifestyles, and patterns of development. Most notably, economic development subjects entire populations to rapid changes in living conditions even though the burden of infections may initially remain harsh (Wells, 2016). Since public health refers to populations rather than individuals, this is exactly the context in which evolutionary theory might offer unique insights, and generate valuable predictions regarding how to reduce disease risk and maximize health.

This chapter aims to set out the rationale for applying evolutionary thinking in public health contexts, and to highlight how it may inform understanding of disease susceptibility. Beyond genetic variability, there is increasing recognition of the importance of other evolutionary processes. One such example is cultural evolution, focusing on the spread of behaviors amongst members of a population. However, the focus of this chapter is phenotypic plasticity, the way in which the body and behavior respond to ecological stimuli and stresses (Wells et al., 2017). It is through this plasticity that variability in disease risk may accrue, but it is also the obvious target of the preventive strategies that are the backbone of public health programs.

The branch of evolutionary theory that is most relevant to phenotypic plasticity within the life course and between generations is life history theory (Stearns, 1992). The chapter will describe life history theory and its application to human plasticity, and show how it can reveal the trade-offs through which disease susceptibility is shaped. It will then explore the inter-generational basis of variability in disease susceptibility, and show how that knowledge can further address the social-determinants of health. In this way, I aim to show that evolutionary theory can play a key role in building healthier societies, whose individual members may themselves have better physical and mental health.

2.1.1 Life history theory

Evolutionary life history theory assumes that organisms are under selective pressure to harvest resources from the environment throughout the life cycle, and divide them between competing functions so as to maximise fitness. The theory is based on the principle of allocation, namely that resources are finite, and if invested in one function are no longer available for other functions (Hill, 1993).

The most important resources are energy and time; hence the theory assumes that organisms making the best use of energy over the life-span will receive the highest fitness pay-offs (Ellison, 2008). At any time point, the optimal pattern of energy investment is determined by the organism's age and condition, and also by the quality of the physical and social environment. Each organism's life history can be summarized as a cumulative series of energy-allocation decisions, represented by a suite of developmental and reproductive traits. These include how fast and large to grow, how to address risks, and defend against threats, and how to schedule reproductive effort (Stearns, 1992).

Initially, life history theory was developed to explain phenotypic variability between species. The diverse selective pressures associated with any given ecological niche favor the emergence of broad species-specific energy allocation strategies. For example, a high risk of extrinsic mortality reduces the pay-off for allocating resources to growth, and instead increases the payoff for early reproduction. To maximize fitness in such environments, organisms should attain only small body size and divert resources to producing large numbers of offspring. At the species level, variability in life histories can be largely attributed to genetic evolution, mediated by such "trade-offs" between competing functions. This led to the terminology of "fast and slow" life histories, contrasting major differences in typical life-span that correlated with many more specific trade-offs (Promislow & Harvey, 1990).

However, life history theory may potentially also be applied to understand phenotypic variability among the members within a species. As with variability between species, some of the life history trade-offs among individuals of a given species may likewise be attributed to genetic factors. Nevertheless, the same theoretical approach can also be applied to phenotypic plasticity, on the assumption that each organism also responds to stimuli and stresses during the life course, differentially allocating resources so as to maximize fitness (Wells, 2018a, Wells, 2016). Such plastic traits are considered to be subject to reaction norms, a term that refers to the spectrum of phenotypes produced by a single genotype across a range of environmental conditions (Stearns, 1992). Life history theory predicts the evolution of reaction norms themselves, while ecological conditions in the broadest sense account for how these norms underpin life history trade-offs in specific environments.

Classically, three competing life history functions were identified: "maintenance," "growth," and "reproduction" (Hill, 1993). Maintenance refers to preserving the body in good condition through diverse homeostatic processes, and thus promotes longevity and future opportunities for reproduc-

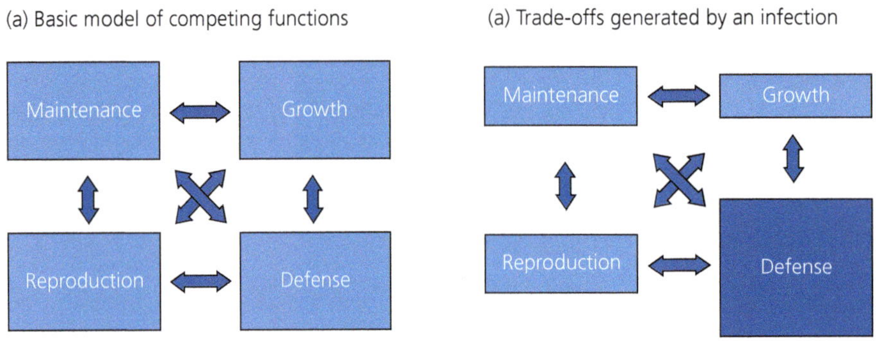

Figure 2.1 Conceptual diagrams showing (a) the basic model of competing life history functions, and (b) alterations in their respective budgets in response to the stress of an infection, reflecting a trade-off.

tion. Growth typically occurs prior to reproduction, especially in mammals, while reproduction diverts resources to the next generation. However, particularly in the context of human medicine, it is helpful to differentiate "defense" as a fourth independent function. During infections, many routine homeostatic processes are curtailed, in order to divert physiological resources to more urgent functions such as immune function. The same scenario is relevant to efforts to avoid predation, when the "fight-or-flight" response is activated. Defense is therefore antagonistic to long-term homeostatic maintenance, and a four-function model is more appropriate for public health application (Wells et al., 2017). Figure 2.1 summarizes the basic model, and provides an illustration of a trade-off.

Phenotypic variability emerges as the product of differential trade-offs between these competing functions, as demonstrated by a wide range of studies on both non-human species and humans. However, in order to obtain the most robust evidence for this principle, it is ideal to change experimentally the availability of energy. Even among wealthy humans with plentiful energy availability, subtle trade-offs are still observed, for example between family size and the growth rate of individual children (Lawson & Mace, 2008). Similarly, although obesity might suggest an excess of stored energy, it is better considered as a state of metabolic disruption, in which perturbed insulin dynamics constrain the optimal allocation of energy to life history traits to maximize fitness (Wells et al., 2017). Nevertheless, observational studies alone may be inadequate to reveal the full extent of trade-offs predicted by the theory, simply because some individuals might have enough energy to invest in all four life history functions simultaneously, while other individuals may have inadequate energy for any of the functions. Only if the supply of energy is experimentally altered within individuals can trade-offs in energy allocation be clearly discerned.

Work on diverse non-human species has employed elegant manipulations of extrinsic mortality risk and resource availability to reveal life history trade-offs. A particularly informative example comes from studies of small freshwater fish called guppies in the mountains of Trinidad. Typically, these streams have waterfalls that restrict predators to the lower reaches. Guppies living downstream, at high risk of predation, have developed very different life history strategies compared to those living upstream. Transplanting "downstream" guppies into the safer upstream environment resulted in slower development and a delayed onset of reproduction, whereas transplanting "upstream" guppies into the more risky downstream environment had the opposite effect (Reznick & Ghalambor, 2005). In this particular example, part of the variability can be attributed to genetic factors, indicating that different life history strategies can evolve in different ecological settings, but other studies have demonstrated similar effects mediated by plastic responses.

Amongst humans, experimental manipulations of environmental factors relevant to health and disease are more difficult to conduct. Most obviously,

Table 2.1 Life history trade-offs relative to public health.

Population	Public policy target	Trade-off	Exposure	Outcome	Reference
UK	Infant mortality	I/M	Mortality variability	Higher infant disease load predicts shorter lifespan	(Barker, 1992)
Europe	Infant mortality	I/G	Mortality decline	Taller surviving adults	(Crimmins & Finch, 2006)
UK	Childhood infections	I/G	Helicobacter pylori	Reduced growth	(Patel et al., 1994)
Guatemala	Childhood infections	I/G	Diarrhoeal diseases	Reduced growth	(Mata, 1992)
Sub-Saharan Africa	Perinatal health	I/R	Maternal malaria	Increased risk of low birth weight offspring	(Guyatt & Snow, 2004)
Senegal	Maternal mortality	R/M	Malaria	Reproduction elevates maternal malaria risk	(Diagne et al., 2000)
Global	Ageing	R/M	Number of offspring	Inverse association of fertility and longevity	(Thomas et al., 2000)
UK aristocracy	Ageing	R/M	Number of offspring	Inverse association with maternal lifespan	(Westendorp & Kirkwood, 1998)
Finland	Ageing	R/M	Bearing twins	Reduced maternal longevity	(Helle et al., 2002)
Sub-Saharan Africa	Family planning	R/G	Maternal fertility	Siblings reduce child growth	(Lawson et al., 2012)
UK	Family planning	R/G	Maternal fertility	Siblings reduce child growth	(Lawson & Mace, 2008)
UK	Teenage pregnancy	R/G	Low birth weight	Earlier reproduction	(Nettle et al., 2010)
Sweden	Infant nutrition	G/M	Post-natal growth	Hypertension if born small	(Leon et al., 1996)
UK	Infant Nutrition	G/M	Childhood growth	Increased cardio-metabolic risk	(Singhal et al., 2004)

M = maintenance (equivalent to survival or longevity); G = growth; R = reproduction; I = immune function

it is not ethical to randomize individuals to malnutrition or infection, or to surplus energy consumption as this would likely result in the development of obesity. This makes it more difficult to manipulate energy allocation in order to produce evidence of the role of trade-offs in differential disease susceptibility. The majority of evidence supporting trade-offs in humans is therefore observational, and hence weaker. Table 2.1 lists a variety of examples of trade-offs, all consistent with life history theory, but nevertheless in most cases not reliably demonstrating causation.

To resolve this issue, we have developed an alternative approach, taking advantage of the way in which many voluntary human sporting activities alter metabolic dynamics, and thus may be used to study potential trade-offs. Consider for example the notion that organs and tissues compete for energy (see Box 2.1), a scenario which has been proposed to contribute to differential susceptibility to type 2 diabetes in contemporary human populations (Hales & Barker, 1992). Evidence that expansion of the metabolically expensive brain may be accompanied by compensatory reductions of other organs and tissues over evolutionary time-scales, or that "brain-sparing" during fetal malnutrition may reduce growth of other vital organs, has been obtained from experiments in animals, but in humans' equivalent evidence is lacking.

To obtain experimental evidence in humans for an energy trade-off between the brain and other tissues, we conducted a study in which young adult men underwent assessments of physical power and cognitive function either separately or simultaneously. This work demonstrated that when subjected to competition for energy, both physical and cognitive function declined, but more so for physical function than cognitive function (Longman et al., 2017). This indicates, as hypothesized, that the brain and muscle tissue compete for energy, and that when this competition is experimentally intensified, brain function is protected at the

Box 2.1 Competition for energy between organs and tissues

Life history theory assumes that energy is allocated in competition between biological functions (Hill, 1993). Since individual organs contribute disproportionately to different functions, we can expect such competition to extend to the growth of tissues and organs during development, as well as their metabolic turnover at any given time point. For humans, this issue has attracted particular interest over the brain, in relation to two complementary questions. First, how did hominins fund the evolutionary expansion of the brain? Second, how can the brain be protected during short-term shortfalls in energy supply?

The brain has obligatory energy requirements, preferentially metabolizing glucose, but also able to process ketones as a "back-up" fuel. During hominin evolution, brain volume expanded from ~ 650 cm^3 in Australopithecines to ~1400 cm^3 in contemporary humans, representing a huge increase in cerebral energy demand. In a classic paper, Aiello and Wheeler proposed that this increase was met in part from a shift to richer diets, but also by an associated decline in the size and hence energy demand of the gut (Aiello & Wheeler, 1995). The same scenario may apply to muscle mass (Leonard et al., 2003).

A very similar logic was invoked in another classic paper by Hales and Barker. Their "thrifty phenotype" hypothesis proposed that fetuses exposed to energy insufficiency *in utero* spare the brain by reducing energy allocation to the growth of other vital organs, such as the pancreas and liver (Hales & Barker, 1992). In the short-term, this would promote survival in the face of energy scarcity, but in the long-term, such individuals would be less able to tolerate energy-dense diets and the metabolic effects of obesity, increasing their susceptibility to type 2 diabetes.

Both of these hypotheses are supported in non-human animals. For example, artificially selecting for larger brain size in fish resulted over time in smaller gut size, especially in males, in association with better performance on tests of cognitive ability (Kotrschal et al., 2013). Likewise, subjecting pregnant rodents to inadequate energy supply results in some organs of the offspring being more reduced in size than others, with the brain again protected (Fortman et al., 2005).

Differential plasticity in organs and tissues also extends to adult life. For example, during prolonged famine, the greatest weight loss occurs in adipose tissue, and the least in the brain, with other organs and tissues ranked hierarchically in between (Rivers, 1988). Finally, we can also observe (see main text) competition for fuel between the brain and muscle on an ongoing basis (Longman et al., 2017).

Competition between organs and tissues for energy can thus be discerned over evolutionary time, during development and in response to a specific energy stress. In each case, we can assume that the relevant energy allocation strategy has been honed by selection in the past to maximize the benefits and minimize the costs. It might be assumed that "immediate competition" might be the least important axis of variability, but it is arguably the most important: only if organs and tissues compete for energy in real time could longer-term trade-offs develop. These trade-offs all have relevance to public health, as summarized in Table 2.2.

Table 2.2 Organ-tissue trade-offs over contrasting timescales, and their public health relevance.

Timescale	Trade-off	Public health relevance	Reference
Hominin evolution	Brain versus gut and muscle mass	Ethnic differences in body composition and physique, associated with varying NCD risk on exposure to obesogenic environments	(Aiello & Wheeler, 1995; Wells, 2012b)
Development	Brain versus pancreas and liver	Within-population variability in NCD risk in association with exposure to variability in nutrition and growth during fetal life and infancy	(Hales & Barker, 1992 ; Li et al., 2015a, Li et al., 2015b)
Adult famine	Brain versus all other organs and tissues, in particular adipose tissue	While much attention is directed to excess adiposity (obesity), inadequate fat stores are associated with increased susceptibility to infection as well as higher mortality from some NCDs	(Rivers, 1988; Wells, 2010a)

expense of muscle. This helps understand both long-term morphological evolution, and developmental pathways to disease vulnerability.

We used a similar approach to investigate changes in the allocation of energy between life-history functions during prolonged energy stress. Endurance sport changes energy dynamics by maximizing the demand for physical activity, predicted in turn to reduce energy allocation to other life-history functions. We showed that participants in an endurance marathon change their energy allocation strategies during the course of the race, reducing investment in markers of maintenance and reproduction, and increasing investment in "survival," demonstrated by increased markers of immune defense (Longman et al., 2018). These adaptive changes were demonstrated by a decrease in salivary testosterone, and reduced "arousal" scores on viewing sexually provocative images, accompanied by significant rises in two markers of immune function, hemolytic complement activity and bacterial killing ability.

Life history theory thus offers a unified theoretical framework within which to consider long-term evolutionary change, developmental adjustments, and short-term plastic responses to ecological stimuli and stresses (Wells, 2017). It provides the core theoretical basis for evolutionary public health, enabling the generation of multiple hypotheses regarding both the causation of disease, and the potential benefits of treatment, which can then be tested empirically (Wells et al., 2017). A key benefit of this approach is that it allows biochemical traits such as hormone concentrations, already linked in numerous studies with disease risk, to be re-evaluated in terms of "life history allocation signals" (Wells et al., 2017). Finally, the theory can be used to elucidate novel targets for interventions, by aiming to change energy allocation strategies in ways that improve health. Increasingly, it is recognized that the complexity of the human life course, with its multiple developmental stages, may not fit well with a dichotomous model of "fast" and "slow" life histories. It may be more helpful simply to refer to trade-offs manifesting at any given life-course stage.

Life history trade-offs are readily tested in the context of infectious diseases, which either tax the organism of energy directly, or impose metabolic costs through eliciting expensive immune responses (Wells, 2010a). However, the unique contribution of life history theory for public health is particularly evident if we turn to non-communicable diseases (NCDs).

The discipline of medicine developed primarily in the context of tackling diverse infectious diseases. The success of germ theory in the 19th century raised expectations that most diseases could be traced to specific microbial organisms. Important public health questions would include: What was the organism? How was the disease transmitted? How could this be prevented? And how could the disease be treated?—This would usually require disposing of the pathogen. Most infections have relatively short incubation periods, making it easy to identify the pathway to disease and develop strategies to prevent or treat disease (Wells, 2016). NCDs pose much greater challenges, developing over entire life courses and indeed across generations. However, this life-course aetiology can be still be addressed through the lens of life history theory, in order to understand how cumulative trade-offs shape metabolic processes and the risk of NCDs.

2.2 Developmental effects

Over the last three decades, biomedical researchers have produced compelling empirical evidence that patterns of nutrition and growth during early life are associated with the risk of diverse components of health and disease susceptibility in later life. This led to the formal emergence of a new research field, targeting the "developmental origins of health and disease" (DOHaD) hypothesis (Barker, 2004). However, developing an adaptive framework for this hypothesis has proven problematic. In particular, why should developmental plasticity, generally considered adaptive from an evolutionary perspective, be associated with ill-health in later life?

Several evolutionary hypotheses have attracted attention in this context. Arguably the first of these was the thrifty phenotype hypothesis (Hales & Barker, 1992), described above. This approach explicitly contrasted with the thrifty genotype hypothesis, which attributed population differences in diabetes susceptibility to differential exposure to cycles of "feast and famine" in ancestral environments, resulting in some populations evolving "thrifty genes" that would raise susceptibility to diabetes in contemporary obesogenic environments (Neel, 1962). Hales and Barker focused on developmental trade-offs rather than past selective pressures, and many now refer to this as a "constraints" hypothesis and consider that developmental constraints undermine optimal phenotype in the long term (Monaghan, 2008).

An alternative hypothesis, developed by Gluckman and colleagues, is known as the predictive adaptive response hypothesis (PAR) (Gluckman & Hanson, 2004). According to this approach, the developing organism seeks "cues" of the quality of the environment during early life, and attempts to develop an adult phenotype in anticipation of reproducing in the same ecological conditions as those experienced during development. If this anticipation fails, the resulting phenotypic "mismatch" is assumed to promote cardio-metabolic disease.

So far, empirical evidence offers minimal support for the PAR hypothesis in humans. Those exposed to energy scarcity in early life are most vulnerable to energy scarcity in later life, suggesting that far from developing adaptive traits, such individuals make trade-offs in early life that predispose to additional trade-offs in subsequent harsh environments (Hayward et al., 2013). This suggests that developmental constraint models offer a more realistic adaptive framework. Nevertheless, the conventional approach in this context also has limitations. Most attention has been directed to severe constraints, equating for example to low birth weight (indicative of *in utero* malnutrition) or markers of psychosocial stress during postnatal life (Painter et al., 2005; Ellis & Garber, 2000). The latter are generally interpreted as a marker of high extrinsic mortality risk. Life history theory predicts that when extrinsic mortality risk is high, development should be accelerated so as to facilitate reproduction before mortality occurs.

Each of these approaches focuses on extreme ecological stresses, and ignores the main range of physiological variability. Moreover, minimal attention is paid to the mother as the source of both ecological information and resources for the developing offspring. I will come back to the second of these points alone, and from now focus on the normal range of variability.

Building on the thrifty phenotype hypothesis, I pay greater attention to the fact that the majority of the association between early growth patterns and later cardio-metabolic risk is evident across the entire spectrum; for example birth weight. Every increasing unit of birth weight is associated with reduced risk of adult disease. This suggests that the underlying mechanism is not overt malnutrition *in utero*, but rather simply variability in early growth patterns (Wells, 2011).

If we look in more detail at neonatal phenotype, we find that a number of physiological traits scale in broadly linear format with birth weight. These traits include lean body mass, nephron number in the kidney, airway size and blood vessel diameter, pancreatic beta cell mass and cardiac structure (Wells, 2011; Wells, 2016). I have collectively termed these traits metabolic capacity, as they tend to track through the life course and therefore remain of great importance for homoeostasis throughout the life course. However, they are strongly contingent on nutritional supply during early life, and hence constitute a generic investment in the life history function of "maintenance" that is strongly correlated with the level of fetal and infant growth. Developmental epigenetic effects can be considered within the same framework. If organ size and structure relates to physiological "hardware," then epigenetic variability may be considered as complementary physiological "software," matching metabolic function to relevant parameters of organ and tissue size and structure.

Weight at birth or in infancy is undoubtedly an imperfect proxy for metabolic capacity (Wells, 2016). For example, early growth faltering *in utero* may induce catch-up prior to birth, such that birth weight does not reliably signal the earlier nutritional insult (Hemachandra & Klebanoff, 2006). Similarly, other components of metabolic capacity may be associated with epigenetic effects that emerged around the time of conception (Fleming et al., 2018). Nevertheless, birth weight has proven a valuable predictor of the risk of non-communicable diseases in middle-/old-age (Li et al., 2015a; Li et al., 2015b), and given its association with diverse physiological traits as highlighted previously, it can be considered a valuable composite marker of meta-

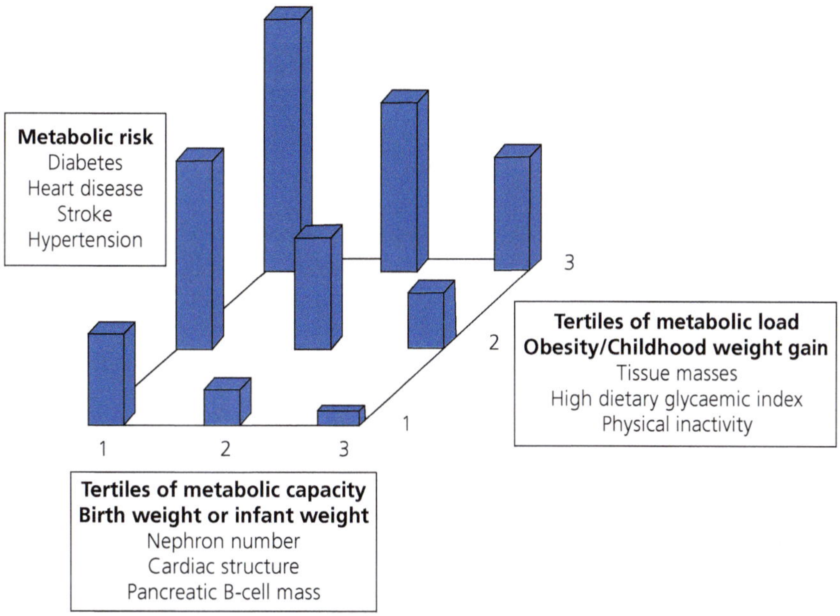

Figure 2.2 The capacity-load model. Generic model of metabolic risk as the product of interactions between metabolic load and metabolic capacity. Metabolic capacity scales directly with birth weight, and is attributable to specific physiological traits with similar scaling effect. Metabolic load scales directly with childhood or adult weight, and is also exacerbated by components of lifestyle and living conditions such as dietary glycemic load and sedentary behavior
Reproduced with permission from Wells (2011).

bolic capacity, and the same approach can be applied to weight in infancy.

What challenges metabolic capacity is a suite of traits that I have termed metabolic load (Wells, 2011). This includes adiposity, especially central truncal fat, unhealthy diet, sedentary behavior, psychosocial stress, toxins and pollutants, and infectious diseases. All of these traits challenge metabolic homeostasis. This simple model allows us to predict long-term NCD risk as a function of the interaction between metabolic capacity and metabolic load (Figure 2.2) (Wells, 2011). For any given level of capacity, NCD risk is predicted to rise with increasing load, while for any given load, NCD risk is predicted to rise with declining capacity. The greatest NCD risk is therefore expected to occur in those who start life with low capacity and go on to accumulate high load. This simple model is well supported by data from recent large cohort studies (Li et al., 2015a; Li et al., 2015b).

This simple model helps explain many characteristics of the ongoing global NCD epidemic. First, it helps explain why NCDs primarily appear in association with nutrition transition and the obesity epidemic. Populations in low and middle-income countries have had relatively low birth weights for centuries, but this reduction in metabolic capacity was not exposed to high load until relatively recently (Wells, 2016). Second, the model can also help explain ethnic differences in NCD risk, as ethnic groups vary in a variety of components of both capacity and load (Wells, 2012b; Wells, 2018b). Third, the model helps understand profound social inequality in NCD risk (Wells, 2016; Wells, 2018b). Poorer social groups tend to have lower birth weights, indicative of reduced metabolic capacity. However in wealthier countries, poor social groups now also demonstrate the highest levels of obesity, indicating higher metabolic load (Wells, 2016).

For cardio-metabolic outcomes, child growth represents a valuable marker of metabolic capacity (Wells, 2016). However, height is itself subject to metabolic trade-offs, and while tall height is associated with lower risk of type 2 diabetes, stroke, and

Box 2.2 Life history trade-offs and the developmental pathway to type 2 diabetes

At the level of physiology, the aetiology of type 2 diabetes can be explained using the conceptual capacity-load model (Wells, 2018b). The primary risk factor is excess circulating fuel and its metabolic correlates, which in the framework of public health relates to obesity (especially truncal adiposity), sedentary behavior and diets with high glycemic load. However, holding constant for these traits, which most commonly provoke diabetes from middle-age onwards, diabetes risk is also increased among those who had low weight at birth or during infancy (Hales and Barker 1992), reflecting their lower metabolic capacity (Wells, 2018b). Life history theory may help understand why individuals may accumulate both low capacity and high load through the life course.

In environments with high extrinsic mortality risk, such as the constant threat of fatal infectious disease, every offspring has a relatively high chance of random death. On this basis, even well-nourished mothers will optimize their fitness by producing greater numbers of smaller offspring, rather than smaller numbers of large robust offspring. Thus populations with higher fertility levels are expected to have higher diabetes risk for any given level of metabolic load in adult life, due to developmental constraints on metabolic capacity (Wells, 2017).

Moreover, in high-risk environments a relatively high proportion of individuals would not live long enough to benefit from investing in homeostasis to an extent that would minimize metabolic deterioration in old age. Instead, fitness would be maximized by investing in reproduction, at the cost of "maintenance," and only a small proportion who by random chance survived past middle-age would pay the long-term costs, for example, by developing diabetes at post-reproductive ages (Wells, 2017). Should energy availability increase at any time in the life course after weaning, it is too late to gain substantial benefits from allocating energy to maintenance, as the most important "critical periods" have already closed. Instead, fitness can be maximized by allocating energy to reproduction (growth and muscle mass in men, peripheral energy stores in women) or immune function (central abdominal fat in both sexes). These traits may promote reproductive fitness but at a cost to health, particularly central abdominal fat which generates chronic inflammation.

This helps understand the "thrifty phenotype" as a life-course developmental strategy, trading off short-term survival and reproduction against longevity (Wells, 2017). Once metabolic capacity is impaired, selection favors diverting subsequent resources to load. Better quality environments with reduced extrinsic mortality risk favor increased parental investment per offspring, and increased allocation of energy to maintenance by each individual offspring in early life. This reduces diabetes risk through the life course, as illustrated in Figure 2.3.

cardiovascular disease, it is also associated with an increased risk of many cancers (Emerging Risk Factors, 2012). Therefore growth and body size may contribute both to capacity and to load, depending on which aspect of homeostasis is being considered.

Whilst the capacity-load model has been used primarily to explore NCD risk, the same framework can be applied to infectious diseases. Here, capacity would represent components of immune function, and load the combination of (a) any direct adverse effects of pathogens on metabolism, and (b) the short-term consequences of immune activation that, while effective at combating pathogens, may themselves challenge homeostasis. Moreover, it is abundantly clear as highlighted in Table 2.1 that activating immune function in response to infections reduces the availability of energy for other life history functions.

More broadly, these associations between developmental experience, metabolic capacity and subsequent NCD risk can be reinterpreted in terms of life history trade-offs. For example, the finding that low birth weight is associated with greater risk of metabolic disease in later life can be re-interpreted in terms of constraint of investment in "maintenance" during early critical windows (Wells, 2018a). Equally, the key role of obesity and excess adiposity in triggering such diseases can be attributed to the broad fact that many components of metabolic load emerge though life history trade-offs that prioritize survival (inflammation, immune function) or in women, reproduction, over long-term maintenance (Box 2.2) (Wells et al., 2016b).

These insights may help provide a robust theoretical framework from which to propose public health interventions, see Figure 2.3. The primary

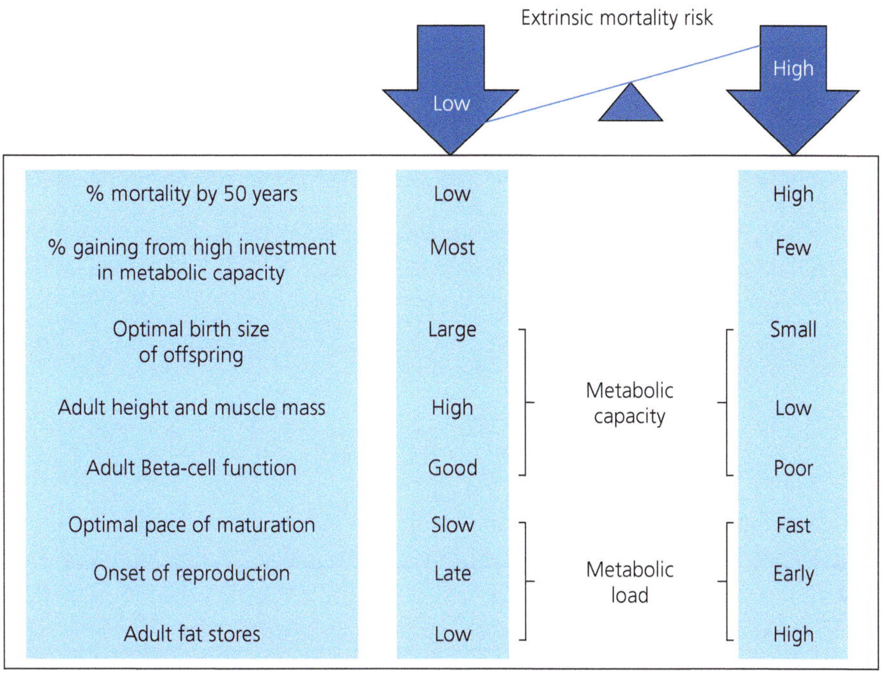

Figure 2.3 Schematic diagram of extrinsic mortality risk, life history strategy and metabolic phenotype
Reproduced with permission from Wells (2017).

public health response to recognition that poor growth in early life predicts NCDs has been to develop nutritional interventions, increasing the supply of either macronutrients or micronutrients. Birth weight or early growth markers have been used as key outcomes in order to assess the success of the intervention. So far, however, results have been disappointing, with neither macro-nutrient nor micro-nutrient supplementation generating long-term improvements in child growth (Kramer, 2000; Devakumar et al., 2016). In contrast, child vaccination programs have been associated with improved child growth outcomes (Anekwe & Kumar, 2012). In this case, reducing the allocation of energy to immune function may itself be a viable strategy for increasing the availability of energy for growth. More generally, the key message is that intervening on trade-offs may represent the best approach for public health nutrition efforts, rather than simply trying to elevate energy supply.

2.3 Intergenerational effects

Once we recognize the importance of growth variability in early life for long-term health outcomes, it becomes clear that we need to focus in more detail on the factors that account for such growth variability. Much attention has been directed to ecological stresses acting during post-natal life, such as psychosocial stresses which might act as a marker of extrinsic mortality risk (Ellis & Garber, 2000).

However, during fetal life, and to a lesser extent during lactation, all ecological stresses are transduced by maternal phenotype (Wells, 2010b). Developmental plasticity must therefore be considered part of a larger intergenerational association between maternal phenotype and that of the offspring. This issue merits its own evolutionary perspective, because energy allocation strategies to maximize maternal fitness are not necessarily those that maximize fitness of the offspring (Trivers, 1974).

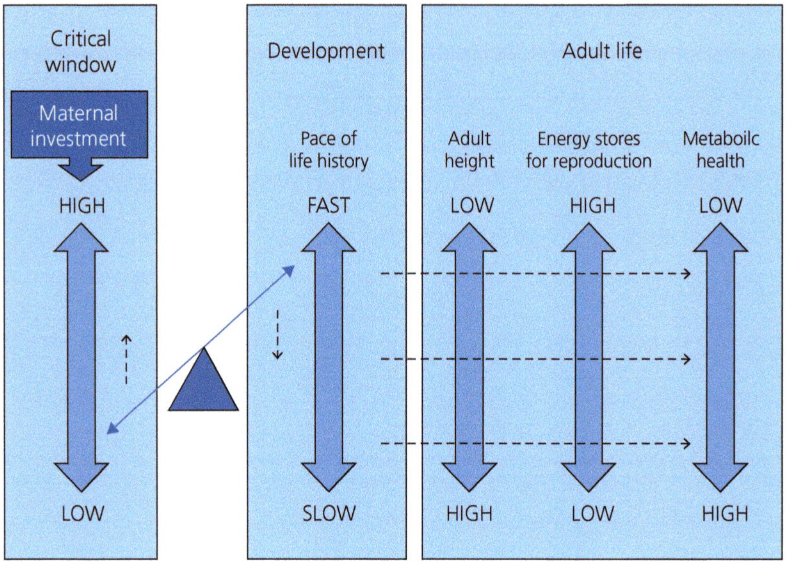
Critical window
Maternal investment
HIGH
LOW
Development
Pace of life history
FAST
SLOW
Adult life
Adult height
LOW
HIGH
Energy stores for reproduction
HIGH
LOW
Metaboilc health
LOW
HIGH

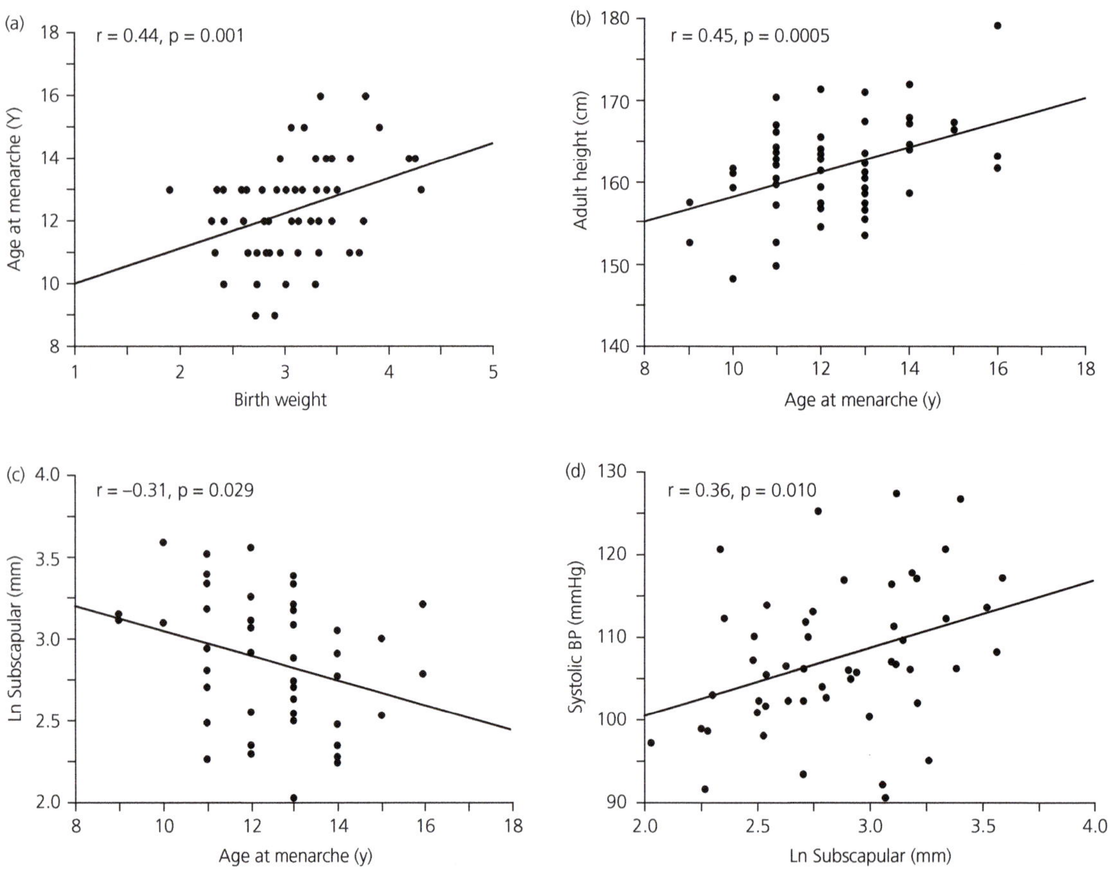
(a)
r = 0.44, p = 0.001
Age at menarche (Y)
Birth weight
(b)
r = 0.45, p = 0.0005
Adult height (cm)
Age at menarche (y)
(c)
r = −0.31, p = 0.029
Ln Subscapular (mm)
Age at menarche (y)
(d)
r = 0.36, p = 0.010
Systolic BP (mmHg)
Ln Subscapular (mm)

As highlighted in Box 2.2, a high extrinsic mortality risk favors diverting energy away from growth and maintenance, since the long-term rewards are reduced, towards survival and earlier reproduction. However, low maternal investment in association with chronic under-nutrition may provoke exactly the same response in the offspring, by constraining fetal energy allocation to "maintenance," and hence reducing the intrinsic "somatic quality" of the offspring (Wells, 2012d; Wells et al., 2016b). Once again, the best response for the offspring is to shunt energy towards reproduction, in order to maximize reproductive fitness before mortality occurs. This hypothesis was supported by a study of South Asian women living in the UK.

Those with low birth weight (indicating low maternal investment during early "critical periods") showed earlier puberty and shorter adult height. In turn, early puberty was associated with higher adiposity, which was associated with higher blood pressure (Figure 2.4) (Wells et al., 2016b). This indicates developmental trade-offs favoring early maturation and storing energy for reproduction, at a cost to growth and homeostasis.

This scenario represents an example of the generic interaction between life histories across two generations, whereby trade-offs manifesting in maternal phenotype steer trade-offs in the offspring's phenotype during its own early development. We can think of two energy-allocation games, one played by the mother, and the other by the offspring (Figure 2.5) (Wells, 2014). The maternal energy allocation game is played across multiple offspring. The more energy she invests in one offspring, the less energy is available for investment in the others. All other things being equal, selection favors a relatively equal allocation of energy between competing offspring, such that each individual offspring receives less maternal investment that would be optimal for its own fitness. This has been demonstrated for example for birth weight, where mean birth weight is somewhat lower than the birth weight at which neonatal mortality rate is minimized (Blurton Jones, 1978).

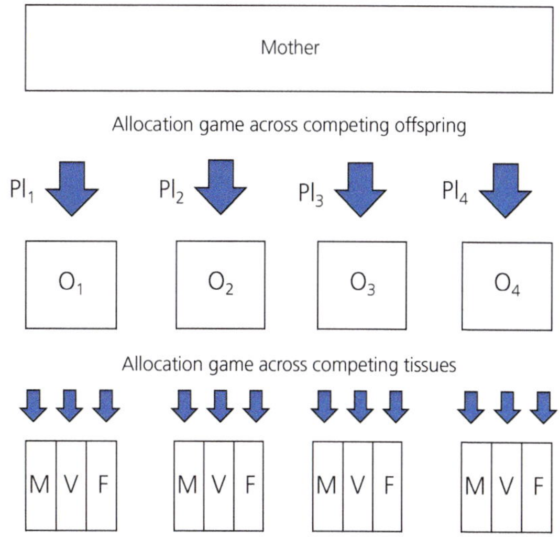

Figure 2.5 Schematic diagram illustrating two different "allocation games" in which each of the mother and the offspring optimize their inclusive fitness. In the first game, the mother optimizes her allocation of parental investment (PI) across competing offspring (O1 to O4). In the second game, which is sensitive to the first game, each offspring optimizes its allocation of that investment between competing phenotypic traits, such as muscle mass (M), vital organ mass (V) and energy stores in fat (F). These traits also represent allocations across life history functions (maintenance, growth, reproduction and immune function)
Reproduced with permission from Wells (2014).

The optimization game played by the offspring is different, for it must allocate its energy income between competing life history traits. If there is an overall decrease in energy supply, the optimal investment pattern may change. This brings the two games into interaction with one another (Wells, 2014).

Figure 2.4 (Opposite) Upper panel: Schematic diagram illustrating how the level of maternal investment in fetal life is associated with the pace of maturation, which then mediates a trade-off between the capacities for reproduction versus maintaining metabolic homeostasis in adult life. Lower panel: Empirical associations between maternal investment (proxied by birth size), maturation rate and adult phenotype in young South Asian women. (a) Birth weight is inversely associated with age at menarche. (b) Earlier menarche is associated with lower adult stature. (c) Earlier menarche is associated with higher adult subscapular skinfold. (d) Subscapular skinfold is positively associated with adult systolic blood pressure
Reproduced with permission from Wells et al. (2016).

Building on the "embodied capital" conceptual model of Kaplan and colleagues (Kaplan et al., 2003), I have emphasized maternal capital as the fundamental niche in which the offspring begins its own development (Wells, 2010b). Paradoxically, developmental plasticity during fetal life does not result in adaptation to external ecological conditions, but rather to the magnitude of maternal capital. This places emphasis on selective pressures acting in early life, rather than adulthood, relating to survival and reproductive potential. Through human evolution, high levels of infant mortality posed a major selective pressure, and to the extent that offspring develop adaptive trade-offs in early life, it is survival in infancy and early childhood that is the likely driving factor (Wells, 2012a).

A variety of components of maternal capital are relevant. These include somatic traits (height, fat-free mass, pelvic dimensions) that are related to the magnitude of fetal growth, nutritional resources (energy stores, micronutrients), and also multiple components of maternal homoeostasis (Wells, 2010b). For example, in contemporary populations, maternal obesity and diabetes challenge maternal homeostasis, and some of the resulting penalties are inflicted on the offspring. Beyond physiological traits, it is also important to consider socio-economic components of maternal capital, such as knowledge and education, social support, and material resources. All of these contribute to the nature of the niche experienced by the developing offspring (Wells, 2010b).

The benefits of applying evolutionary theory to public health interventions are nowhere better illustrated than in this context. Given the fundamental connection between low birth weight and maternal chronic under-nutrition, interventions targeting maternal nutritional status clearly offer an opportunity to improve the health of both mothers and their offspring. However, human metabolism has evolved to maximize fitness rather than health, and simply changing the availability of energy may not produce better health outcomes. This was demonstrated by analysis of an intervention that aimed to improve maternal nutritional status in Ethiopian villages. Traditionally, village women traveled lengthy distances on a daily basis to collect water, raising their energy expenditure. The installation of water supplies aimed to reduce this expenditure, thereby allowing greater energy allocation to fetal growth. However, the consequences were rather different: the intervention promoted maternal fertility and increased the birth rate in comparison with control villages, while the nutritional status of the individual children declined. Increasing the availability of energy had therefore promoted maternal fitness, at a cost to the health of individual offspring. The researchers who conducted this analysis concluded that the intervention should have incorporated a family planning component in order to counter the fertility effect (Gibson & Mace, 2006).

While this study illustrates the benefit of applying an evolutionary perspective to public health interventions, preferably in future at the planning stage, it also highlights the fact that human metabolism has been shaped to maximize fitness over individual health (Nesse & Williams, 1994). Developmental trade-offs shape the long-term predisposition for NCDs, in particular because later trade-offs occur in the context of those made earlier. For example, allocating energy to fat in adult life may be a broadly fitness-enhancing strategy, but it may be also associated with elevated health costs if it follows another trade-off, the diversion of energy from pancreatic growth to brain development (the thrifty phenotype) in fetal life (Wells, 2016) (Figure 2.6). Among the physiological mechanisms that maximize fitness through plastic responses, it is particularly useful to consider hormones as these are already widely investigated in biomedical research on disease aetiology.

2.4 Mechanisms of plasticity

Life history plasticity is achieved through multiple biological mechanisms, which cannot be considered in detail here. Relevant pathways include epigenetic effects, differential development of the gut biota, differential development of different organs and tissues, and diverse components of hormonal regulatory systems (Wells, 2016). The latter are worth discussing briefly, because hormones are already strongly implicated in differential risk of NCDs.

Hormones allow organisms to respond to a variety of endogenous and exogenous environmental

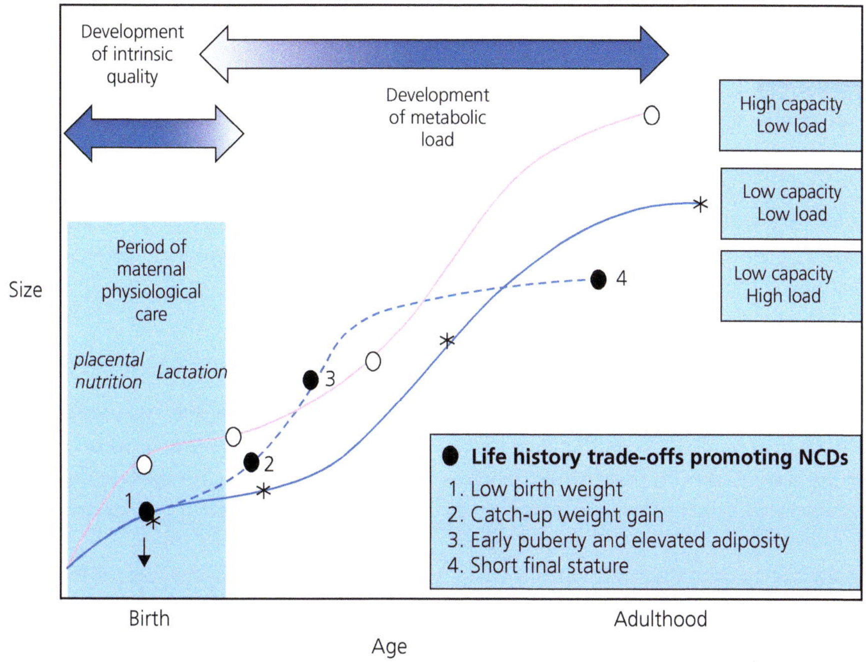

Figure 2.6 Conceptual diagram illustrating life history trajectories and their associated risk of Non-Communicable Diseases (NCDs). Each trajectory demonstrates different trade-offs, shaping both metabolic capacity and load (traits relevant to the life-course emergence of NCD risk) and the somatic traits that underlie partitioning of maternal investment between pregnancy and lactation

○: high maternal investment in pregnancy favours energy allocation to growth/maintenance, promoting longevity and health.

*: low maternal investment in pregnancy is followed by post-natal energy constraint, preventing catch-up. This leads to small adult size but low NCD risk.

●: low maternal investment in pregnancy is followed by catch-up during lactation, continuing into childhood. The extra energy accelerates maturation and increases adiposity without benefitting adult size, thus increasing NCD risk. Each of these trajectories produces in adult women the traits that favor the same trajectory in the next generation

Reproduced with permission from Wells (2018a).

factors by modifying diverse cell functions on a tissue- and organ-specific basis (Finch & Rose, 1995). Hormones generate multiple physiological effects, a scenario known as pleiotropy (Finch & Rose, 1995), and the functions emphasized by biomedical researchers are often only a narrow part of the full spectrum (Wells et al., 2017).

Conventionally, for example, clinicians think of insulin as responsible for regulating blood glucose, and variability in its production or activity is well understood to be central to the constellation of diseases grouped as diabetes. However, the total metabolic profile of insulin is far more complex, and it affects diverse functions in tissue-specific ways (Watve, 2013). Insulin modulates the regulation of peripheral metabolism, including appetite, reproductive function, thermoregulation, and adiposity, via receptors in the brain (Watve, 2013; Kleinridders et al., 2014). Within the brain, insulin also regulates cognitive functions such as learning and memory (McGregor et al., 2015). Although muscle insulin resistance increases the risk of diabetes, this can also be considered from an adaptive perspective. Diverting energy away from muscle may on the one hand reduce oxidative stress and protect mitochondrial DNA (Hoehn et al., 2009), while on the other hand it also allows the diversion of fuel to other tissues (Watve, 2013).

A similar approach can be applied to leptin, which is secreted by adipose tissue. This hormone signals the magnitude of energy stores to the brain, but it also has broader functions, contributing to the regulation of reproduction, cognitive function, and immune function (Schneider, 2004; Li et al., 2006). Indeed, many of the inflammatory markers that respond to leptin contribute not only to immune function but also to the regulation of energy balance, emphasizing the complex molecular machinery that orchestrates energy allocation strategies and their trade-offs between functions. Since adipose tissue represents a "savings bank" of stored energy, it can be considered to reduce the obligation to allocate all energy on a moment-by-moment basis between the four main life history functions (Wells, 2012c). Leptin signals the "account balance" to the brain, and it is no surprise that it and insulin are two of the hormones fundamentally implicated both in life history trade-offs and in the aetiology of NCDs (Wells, 2016).

Another influential hormone is cortisol, produced by the adrenal glands in response to diverse types of stress including illness, trauma, fear, pain, and psychosocial stress. Although it is tempting to consider cortisol as the archetypal "stress hormone," this results in a circular argument: stresses raise cortisol, and a stress is then defined as anything that raises cortisol. It is more appropriate to consider cortisol as another energy allocation hormone, playing a similar role to insulin but stimulated by a different class of ecological signals. Like insulin, cortisol affects diverse metabolic activities, for example suppressing immune function while increasing blood pressure and blood glucose (Desborough, 2000).

In each case, therefore, these hormones implement the energy–allocation trade-offs between life history functions. Although such plasticity has adaptive components, especially in the short term, there are fundamental trade-offs between survival, fitness, and health. As environments become increasingly harsh, the trade-offs that promote survival and reproduction become increasingly detrimental to health (Wells et al., 2017). From this perspective, evolutionary theory provides another key insight for public health policies: reducing environmental harshness may play a key role in improving health (Wells et al., 2017). This insight was well recognized in the pioneering Victorian era, but has since been replaced by a new focus on "individual responsibility".

More generally, improved understanding of the twin roles of hormonal mechanisms in life history trade-offs and in the pathway to disease will help develop interventions that change phenotype in ways that favor long-term health over immediate reproductive fitness.

2.5 Opportunities for intervention

If evolutionary public health is to offer benefits, it must guide future interventions. The major insight generated from this perspective is that rather than physiological traits, it is the *trade-offs* between biological functions that are the ideal target for public health dimensions. A number of the trade-offs highlighted in Table 2.1 can be targeted in this way, one example being that efforts to reduce infections may benefit child growth (Martins et al., 2017; Anekwe & Kumar, 2012).

Another relevant trade-off is between linear growth and adiposity. In many populations, short stature is associated with an elevated risk of overweight and obesity (Martins et al., 2004), mediated by differences in fat oxidation rate (Hoffman et al., 2000), and shorter adults also have elevated NCD risk (Wells et al., 2016a; Paajanen et al., 2010). Simply increasing energy supply may not necessarily resolve poor growth, as it may simply promote adiposity. Two recent studies suggest how this adverse outcome can be avoided. First, high-energy diets that are low in protein and micronutrients may signal harsh ecological conditions, and promote trade-offs favoring survival and reproduction over-growth and maintenance (Penuelas et al., 2017). Consistent with that, the magnitude of secular trend in, height correlates across populations with the protein quality of the diet. Second, combining nutritional therapy for acute malnutrition with efforts to cut the burden of infection likewise not only promotes growth in under-nourished children, but also reduces a range of NCD risk markers (Martins et al., 2017). Collectively, these interventions increase the potential pay-offs from investing in maintenance and growth, by making the long-term future look more "attainable". The body responds by promoting linear growth rather than adiposity, whilst restoring metabolism to lower levels of NCD risk.

A third example involved providing relaxation therapy to first-time mothers during lactation. Maternal anxiety may consume energy that would otherwise be available for the offspring. Consistent with that, our intervention to reduce maternal stress altered hormonal signals and significantly increased weight gain in the infant (Shukri et al., 2016).

All of these examples illustrate the point made previously, that interventions may be most effective if they change hormonal signals that simultaneously underpin life history trade-offs and shape the predisposition for health or disease.

2.6 Conclusion

This chapter has aimed to show that an evolutionary perspective provides new understanding of the aetiology of human disease, in particular NCDs which are now the primary source of morbidity and premature mortality worldwide. At the core of the approach is the use of life history theory to highlight trade-offs between competing life history functions. Interventions could potentially target those trade-offs, increasing the allocation of energy to "maintenance" which is closely associated with better health. Such interventions might be very different to those addressing acute illness, which would prioritise immediate survival over long-term disease risk.

Life history theory remains a relatively crude conceptual model, treating energy as a single common currency for energy allocation. Such a broad integrative model can be extremely valuable due to its very wide applicability. However, in order to apply it successfully to promote human health, we need to improve our understanding of how energy allocation strategies are mediated by a huge host of individual signaling molecules. This will require stepping back form the conventional biomedical approach, and considering the wider pleiotropic functions of diverse hormones.

Linking life history theory with the developmental origins of adult health and disease hypothesis sheds new light on *why* experience in early life is so important for long-term health. Not only does poor nutrition in early life constrain metabolic capacity, but it may also underpin subsequent trade-offs that promote survival and reproduction over maintenance and growth. Whilst these may benefit genetic fitness, helping understand why they have been favoured by selection, they may reduce health. An intergenerational perspective indicates that the trade-offs occurring within a generation are relevant to parental as well as offspring fitness. These insights may help develop interventions that promote health of both mothers and their offspring, increasing energy allocation to maintenance in both generations.

An evolutionary approach may be particularly valuable for addressing variability within and between populations in disease risk. Trade-offs may vary substantially between ethnic groups, in association with adaptation to contrasting ancestral environments, or simply due to neutral evolution. Trade-offs may also vary substantially between groups occupying contrasting living conditions, and recognition of this scenario may stimulate new public health policies targeting the social determinants of health. For example, social inequality is associated on the one hand with differential acquisition of metabolic capacity and load through the life course, contributing to contrasting NCD risk, but it is also associated with differential life history trade-offs. Interventions targeting those trade-offs may therefore be most successful in reducing social inequality in health outcomes.

References

Aiello, L. C. & Wheeler, P. (1995). The expensive-tissue hypothesis—the brain and the digestive-system in human and primate evolution. *Curr Anthropol* 36: 199–221.

Anekwe, T. D. & Kumar, S. (2012). The effect of a vaccination program on child anthropometry: Evidence from India's Universal Immunization Program. *J Public Health (Oxf)*, 34: 489–97.

Arias, C. A. & Murray, B. E. (2009). Antibiotic-resistant bugs in the 21st century—a clinical super-challenge. *N Engl J Med*, 360: 439–43.

Barker, D. J. (1992). *Fetal and Infant Origins of Adult Disease*. London: British Medical Journal.

Barker, D. J. (2004). The developmental origins of adult disease. *J Am Coll Nutr*, 23: 588S–595S.

Blurton Jones, N. (1978). Natural selection and birth-weight. *Ann Hum Biol*, 5: 487–9.

Cohen, J. C., Kiss, R. S., Pertsemlidis, A., Marcel, Y. L., McPherson, R., & Hobbs, H. H. (2004). Multiple rare alleles contribute to low plasma levels of HDL cholesterol. *Science*, 305: 869–72.

Crimmins, E. M. & Finch, C. E. (2006). Infection, inflammation, height, and longevity. *Proc Natl Acad Sci U S A*, 103: 498–503.

Darwin, C. (1859). *On the Origins of Species by Means Of Natural Selection*. London: Murray.

Desborough, J. P. (2000). The stress response to trauma and surgery. *Br J Anaesth*, 85: 109–17.

Devakumar, D., Fall, C. H., Sachdev, H. S., Margetts, B. M., Osmond, C., Wells, J. C., Costello, A., & Osrin, D. (2016). Maternal antenatal multiple micronutrient supplementation for long-term health benefits in children: A systematic review and meta-analysis. *BMC Med*, 14: 90.

Diagne, N., Rogier, C., Sokhna, C. S., Tall, A., Fontenille, D., Roussilhon, C., Spiegel, A., & Trape, J. F. (2000). Increased susceptibility to malaria during the early postpartum period. *N Engl J Med*, 343: 598–603.

Dwork, D. (1987). *War is Good for Babies and Other Young Children: A History of the Infant and Child Welfare Movement in England 1898–1918*. London: Tavistock Publications.

Ellis, B. J. & Garber, J. (2000). Psychosocial antecedents of variation in girls' pubertal timing: Maternal depression, stepfather presence, and marital and family stress. *Child Dev*, 71: 485–501.

Ellison, P. T. (2008). Energetics, reproductive ecology, and human evolution. *PaleoAnthropol*, 2008: 172–200.

Emerging Risk Factors, C. (2012). Adult height and the risk of cause-specific death and vascular morbidity in 1 million people: Individual participant meta-analysis. *Int J Epidemiol*, 41: 1419–33.

Finch, C. E. & Rose, M. R. (1995). Hormones and the physiological architecture of life history evolution. *Q Rev Biol*, 70: 1–52.

Fleming, T. P., Watkins, A. J., Velazquez, M. A., Mathers, J. C., Prentice, A. M., Stephenson, J., Barker, M., Saffery, R., Yajnik, C. S., Eckert, J. J., Hanson, M. A., Forrester, T., Gluckman, P. D., & Godfrey, K. M. (2018). Origins of lifetime health around the time of conception: Causes and consequences. *Lancet*, 391: 1842–52.

Fortman, J. K., 2nd, Reichling, T., & German, R. Z., (2005). The impact of maternal protein malnutrition on preweaning skeletal and visceral organ growth in neonatal offspring of Rattus norvegicus. *Growth Dev Aging*, 69: 39–52.

Frazer, K. A., Murray, S. S., Schork, N. J., & Topol, E. J. (2009). Human genetic variation and its contribution to complex traits. *Nat Rev Genet*, 10: 241–51.

Gandon, S., Mackinnon, M. J., Nee, S, & Read, A. F. (2001). Imperfect vaccines and the evolution of pathogen virulence. *Nature*, 414: 751–6.

Gibson, M. A. & Mace, R. (2006). An energy-saving development initiative increases birth rate and childhood malnutrition in rural Ethiopia. *PLoS Med*, 3: e87.

Gluckman, P. & Hanson, M. (2004). *The Fetal Matrix: Evolution, Development and Disease*. Cambridge: Cambridge University Press.

Guyatt, H. L. & Snow, R. W. (2004). Impact of malaria during pregnancy on low birth weight in sub-Saharan Africa. *Clin Microbiol Rev*, 17: 760–9, table of contents.

Hales, C. N. & Barker, D. J. (1992). Type 2 (non-insulin-dependent) diabetes mellitus: The thrifty phenotype hypothesis. *Diabetologia*, 35: 595–601.

Hancock, A. M., Witonsky, D. B., Ehler, E., Alkorta-Aranburu, G., Beall, C., Gebremedhin, A., Sukernik, R., Utermann, G., Pritchard, J., Coop, G., & Di Rienzo, A. (2010). Human adaptations to diet, subsistence, and ecoregion are due to subtle shifts in allele frequency. *Proc Natl Acad Sci U S A*, 107 Suppl 2: 8924–30.

Hayward, A. D., Rickard, I. J,. & Lummaa, V. (2013). Influence of early-life nutrition on mortality and reproductive success during a subsequent famine in a preindustrial population. *Proc Natl Acad Sci U S A*, 110: 13886–91.

Helle, S., Lummaa, V., & Jokela, J. (2002). Sons reduced maternal longevity in preindustrial humans. *Science*, 296: 1085.

Hemachandra, A. H. & Klebanoff, M. A. (2006). Use of serial ultrasound to identify periods of fetal growth restriction in relation to neonatal anthropometry. *Am J Hum Biol*, 18: 791–7

Hill, K. (1993). Life history theory and evolutionary anthropology *Evol Anthropol*, 2: 78–89.

Hoehn, K. L., Salmon, A. B., Hohnen-Behrens, C., Turner, N., Hoy, A. J., Maghzal, G. J., Stocker, R., Van Remmen, H., Kraegen, E. W., Cooney, G. J., Richardson, A. R., & James, D. E. (2009). Insulin resistance is a cellular antioxidant defense mechanism. *Proc Natl Acad Sci U S A*, 106(42): 17787–92.

Hoffman, D. J., Sawaya, A. L., Verreschi, I., Tucker, K. L., & Roberts, S. B. (2000). Why are nutritionally stunted children at increased risk of obesity? Studies of metabolic rate and fat oxidation in shantytown children from Sao Paulo, Brazil. *Am J Clin Nutr*, 72(3): 702–7.

Jørgensen, C., Enberg, K., Dunlop, E. S., Arlinghaus, R., Boukal, D. S., Brander, K., Ernande, B., Gardmark, A.,

Johnston, F., Matsumura, S., Pardoe, H., Raab, K., Silva, A., Vainikka, A., Dieckmann, U., Heino, M., & Rijnsdorp, A. D. (2007). Ecology: Managing evolving fish stocks. *Science*, 318: 1247–8.

Kaplan, H., Lancaster, J., & Robson, A. (2003). Embodied capital and the evolutionary economics of the human life span. In: Carey, J. R., & Tuljapurkar, S. (eds.) *Life Span: Evolutionary, Ecological, and Demographic Perspectives*. New York: Population Council.

Kleinridders, A., Ferris, H. A., Cai, W., & Kahn, C. R. (2014). Insulin action in brain regulates systemic metabolism and brain function. *Diabetes*, 63(7): 2232–43.

Kotrschal, A., Rogell, B., Bundsen, A., Svensson, B., Zajitschek, S., Brannstrom, I., Immler, S., Maklakov, A. A., & Kolm, N. (2013). Artificial selection on relative brain size in the guppy reveals costs and benefits of evolving a larger brain. *Curr Biol*, 23: 168–71.

Kramer, M. S. (2000). Balanced protein/energy supplementation in pregnancy. *Cochrane Database Syst Rev*, 2(2): CD000032.

Kuper, A. (1999). *Culture: The Anthropologists' Account*. Cambridge, MA, Harvard University Press.

Lawson, D. W., Alvergne, A., & Gibson, M. A. (2012). The life-history trade-off between fertility and child survival. *Proc Biol Sci*, 279(1748).

Lawson, D. W. & Mace, R. (2008). Sibling configuration and childhood growth in contemporary British families. *Int J Epidemiol*, 37(6): 1408–21.

Leon, D. A., Koupilova, I., Lithell, H. O., Berglund, L., Mohsen, R., Vagero, D., Lithell, U. B., & Mckeigue, P. M., (1996). Failure to realise growth potential in utero and adult obesity in relation to blood pressure in 50 year old Swedish men. *BMJ*, 312(7153): 401–06.

Leonard, W. R., Robertson, M. L., Snodgrass, J. J., & Kuzawa, C. W. (2003). Metabolic correlates of hominid brain evolution. *Comp Biochem. Physiol A Mol Integr Physiol*, 136: 5–15.

Li, J., Li, F., & Zhao, A. (2006). Inflammation and leptin. *Drug Discovery Today: Dis Mech*, 3: 387–93.

Li, Y., Ley, S. H., Tobias, D. K., Chiuve, S. E., Vanderweele, T. J., Rich-Edwards, J. W., Curhan, G. C., Willett, W. C., Manson, J. E., Hu, F. B., & Qi, L. (2015a). Birth weight and later life adherence to unhealthy lifestyles in predicting type 2 diabetes: Prospective cohort study. *BMJ*, 351: h3672.

Li, Y., Ley, S. H., Vanderweele, T. J., Curhan, G. C., Rich-Edwards, J. W., Willett, W. C., Forman, J. P., Hu, F. B., & Qi, L. (2015b). Joint association between birth weight at term and later life adherence to a healthy lifestyle with risk of hypertension: A prospective cohort study. *BMC Med*, 13: 175.

Longman, D., Stock, J. T., & Wells, J. C. K. (2017). A trade-off between cognitive and physical performance, with relative preservation of brain function. *Sci Rep*, 7: 13709.

Longman, D. P., Prall, S. P., Shattuck, E. C., Stephen, I. D., Stock, J. T., Wells, J. C. K., & Muehlenbein, M. P. (2018). Short-term resource allocation during extensive athletic competition. *Am J Hum Biol*. 2018 Jan;30(1). doi: 10.1002/ajhb.23052.

Martins, P. A., Hoffman, D. J., Fernandes, M. T., Nascimento, C. R., Roberts, S. B., Sesso, R., & Sawaya, A. L. (2004). Stunted children gain less lean body mass and more fat mass than their non-stunted counterparts: A prospective study. *Br J Nutr*, 92: 819–25.

Martins, V. J., De Albuquerque, M. P., & Sawaya, A. L. (2017). Endocrine changes in undernutrition, metabolic programming, and nutritional recovery. In Preedy, V. R. & Patel, V. B. (eds.) *Handbook of Famine, Starvation, and Nutritional Deprivation*, Switzerland: Springer International Publishing AG.

Mata, L. (1992). Diarrheal disease as a cause of malnutrition. *Am J Trop Med Hyg*, 47: 16–27.

Mcgregor, G., Malekizadeh, Y., & Harvey, J. (2015). Minireview: Food for thought: Regulation of synaptic function by metabolic hormones. *Mol Endocrinol*, 29: 3–13.

Monaghan, P. (2008). Early growth conditions, phenotypic development and environmental change. *Philos Trans R Soc Lond B Biol Sci*, 363: 1635–45.

Neel, V. (1962). Diabetes mellitus: A "thrifty" genotype rendered detrimental by "progress"? *Am J Hum Genet*, 14: 353–62.

Nesse, R. M. & Williams, G. C. (1994). *Why we Get Sick: The New Science of Darwinian Medicine*. New York: Times Books.

Nettle, D., Coall, D. A., & Dickins, T. E. (2010). Birthweight and paternal involvement predict early reproduction in British women: Evidence from the National Child Development Study. *Am J Hum Biol*, 22(2): 172–9.

Paajanen, T. A., Oksala, N. K., Kuukasjarvi, P., & Karhunen, P. J. (2010). Short stature is associated with coronary heart disease: A systematic review of the literature and a meta-analysis. *Eur Heart J*, 31: 1802–9.

Painter, R. C., Roseboom, T. J., & Bleker, O. P. (2005). Prenatal exposure to the Dutch famine and disease in later life: An overview. *Reprod Toxicol*, 20(3): 345–52.

Patel, P., Mendall, M. A., Khulusi, S., Northfield, T. C., & Strachan, D. P. (1994). Helicobacter pylori infection in childhood: Risk factors and effect on growth. *BMJ*, 309: 1119–23.

Penuelas, J., Janssens, I. A., Ciais, P., Obersteiner, M., Krisztin, T., Piao, S., & Sardans, J. (2017). Increasing gap in human height between rich and poor countries associated to their different intakes of N and P. *Sci Rep*, 7(1): 17671.

Promislow, D. E. & Harvey, P. H. (1990). Living fast and dying young: A comparative analysis of life-history variation among mammals. *J Zool*, 220: 417–37.

Reznick, D. N. & Ghalambor, C. K. (2005). Selection in nature: Experimental manipulations of natural populations. *Integr Comp Biol*, 45: 456–62.

Rivers, J. P. (1988). The nutritional biology of famine. In Harrison, G. A. (ed.). *Famine*. Oxford: Oxford University Press.

Schneider, J. E. (2004). Energy balance and reproduction. *Physiol Behav*, 81: 289–317.

Mohd Shukri, N. H., Wells J., Eaton S., Mukhtar F., Petelin A., Jenko-Pražnikar Z., & Fewtrell, M. Randomized controlled trial investigating the effects of a breastfeeding relaxation intervention on maternal psychological state, breast milk outcomes, and infant behavior and growth. Am J Clin Nutr. 2019 Jun 4. pii: nqz033. doi: 10.1093/ajcn/nqz033.

Singhal, A., Cole, T. J., Fewtrell, M., Deanfield, J., & Lucas, A. (2004). Is slower early growth beneficial for long-term cardiovascular health? *Circulation*, 109(9): 1108–13.

Stearns, S. C. (1992). *The Evolution of Life Histories*. Oxford: Oxford University Press.

Thomas, F., Teriokhin, A., Renaud, F., De Meeüs, T., & Guégan, J. F. (2000). Human longevity at the cost of reproductive success: Evidence from global data. *J Evol Biol*, 13: 409–14.

Toubas, P. L. (2007). Dr. Pierre Budin: Promoter of breastfeeding in 19th century France. *Breastfeeding Med*, 21: 45–9.

Trivers, R. L. (1974). Parent-offspring conflict. *Am Zool*, 14: 249–64.

Watve, M. (2013). *Doves, Diplomats and Diabetes: A Darwinian Interpretation of Type 2 Diabetes and Related Disorders*. New York: Springer.

Wells, J. C. (2010a). *The Evolutionary Biology of Human Body Fat: Thrift and Control*. Cambridge: Cambridge University Press.

Wells, J. C. (2010b). Maternal capital and the metabolic ghetto: An evolutionary perspective on the transgenerational basis of health inequalities. *Am J Hum Biol*, 22(1): 1–17.

Wells, J. C. (2011). The thrifty phenotype: An adaptation in growth or metabolism? *Am J Hum Biol*, 23: 65–75.

Wells, J. C. (2012a). A critical appraisal of the predictive adaptive response hypothesis. *Int J Epidemiol*, 41: 229–35.

Wells, J. C. (2012b). Ethnic variability in adiposity, thrifty phenotypes and cardiometabolic risk: Addressing the full range of ethnicity, including those of mixed ethnicity. *Obes Rev*, 13 Suppl 2: 14–29.

Wells, J. C. (2012c). The evolution of human adiposity and obesity: Where did it all go wrong? *Dis Model Mech*, 5: 595–607.

Wells, J. C. (2012d). Obesity as malnutrition: The role of capitalism in the obesity global epidemic. *Am J Hum Biol*, 24(3): 261–76.

Wells, J. C. (2014). Adaptive variability in the duration of critical windows of plasticity: Implications for the programming of obesity. *Evol Med Public Health*, 83(197): 109–21.

Wells, J. C. (2016). *The Metabolic Ghetto: An Evolutionary Perspective on Nutrition, Power Relations and Chronic Disease*. Cambridge: Cambridge University Press.

Wells, J. C. (2018a). Life history strategy and the partitioning of maternal investment: Implications for health of mothers and offspring. *Evol Med Public Health*, 2018 Aug 16;2018(1):153–166.

Wells, J. C., Pomeroy, E., Walimbe, S. R., Popkin, B. M., & Yajnik, C. S. (2016a). The elevated susceptibility to diabetes in India: An evolutionary perspective. *Front Public Health*, 4: 145.

Wells, J. C., Yao, P., Williams, J. E., & Gayner, R. (2016b). Maternal investment, life-history strategy of the offspring and adult chronic disease risk in South Asian women in the UK. *Evol Med Public Health*,1: 133–45.

Wells, J. C. K. (2017). Body composition and susceptibility to type 2 diabetes: An evolutionary perspective. *Eur J Clin Nutr*, 71: 881–9.

Wells, J. C. K. (2018b). The capacity-load model of non-communicable disease risk: Understanding the effects of child malnutrition, ethnicity and the social determinants of health. *Eur J Clin Nutr*, 72: 688–97.

Wells, J. C. K., Nesse, R. M., Sear, R., Johnstone, R. A., & Stearns, S. C. (2017). Evolutionary public health: Introducing the concept. *Lancet*, 390: 500–09.

Westendorp, R. G. & Kirkwood, T. B. (1998). Human longevity at the cost of reproductive success. *Nature*, 396: 743–6.

Whittles, L. K. & Didelot, X. (2016). Epidemiological analysis of the Eyam plague outbreak of 1665–1666. *Proc Biol Sci*, 283(1830): 20160618.

Willyard, C. (2018). New human gene tally reignites debate. *Nature*, 558: 354–5.

CHAPTER 3

Evolutionary Medicine and Women's Reproductive Health

Wenda R. Trevathan and Karen R. Rosenberg

3.1 Introduction/Overview

Reproduction is the currency of evolution. A person with a single child who lives to reproduce is evolutionarily more successful than one who has a dozen children none of whom reproduce. What really counts in evolution is getting as many of your genes into subsequent generations as possible, so having a child who shares 50% of your genes is a good way to do that. Of course, this is an oversimplification because one's genes survive in all of one's relatives, to differing degrees, depending on the relationship. Successfully passing genes on to the next generation is what "counts," emphasizing the critical role of reproduction (rather than survival) in the evolutionary process.

A number of things can go wrong with the human reproductive process, some of which may require clinical intervention to resolve. However many conditions that modern medicine considers "problems" can be considered as healthy responses worthy of attention and explanation from an evolutionary medicine perspective. This chapter will focus on the evolution of human female reproductive processes and the ways in which some of those processes may be at odds with some of the environmental contexts within which reproduction occurs today. We note that natural selection may favor physiological processes that seem maladaptive at first glance, but may actually mean that the body is acting in ways that enhance long-term reproductive success. This understanding may provide the basis for making better medical decisions about reproductive health in contemporary settings.

From an anatomical perspective, human female reproductive organs include the vagina, cervix, uterus, ovaries, fallopian tubes, and birth canal; peripheral to these organs, but equally important for reproduction, are the mammary glands. Hormones such as progesterone and estrogen coordinate reproductive processes of ovulation, menstruation, conception, implantation, gestation, labor and delivery, and lactation. Reproduction is a complex process that may benefit from medical assistance in some cases, but a glance at world population data tells us that humans have clearly been successfully reproducing for hundreds of thousands of years with minimal medical assistance and they continue to do so today. In contrast, there are well-known examples of failures in the reproductive process that may be caused or exacerbated by medical intervention. Some complications of human childbirth are due to mismatches between our evolved biology of reproduction and the medical and social systems that deal with birth. In our view, these mismatches are responsible for a number of contemporary problems that may be

Trevathan, W. R. and Rosenberg, K. R., *Evolutionary Medicine and Women's Reproductive Health* In: *Integrating Evolutionary Biology into Medical Education—for maternal and child healthcare students, clinicians, and scientists.* Edited by: Jay Schulkin and Michael L. Power, Oxford University Press (2020).
DOI: 10.1093/oso/9780198814153.003.0004

partially resolved if lessons from evolutionary medicine are incorporated into contemporary obstetrics and midwifery.

In this chapter we will consider the healthy human female reproductive system, shaped by millions of years of evolution and developed over individual lifetimes in various environmental contexts to enable conceiving, bearing, and raising offspring. To begin the discussion, we must understand the characteristics of the "normal," healthy reproductive function. Although this understanding is based on substantial clinical research, most of the research has been conducted in modern industrialized settings based on women whose lives are very different from the lives of their ancestors in fundamental ways (Trevathan, 2010). Today's medical understanding of "normal" is typically based on averages reported for well-nourished women from resource-rich populations. By contrast, the perspective of evolutionary medicine (Williams & Nesse, 1991) argues that for any trait or behavior, there is a broad range within which development occurs and within which reproduction occurs. What may be seen as "abnormal" in one setting may actually represent healthful responses to local environments. Recognition of this variation by clinicians and other health practitioners may improve health care by guiding medical intervention based on variation and context rather than on averages.

3.2 Overview of life history

An evolutionary perspective on reproductive health invokes life history theory. At the most basic level, living, growing, and reproducing require energy. Energy, obtained from foods consumed, is limited and its availability varies throughout the year and throughout the life course. Because it is limited, natural selection has shaped the allocation of energy throughout life to optimize its use and to maximize individual reproductive success. The reproductive life course from menarche to menopause can be viewed as a series of trade-offs between various aspects of life history, such as age of onset of reproductive function, length of gestation, number of offspring, length of breastfeeding, and length of reproductive life span.

For example, energy is required both for growth and for carrying a pregnancy to term, so unless calories are abundantly available, it is usually to her fitness advantage for a young woman to delay having children until she has completed growing. Likewise, when she is nursing a young infant, a woman may not have reserve calories for another pregnancy, and ovulation is often inhibited in the first few months of lactation. On the other hand, if she lives in an environment of abundant calories, she may have sufficient reserves to both lactate and gestate. This partially explains why breastfeeding does not work to inhibit ovulation in well-nourished (even overly nourished) women (Valeggia & Ellison, 2003). Although breastfeeding an infant for several years may enhance infant survival (Dieterich et al., 2013) it can also delay the next pregnancy for the mother. In an environment in which early weaning compromises infant survival, it may be advantageous to continue to breastfeed to enhance survival of a child rather than wean and become pregnant again. If she weans earlier and becomes pregnant sooner, both her weanling and her newborn infant may die. The mother who breastfeeds for longer (e.g., 3 years) has "traded" higher numbers of less healthy young for fewer, but healthier, infants and the payoff may be greater lifetime reproductive success.

For humans, who give birth to costly offspring (Trevathan & Rosenberg, 2016; see Stuebe & Tully, Chapter 4, this volume), quantity may not be as important as quality. By "costly," we mean that because they are born in an immature and highly dependent state, with expensive brains still growing at fetal rates of growth, human infants require inordinate investments of time and energy to reach maturity. Humans grow very slowly, requiring longer to reach maturity than any other primate. Kaplan (1994) estimates that an investment of 13 million calories is needed to raise a child to maturity (approximately age 18), almost all of which must come from someone other than the child himself. (In the United States in 2017, the financial cost was almost a quarter of a million dollars to raise a middle-income child to age 18.) Fortunately, humans have cultural behavior that enables them to have several dependent offspring at a time: cooperative childcare or the practice of sharing of resources and child care (Hrdy, 1999, 2009). As anthropologist Bentley notes, humans have the ability to "maximize reproduction through cultural interventions," which affects life history theory in ways not known for other animals (Bentley, 1999). Among those cultural interventions are medical and scientific discoveries that alter reproductive lives in ways that often modify how natural selection acts. For more on life history theory, see Stuebe & Tully, Chapter 4, this volume; Wells, Chapter 2, this volume.

3.3 Ovarian function

3.3.1 Hormonal regulation of reproductive function

Human reproduction in women is regulated via the hypothalamus, pituitary, and gonadal glands. With regard to reproduction, the primary endocrine glands of significance include the ovaries in women and the testes in men. Gonadotropins, luteinizing hormone (LH) and follicle stimulating hormone (FSH), released by the anterior pituitary stimulate production of the hormones, estrogen, progesterone, and testosterone, which, in turn, influence ovulation and sperm production.

In general, hormones can be classified as steroid (e.g., estrogen, testosterone) or peptide (e.g., prolactin, oxytocin) depending on their molecular structure. Peptide hormones are proteins, whereas steroid hormones are lipids, synthesized from cholesterol. One distinction between the two classes of hormones that is particularly important for field (non-clinical) studies of reproduction is that peptides are detectable only in blood whereas steroids can be measured in other bodily fluids such as saliva and urine, in addition to blood. Field studies are our most important source of information about the range of healthy reproductive functions. Because they can be assessed from urine and saliva, steroid hormones are usually easy to collect where refrigeration and other equipment are unavailable (Ellison, 1988). Additionally, people are usually more willing to donate saliva or urine than they are to donate blood. Fortunately, because peptide hormones play important roles in reproduction, reproductive biologists and anthropologists have developed techniques to measure them in blood spots that can be collected with only a finger prick (Worthman & Stallings, 1994, 1997).

Estrogens are often referred to as "female hormones" and testosterone as "male hormones" but both males and females produce both hormones, although in very different quantities. Estrogens exist in three forms: estriol, estrone, and estradiol. When we discuss these hormones we will use the collective term except when we are referring to studies that focus on one of the specific types. Estrogens are produced in the ovaries, the corpus luteum, the placenta, and in other organs such as liver, adrenals, breast, and adipose tissue. Another hormone that is usually thought of as a "female hormone" is progesterone (literally, "in favor of gestation"). This is another steroid hormone that is produced by the ovaries, corpus luteum, and the placenta and is important in the ovarian/menstrual cycle, in implantation and in maintaining a pregnancy. It is also an immunosuppressant. Testosterone is another steroid hormone, produced by the testes (the Leydig cells) and the adrenals; in women it is produced in the ovaries and the adrenals.

Oxytocin is an important peptide hormone that plays a role in almost every aspect of reproduction: sexual response, labor and delivery, mother-infant bonding, and breastfeeding. Oxytocin appears to be involved in orgasmic response and may facilitate sperm transport (Kunz et al., 2007). Although it may not be the trigger, oxytocin is responsible for stimulating and maintaining uterine contractions and for clamping down the uterus after birth, thus preventing postpartum hemorrhage (PPH). It is also responsible for the let-down reflex that enables milk to flow into the mammary ducts when the infant suckles. Synthetic forms are often used to speed labor progression and to avert PPH. Excessive reliance on synthetic oxytocin has been linked to neonatal jaundice and may interfere with mother-infant bonding, breastfeeding, and infant behavioral adaptation (Bell et al., 2014; Schiefenhövel & Trevathan, 2019).

3.4 Development of the reproductive system in females

Development of the female reproductive system occurs throughout the life course, accelerating in the mid-teens with the onset of puberty marked by enlargement of the breasts, growth of axial and pubic hair, and onset of menstrual cycles, albeit ones that may be highly irregular and often not ovulatory. This phase of sexual maturity is known as menarche and is often marked around the world by ritual and change in social status. Historically and in many contemporary foraging and horticultural populations, menarche occurred in the late teens, approximately age 17. A few years of irregular cycling ensued following menarche so that the first pregnancy typically occurred in the early 20s when girls had completed most of their skeletal growth and were socially and economically mature enough to marry and begin a family. In the past century and a half, however, the age of onset of menarche in developed countries has steadily declined, a phenomenon known as a "secular trend" (Tanner, 1987) (Figure 3.1). What this means for many families today is that girls are reaching sexual maturity as young as 11 or 12 years and, if they are sexually active, they may become pregnant as young as 14 or 15 years. Thus, they are becoming pregnant and having children when they themselves have not finished growing and they are rarely socially adults or economically independent. This is often regarded as a societal problem in many industrialized populations where young (frequently unmarried) mothers are seen as an economic burden.

A number of factors have contributed to the declining age of menarche in the past 100 years; most are related to changes in diet and activity levels associated with modernization. Considering the early onset of reproductive function through an evolutionary lens reminds us that natural selection has favored high reproductive output when resources are sufficient. In times of dietary scarcity, ovarian function is dampened in such a way that likelihood of conception is reduced. In these situations, becoming pregnant, giving birth, and nursing an infant would take a toll on a woman's body in ways that might interfere with future reproduction and would be a waste of valuable resources. Furthermore, infant health may also be compromised, potentially having lifelong effects. Under these

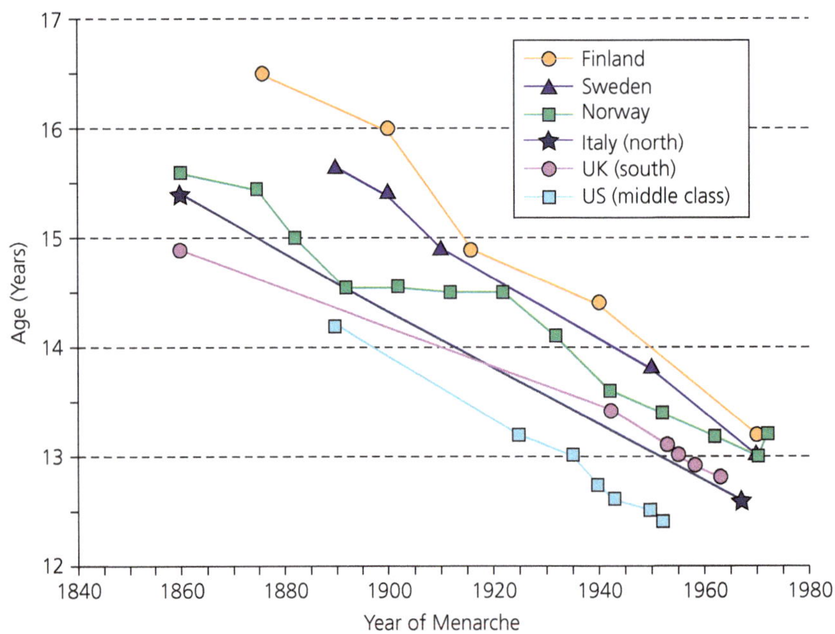

Figure 3.1 Secular trend of a declining age at menarche in European Countries and the United States Trevathan, (2010).

conditions, delaying the onset of reproduction (menarche) until more favorable times could potentially lead to greater reproductive success for the woman and her offspring.

What signals a time favorable for reproduction in adolescent females? As a girl's body is growing, most calories that she consumes are given over to growth and physical activity. When she has completed growth, calories that are not needed for physical activity can then be invested in reproduction. This convergence of growth completion and adequate calorie consumption leads to onset of reproductive function at a time that can be considered "good for" reproduction. Many contemporary diets are, however, not merely *adequate* for growth and physical activity for young girls, they are *excessive* in calories, especially those that can be readily stored as fat. Although identifying a minimum or "critical" fat level for reproduction has been elusive (Frisch & Revelle, 1970; Frisch & McArthur, 1974), it is a safe generalization that successful reproduction cannot usually occur unless the mother consumes calories beyond what are needed for maintenance of her life. Today, these extra calories are readily available in most modern diets and young girls begin storing surplus energy as fat, signaling to the body that it is a "good time" to reproduce. This can happen as young as 12 years in some populations where caloric intake is high and exercise levels are low.

3.5 Understanding ovulation and menstruation

The typical ovulatory cycle in human females lasts approximately 28 days on average, but following a theme in this chapter, what characterizes a "normal" menstrual cycle is highly variable. The first part of the cycle is the follicular phase (also known as the proliferative phase) and is the period in which the oocyte is preparing for ovulation under the influence of estrogen/estradiol (see Figure 3.2). It takes 9 to 12 days for this preparation, followed by a surge in the hormones luteinizing hormone (LH) and follicle stimulating hormone (FSH), which stimulates ovulation. Following ovulation, levels of the hormone progesterone rise, facilitating the development of the endometrium for implantation of the egg if fertilization occurs. This process takes about 12 days and is called the luteal phase. Progesterone (and some estrogen) production during this phase is primarily by the follicle that formerly housed the egg, known as the corpus luteum. If conception does not occur, the corpus luteum dies/regresses and progesterone production ceases. At this point, the endometrium disintegrates, and is shed with the unfertilized egg.

To tell the story of the ovarian cycle from the point of view of the egg, we would have to go all the way back to a woman's 20th week of gestation, when she, herself, was a fetus. At this time her

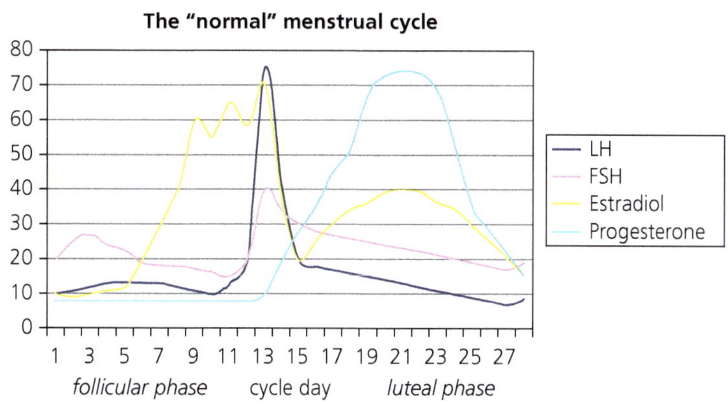

Figure 3.2 Fluctuations in hormones over the course of the "normal" menstrual cycle

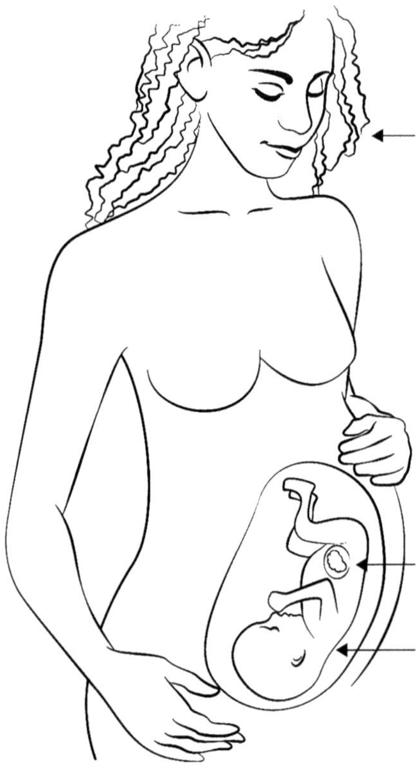

Figure 3.3 When a woman is pregnant with a female fetus, the oocytes of her grandchild are already present

ovary held about 4 to 6 million "pre-eggs" known as oogonia, each containing 46 chromosomes (that is, 23 pairs). A few weeks later (still during fetal development), these oogonia undergo meiosis and become primary oocytes with 23 single chromosomes each. They are arrested at this stage and develop no further until puberty. The number of oocytes at this stage is at the lifetime maximum and they will decline until their number reaches near zero at menopause. In other words, there will never be additional oocytes following gestation so that any stress experienced in utero or at other times in the life course will have an impact on the oocytes that can never be corrected by replacement (but see White et al., 2012). (This is in contrast to sperm, which are continually produced anew from puberty until death in men, although production seems to decrease with increasing age). Furthermore, a woman's health during pregnancy has an impact not only on her developing fetus, but also, if it is a girl, on her grandchildren whose potential existence lies in the oocytes developing in her female fetus (see Figure 3.3). This process is perhaps the ultimate and extreme form of the developmental origins of health and disease (see Gluckman et al., Chapter 1, this volume) in which a woman's pregnancy affects the risk for disease of her grandchildren.

For most women living in industrialized countries, the most obvious aspect of their biology during the reproductive years is menstruation. If a woman is using contraception or not engaging in behavior that can lead to conception, she typically has about 13 menstrual cycles a year, potentially 450 between puberty and menopause. Thus, the expected manifestation of her biology for many women today is menstrual cycling. For women in many foraging and horticultural societies today and most likely for all women a hundred or more years ago, the expected biological state is more often non-cycling because so much of the reprod-

Table 3.1 Reproductive Milestones in Foraging Populations compared to U.S. population.

	Foraging Populations	Americans
Age at menarche	16.1	12.5
Age at first birth	19.5	24.0
Years between menarche and first birth	3.4	11.5
Years of lactation per birth	2.9	0.25
Completed family size	5.9	1.8
Total years of lactation (approx.)	17	0.5
Age at menopause	47	50.5
Estimated total lifetime ovulations	160	450

Eaton & Eaton (1999).

uctive years are spent in pregnancy and nursing children, both behaviors that interrupt menstrual cycling. In other words, the "normal" and expected manifestation of female biology is not menstruation; rather it is pregnancy and breastfeeding. In fact, it has been estimated that women who don't use contraception and who nurse their children for 2 to 3 years have as few as 40 menstrual cycles over their lifetimes, as has been reported for women in the New Guinea Highlands (Vitzthum et al., 2001), see Table 3.1.

In addition to lifetime frequency and variation in cycle length, there are other aspects of the menstrual cycle that invite inquiry about what is "normal." Quantity of blood shed during menstruation appears to vary both by population and among women in a number of ways. It appears, for example, that the amount of bleeding that is reported in healthy women from resource-rich populations, and the amount deemed "normal," is seen as excessive in women who live in environments of low resource availability and who perform high levels of physical work during their reproductive years. Furthermore, the number of days of menstrual flow may also be longer in women from industrialized societies than for women in resource-poor societies. Reported median days of flow in the former is about 6 days, whereas the number for women in the latter may be about 3.5 days, based on a sample of Andean women studied by Vitzthum and her colleagues (Vitzthum et al., 2001). So what is "normal"? Given that women with a wide range of menstrual cycle characteristics conceive, maintain pregnancies, and have children, it may be that the definition of normal is broader than textbooks based on well-nourished, inactive populations might indicate. On the other hand, populations with women who report more days of menstruation and greater volume of flows have higher rates of reproductive cancers, suggesting that shorter flows and lower volume may be healthier in the long-term, a suggestion to which we will return later. Furthermore, for women living at high altitudes with low oxygen levels (e.g., the Andean women cited previously), minimizing monthly blood loss may be adaptive (Power, personal communication, 2018).

One example of how an understanding of menstrual cycle characteristics may affect health more directly is in the development of oral contraceptives (Lovett et al., 2017). Most research and work on oral contraceptives has been conducted in resource-rich populations where women have frequent cycles with 5–6 days of menstruation and relatively high volume of blood flow. When a cycle of contraceptives is developed to mimic that "normal" cycle, it may not match the expectations of women from resource-poor populations who begin a course of oral contraceptives and are alarmed when they have what they perceive to be longer cycles, more days of menstruation, and heavier flows than they are familiar with. They may perceive this as abnormal and cease taking oral contraceptives, leaving them with more limited and less reliable ways of controlling reproduction (Vitzthum et al., 2000). This is an example of the challenges faced when industrialized medicine is exported to populations in which the original medical research and development did not occur. Similarly, many American women were unhappy taking the pill that eliminates or reduces the frequencies of periods because they said that it was "unnatural" (Sievert, 2008) although less frequent menstrual periods than we see in developed countries are actually closer to the typical human norm.

When one examines epidemiological data on reproductive cancers (endometrial, ovarian, and breast) striking correlations are seen with highly frequent menstrual cycling and high levels of

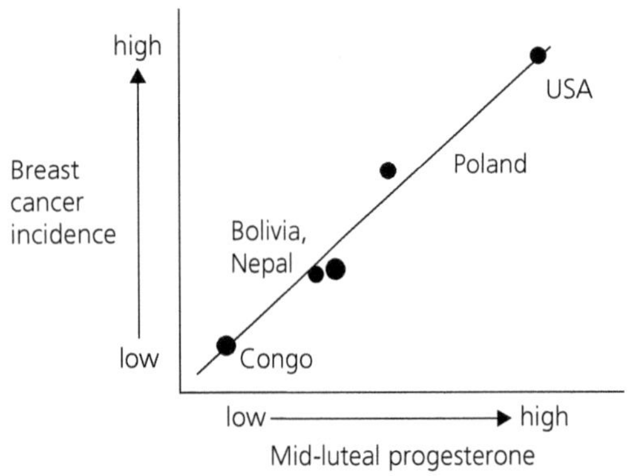

Figure 3.4 Breast cancer rates and mid-luteal progesterone
Trevathan, (2010 from Jasieńska & Thune, 2001).

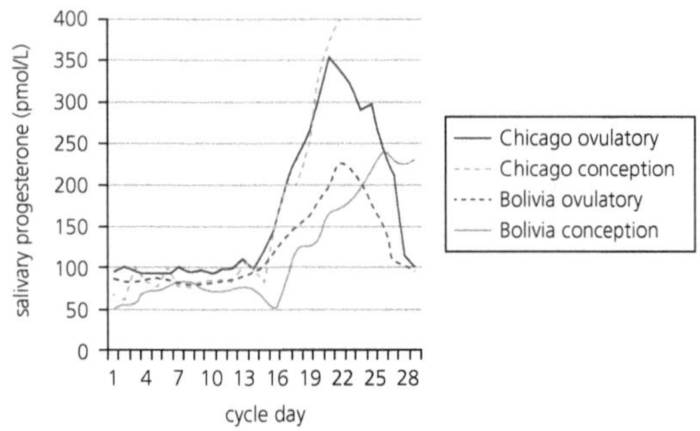

Figure 3.5 Progesterone levels in conception and non-conception cycles in two populations
Trevathan, (2010, from Vitzthum et al., 2004).

reproductive hormones (Aktipis et al., 2015; Hidaka & Boddy, 2016). Although correlations are not necessarily indicative of causes, investigation into these associations is warranted as we explore ways of reducing the incidence of reproductive cancers worldwide. As an example, Figure 3.4 shows breast cancer rates and levels of mid-luteal progesterone in five samples: Congolese, Nepalese, Bolivians, Poles, and Americans. In those countries in which progesterone levels are high, so are breast cancer rates (Jasieńska & Thune, 2001).

3.6 Levels of reproductive hormones

Many of the factors of concern related to menstrual cycling may trace their underlying causes to the levels of hormones produced by women in the course of their reproductive lives. As an example, Figure 3.5 illustrates the levels of progesterone observed in a single cycle of women from four populations: Chicago ovulatory cycle with no conception, Chicago cycle with conception, Bolivia ovulatory cycle with no conception, and Bolivia cycle with conception (Vitzthum, et al., 2004). It is clear that the absolute levels of progesterone produced in these four samples are bimodally distributed, with far higher levels in both Chicago samples than in the two Bolivian samples. In fact, the levels observed in the Bolivian sample would be perceived as abnormally low and not likely to lead to conception if they were reported in American or European women. It may be true that the levels would cause concern in clinical samples, but the Bolivian women seem to have gotten pregnant as easily as the Chicago women. This highlights the importance of context and illustrates the folly of trying to find optimal levels of hormones that apply to all women. The take-home message for clinicians is that what may be of concern as too low in one woman may not be too low in another, especially if they come from populations with very different dietary and physical lifeways.

As noted previously with regard to the number of menstrual cycles in a lifetime, the high amounts of reproductive hormones to which women can be exposed in their lives may account for other health concerns related to reproduction: Premenstrual syndrome (PMS), postpartum depression, and physical and emotional challenges of the menopausal years. These are almost routine and expected (although generally unpleasant) in some populations and rare to absent in others. What accounts for this variation?

Premenstrual syndrome or PMS has rarely been reported in non-Western populations, leading some scholars to refer to it as a "disease/disorder of civilization" and a "culture-bound syndrome" (Johnson, 1987). Although PMS is not easy to define and experiences with it vary widely between individuals, reported symptoms include weight gain, water retention, breast pain, dizziness, headaches, sleep problems, mood swings, irritability, anxiety, and mild depression (Martin, 1988). These symptoms are usually reported to occur 2 to 3 days before menstruation and they are often relieved when menstruation begins, sometimes to be replaced by other problems such as abdominal cramping and lower back pain. Although PMS may be unpleasant, it only rarely reaches the level of medical concern, at which point it becomes not merely a nuisance, but perhaps a debilitating disorder that compromises quality of life (Taylor, 2006). In these cases medical intervention may be warranted.

As noted previously, PMS is not usually cited as a routine and expected phenomenon in horticultural and hunter-gatherer societies (although one possible explanation may be that researchers have not asked about it). Perhaps a reason for this is that when women spend most of their lives in non-cycling states (pregnancy and lactation), PMS is as rare as menstruation and is, therefore, rarely remarked upon. It is also possible that the high levels of reproductive hormones reported in industrialized populations lead to greater expression of symptoms during the premenstrual phase. The "fall" from the high levels at ovulation is more precipitous, perhaps resulting in more symptoms, especially those related to moods (for another view, see Reiber, 2008). As we will see, this may also help explain why the postpartum and menopausal periods are often associated with depressed mood in industrialized populations, whereas they are less often reported in foraging and horticultural populations.

Another hypothesis to explain PMS from an evolutionary medicine perspective considers the immunosuppression that characterizes the late luteal phase when implantation of a fertilized egg typically occurs. A fertilized egg is genetically different from the mother (in 50% of its genes) and, as such, expresses antigens that can trigger a maternal immune response leading to rejection of the zygote before implantation. Progesterone in the luteal phase serves to suppress her immune function enabling implantation, but it can also increase her susceptibility to infectious agents, some of which may be related to symptoms perceived as PMS (e.g., fatigue, headaches, gastrointestinal upset). Caroline Doyle and her colleagues have suggested that PMS could better be treated with anti-virals rather than targeting symptoms (Doyle et al., 2008).

3.7 Fecundity, fertility, and infertility

The terms fecundity and fertility are often used interchangeably, although they refer to different phenomena (Leslie, 2018). Fecundity means the ability to have offspring whereas fertility refers to the actual bearing of offspring. Women between puberty and menopause are usually fecund, but only if they have children would they be referred to as fertile. The distinction is important because the factors that affect fecundity and fertility, while similar, differ in important ways. Clearly fecundity and fertility are linked and factors causing what is referred to as "subfecundity" also have effects on fertility. For example, intense physical activity often inhibits ovulation and athletes such as marathon runners and ballerinas often cease ovulating while they are training. This state of subfecundity is usually reversed when levels of physical activity decline. Extreme food restriction as may occur in severe dieting or fasting and economic-induced starvation can also interfere with ovulatory cycling. In these cases, individual cycles, when they occur, may be non-ovulatory and therefore, a woman would not get pregnant (i.e., not be fertile) even if she engages in intercourse.

Another example is seen in what is known as "luteal phase defect" (LPD) or "luteal phase insufficiency" (LPI) in which ovulation occurs, but the luteal phase is not long enough for implantation to take place, even when conception occurs. Extreme physical activity and food deprivation have been associated with luteal phase defect. In these cases, intervention to improve likelihood of conception is usually fairly simple. Luteal phase defect may also refer to levels of progesterone that are too low to enable implantation. Variability in baseline levels must be considered here, however, because what is too low for one woman may be sufficient for another as observed in the women from Bolivia and Chicago previously discussed. In other words, treatment regimens, which may consist of hormone injections, must take into consideration baseline levels in populations where "normal" levels of all ovarian hormones are lower than seen in women from resource-rich populations. Hormone levels alone are not good indicators of LPD/LPI and should be confirmed by additional investigation.

Many causes of infertility or subfecundity can be considered pathological, requiring medical intervention, but in some cases, their causes may be seen through the lens of evolutionary medicine as resulting from conditions that are not suitable for pregnancy. For example, a woman who engages in extreme dieting or training for a marathon often has so little body fat that supporting a healthy pregnancy would be difficult, if not impossible. There is a physiological trade-off between her increased ability to run long distance and her ability to conceive. If she is trying to become pregnant, a solution to her failure to conceive would be to modify her dieting or training regimens to decrease energetic stress on her body. In other words, she needs to divert calories away from keeping herself alive or athletic toward storage of sufficient body fat for ovulation and gestation. This would be an evolutionary medicine "solution" to a problem of infertility that does not require medical intervention such as hormone injections to stimulate ovulation.

On the other hand, there are thousands of examples of women living in impoverished circumstances who conceive, gestate, birth, and nurse infants under conditions that would likely inhibit ovulation in women from resource-rich populations. In these cases, the bodies of impoverished women seem to have adapted to chronically low caloric intake so that the HPG reproductive axis has readjusted enabling reproduction to proceed. For women in resource-rich populations who go on crash diets or begin a strenuous exercise routine, it may take a few months for their bodies to adjust to the new demands and they may have trouble getting pregnant.

With a world population exceeding seven billion, it seems that getting pregnant must be relatively easy. This is true for many women, but, sadly, it is difficult and sometimes impossible for others. Some of the reasons for failure of conception or implantation fall under the purview of clinical medicine and will not be discussed in much detail here. Examples include anatomical disorders of the ovaries, uterus, fallopian tubes, and vagina; hormonal disorders of the endocrine glands or of receptors; presence of diseases such as sexually transmitted diseases (STDs) and malaria; endometriosis; premature ovarian failure; and of course, male infertility. Causes of infertility that have potential solutions via public health activities include smoking, obesity, some environmental and occupational hazards, and to some extent, STDs.

Societal norms and expectations likely play a role in patterns of reduced fertility in some populations. The optimal ages for conception and pregnancy, likely shaped by natural selection, are the early twenties (Liu et al., 2011). In many populations, however, having children is postponed until schooling is finished, careers have become established, or economic stability has been achieved (among other reasons), milestones that are typically reached in the early- to mid-30s, when fecundity has usually decreased. Thus, in many contemporary populations, there is a mismatch between the biological and sociocultural optimal times for childbearing. For example, in the United States, first births to women over 30 increased from 5% to 24% of the population between 1975 and 2006 (Macaluso et al., 2008) as women moved into the workforce and professional careers.

A frequent cause of female infertility is polycystic ovarian disease/syndrome (PCOD or PCOS), a disorder that is increasing with rising rates of obesity and metabolic syndrome. The word "polycystic" refers to the fact that women with PCOS/D produce an excess of smaller-than-normal follicles ("cysts") in their ovaries that interfere with hormone balance and disrupt ovulation (Gleicher & Barad, 2006). PCOS/D can sometimes be treated with weight reduction, suggesting that some of the causes may be environmentally induced. Women who migrate from environments of low resource availability and high energy expenditure to environments of plenty are at greater risk for obesity and insulin disorders and, thus, for PCOS and associated infertility (Pollard & Unwin, 2008). High androgen and insulin levels are also implicated.

A view from evolutionary medicine suggests that PCOS may be a by-product of the elaborate endometrium needed to support a large-brained human fetus during gestation (Barnett & Abbott, 2003). The human endometrium requires a prolonged follicular phase and greater hormone production for sufficient build-up to occur for implantation. Unfortunately, excess hormone production can also disrupt ovulation and it may predispose women to endometrial cancer. PCOS may be an example of a "disease of civilization" that results when a system designed to make the most of every calorie meets resource abundance.

Most women are eventually able to conceive and if all goes well, pregnancy proceeds. Unfortunately, sometimes the zygote fails to implant and may require medical assistance. In some cases failure to implant is due to abnormalities in the zygote or in the maternal reproductive system, such that conceptions that would develop unsuccessfully are weeded out before too much energy (caloric and emotional) is invested. An evolutionary medicine perspective suggests that many cases of failed conceptions result from the reproductive system doing what it evolved to do—avoid wasting reproductive effort on offspring that are not viable or are otherwise unable to increase reproductive fitness. One cause of failure to implant is rejection by the maternal immune system, a factor that is discussed in Chapter 5 by Muglia and Pavlicev, this volume.

In some cases, an implanted embryo may be rejected due to chromosomal abnormalities in the fetus (Guerneri et al., 1987; Goddijn & Leschot, 2000; Macklon et al., 2002), an action that makes "evolutionary sense" by terminating a pregnancy that would lead to a nonviable infant after nine months of gestation or an individual incapable of reproducing after years of being breastfed and cared for. The evolutionary argument is that that energy would be better spent on a reproductively viable future offspring and that terminating a pregnancy with poor prospects would benefit a woman's long-term fitness (reproductive success). Although clinicians and families may see every failed conception as a problem, perhaps even a personal tragedy, in some cases it is an evolutionary success story.

3.8 Successful conceptions: Pregnancy

Once implantation has taken place, the risk of early pregnancy loss declines and most gestations proceed with few challenges. The placenta performs critical functions as it produces reproductive hormones, transmits gasses and nutrients to and from the fetus, and shields the fetus from toxins and other dangerous substances in the mother's environment. It is not a perfect filter—some drugs, alcohol, and other toxins are able to cross from the maternal to the fetal circulation, where they may cause damage to fetal cells.

Female physiology becomes more efficient at metabolizing food so that a pregnant woman gets more nutrients from the same amount of food than she did before she became pregnant (Cunningham et al., 2014). In this sense, her body is becoming more efficient with respect to dietary intake, so that even if she doesn't increase the volume of the food she consumes, she can still gain the weight needed to support her growing fetus. There are obvious advantages to improved metabolism for women living in marginal environments, but in places where food availability is high, greater potential problems emerge when women eat more food at the same time that they reduce activity levels, behaviors that may lead to excessive weight gain during pregnancy. This can lead to gestational diabetes, which may predispose the mother to regular diabetes later in life. Furthermore, weight gain in pregnancy is usually not easy to reverse after delivery, especially in the absence of breastfeeding.

During the first trimester of pregnancy when fetal organ systems are being developed, environmental factors to which the fetus may be exposed are especially important. These include air and water pollution, food, drug, and alcohol intake, and other toxins. In addition, nutrient deficiencies during this period can compromise skeletal, neurological, and motor development. This especially vulnerable time occurs when some women are not even aware that they are pregnant, but their awareness may be boosted by "morning sickness," also known as nausea and vomiting of pregnancy (NVP). NVP usually results in a woman limiting her food intake and is so common that it may not cause concern unless it becomes extreme in the context of marginal food resources (see Pike, 2000, for examples). Although this "sickness" is usually unpleasant, a somewhat speculative evolutionary proposal is that it may be beneficial to the health of both the mother and her fetus if it leads her to avoid exposure to toxins that may interfere with fetal development. Overripe, spicy, fermented, and

Table 3.2 Neonatal brain size relative to adult brain size for selected primate species.

Species	Brain size at birth	Adult brain size	Brain size at birth as a percentage of adult brain size
Saimiri sciureus	15.3	26.2	58.4%
Cercocebus atys	57.5	107.9	53.3%
Macaca fuscata	69.4	109.7	63.3%
Macaca mulatta	58.3	91.1	64.4%
Macaca nemestrina	61.8	104.6	59.1%
Papio anubis	82.9	161.4	51.4%
Pan troglodytes	150.9	381.7	39.5%
Homo sapiens	381.8	1336.1	28.6%
Homo sapiens	367.7	1306.9	28.1%

Data from DeSilva & Lesnik, 2008.

smoked foods are often those that women find especially noxious early in pregnancy and many of these have been found to have contents that are harmful to the developing fetus. Fortunately, low food intake (as long as quality of food is not compromised) is not problematic in the first trimester when fetal weight gain is not of concern. The nausea usually disappears in the second trimester when weight gain is the primary goal of a fetus. When maternal appetite is restored, she can begin the period of "eating for two."

In the third trimester fetal growth accelerates as the fetus prepares for life outside the womb. Fat accumulation during this period is particularly important for humans who are fatter than the infants of many, if not most, mammalian species—fat provides resources for the rapidly growing energetically demanding fetal brain (Kuzawa, 1998). Babies born to mothers who had low nutrient intake during pregnancy can be at risk for compromised health if they are born with lower fat reserves.

3.9 Gestation length

The healthiest human infants are born at the end of 37–42 weeks of gestation (Liao et al. 2005), weighing about 2.5–4 kilograms (6–9 pounds), although there is variation in what is seen as a healthy neonatal weight. Babies born before 37 weeks of gestation are usually identified as "premature" and many require intensive neonatal care to improve their chances of survival (Phillips et al., 2015). Infants born before 35 weeks of gestation in the past, before modern medical treatment options were developed, most likely died. Death of a newborn, after months of investment through gestation, is a blow to a woman's reproductive effort (not to mention her emotions), suggesting that natural selection has probably favored mechanisms for maintaining a pregnancy sufficiently long enough for the infant to be viable without clinical intervention.

Constraints on the length of gestation come from several sources: limits on maternal metabolism in provision of energy to the developing fetus with its hungry brain; limits on the placental ability to support the fetus; limits on the size of the neonate that can successfully traverse the birth canal; and, at the lower end, limits on the fetal respiratory system for survival outside the womb. Humans give birth to relatively large infants after relatively long gestations, but both measures fall in line with most other primates. One neonatal measure that falls outside the parameters of the other primates is neonatal brain size as a percentage of adult brain size. Table 3.2 shows neonatal brain size as a percentage of adult brain size for selected primate species for which we have data (data taken from DeSilva & Lesnik, 2008). For humans, that measure is notably smaller, suggesting that brain size, in particular, may be important in understanding why human gestation lasts approximately 38 weeks.

The brain is the most energetically expensive organ in the body and as the fetal brain increases in size, it draws more and more of the available energy, in the form of glucose, from the mother, eventually reaching a point where she can no longer meet the needs of herself and her growing fetus. Ellison (2001), referring to this as the "metabolic crossover" point (Figure 3.6), argues that this is one of the major determinants of gestation length and that fetal stress as it begins to "starve" is one of the triggers for the onset of labor (see also, Dunsworth et al., 2012; Dunsworth & Eccleston, 2015; Epstein, 1973; Martin, 1983, 2007). One piece of evidence of this is that fetuses begin to lose weight and

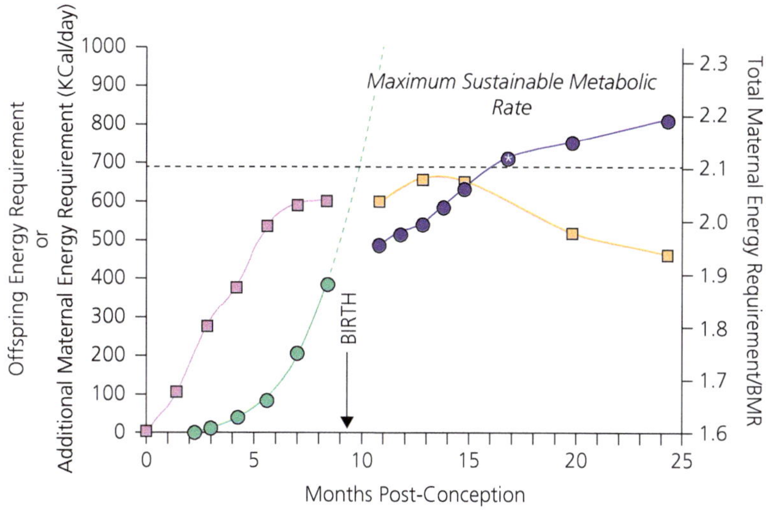

Figure 3.6 Showing metabolic constraint on gestation length and fetal size. Throughout pregnancy, as the fetus grows, its energy needs (circles, kcal/d) increase exponentially. By contrast maternal energy expenditure (squares) rises with fetal needs in the first two trimesters but reaches a metabolic limit late in pregnancy
Graph from Dunsworth et al. (2012).

body fat if they remain in the womb beyond the expected due date. An evolutionary cost of this expensive brain is the birth of an infant with less neurological development and a greater degree of helplessness than observed for other primates, both of which require extensive parental care.

Metabolic costs of fetal growth in the last trimester of pregnancy average about 300 calories per day, but the costs of lactation exceed 500 calories a day, so it could be argued that keeping the fetus in the womb for a bit longer would be an energetic cost saving measure for the mother. Although larger babies are generally healthier at birth, if a baby is too large, difficulties navigating the birth canal may increase birth complications, as will be discussed next. This is the basis of the argument that anatomical constraints also impose a limit on gestation length.

The high metabolic costs of pregnancy can also help explain the existence of an especially problematic clinical disorder of pregnancy, preeclampsia and its more extreme form, eclampsia. Preeclampsia and eclampsia are common worldwide in both resource-rich and resource-poor populations. The primary manifestation is maternal hypertension that, when untreated, can lead to liver, kidney, and brain damage and even death. The best way to cure this disorder is to deliver the infant, which often means doing so prematurely. In general, mortality is higher in resource-poor populations than in those with access to medical care and surgery. This condition affects 2.5–3.0% of women worldwide (Redman & Sargent, 2005).

A view from evolutionary medicine on preeclampsia and eclampsia is that it is related to parent-infant conflict that it is an inevitable aspect of the fact that maternal and fetal needs and fitness goals are not fully aligned and that what may benefit the fetus (e.g., more resources) may be at the expense of maternal physiology and health (Hollegaard et al., 2017). Another view relates to the deep invasion of the placenta into the uterine wall (known as the secondary invasion of the trophoblast) necessary to maintain the energetically expensive fetal brain. If this invasion is not complete, there comes a point in gestation (third trimester) when it becomes difficult to maintain nutritional and oxygen support of the fetus as maternal blood pressure rises to compensate. As noted, the only solution to this disorder is delivery of the fetus. When this occurs in settings where medical technology is available to care for the premature neonate, that solution may work, but in other contexts, death is the more likely outcome for both mother and infant. The World Health Organization estimates that as many as 70,000 women die from preeclampsia/eclampsia each year. Furthermore, there is increasing evidence that exposure to even mild pre-eclampsia late in pregnancy may compromise later-life health for children, most especially for endocrine, nutritional, and metabolic diseases (Wu et al., 2009; Hollegaard et al., 2017).

Advanced pregnancy places unusual stress on the upright, bipedal human. Unlike quadrupedal mammals that carry the weight of the fetus in a "sling," evenly distributed along the vertebral column, the human vertebral column supports increasing weight from top to bottom. This is challenging enough for all humans, as attested to by the lower back problems that occur very frequently with age, but becomes increasingly difficult as pregnancy advances. The weight of the fetus throws off a pregnant woman's center of gravity making walking increasingly difficult and uncomfortable. Natural selection has favored alteration of the vertebral column to accommodate this challenge to bipedal walking in women, but it puts them at greater risks for lower back pain and slipped discs (Whitcome et al., 2007).

So far we have focused on the physical challenges of pregnancy, but psychosocial stresses also play major roles in affecting outcomes. Stress hormones such as cortisol are associated with compromised immune function and developmental abnormalities of metabolic, pancreatic, cardiovascular, and liver functions (Seckl et al., 2005). Furthermore, stress during pregnancy has been linked to behavioral disorders in childhood (Davis et al., 2005), suggesting that reduction of stress during pregnancy will likely benefit both mother and infant. Social support has been shown to be especially effective in reducing stress during pregnancy by buffering the effects of psychological stress hormones (Giesbrecht et al., 2013). These effects are not surprising, given the dense social and kin networks in which pregnancy was experienced in the past and in many communities today.

3.10 Labor and delivery

Complications related to childbirth are the primary source of morbidity and mortality for women of reproductive age throughout the world, with an estimated 830 women dying each day (World Health Office (WHO) Facts Sheet). Although mortality rates have steadily declined since 1980, childbirth is still a formidable force of natural selection, especially in developing nations where 99% of maternal deaths occur. Rates range from a high of 239 per 100,000 live births to a low of 12 per 100,000, with considerable variation within nations. Adolescent girls are at particular risk. Leading causes of death are postpartum hemorrhage, sepsis, pre-eclampsia and eclampsia, delivery complications, and unsafe abortion. Some of these causes of disability and death have probably been factors for hundreds of thousands of years, although their frequency and severity have likely increased since the origins of agriculture, as is true for many diseases (Larsen, 2006).

3.11 Childbirth

Human childbirth is different from birth in our closest living relatives in several important ways. The differences are the result of two significant human adaptations that have arisen over the last 4 million years: 1) bipedalism (walking on two legs rather than on all fours), which involved a major reshaping of the muscles and bones of the pelvic girdle to accommodate upright posture and locomotion, and 2) encephalization (the evolution of larger brains). Although both of these adaptations significantly affect how we give birth, they occurred at very different times in our evolutionary history. Evidence from the human fossil record shows that bipedalism evolved by about 4 million years ago while significant brain size expansion began around 2.5 million years ago. Nevertheless, our current mode of childbirth results from these two evolutionary changes and in particular the interaction between them as selection for locomotion favors a less spacious pelvis whereas selection for easier birth favors a spacious birth canal.

In humans, there is a tight fit between the relatively large neonate (in particular the head and

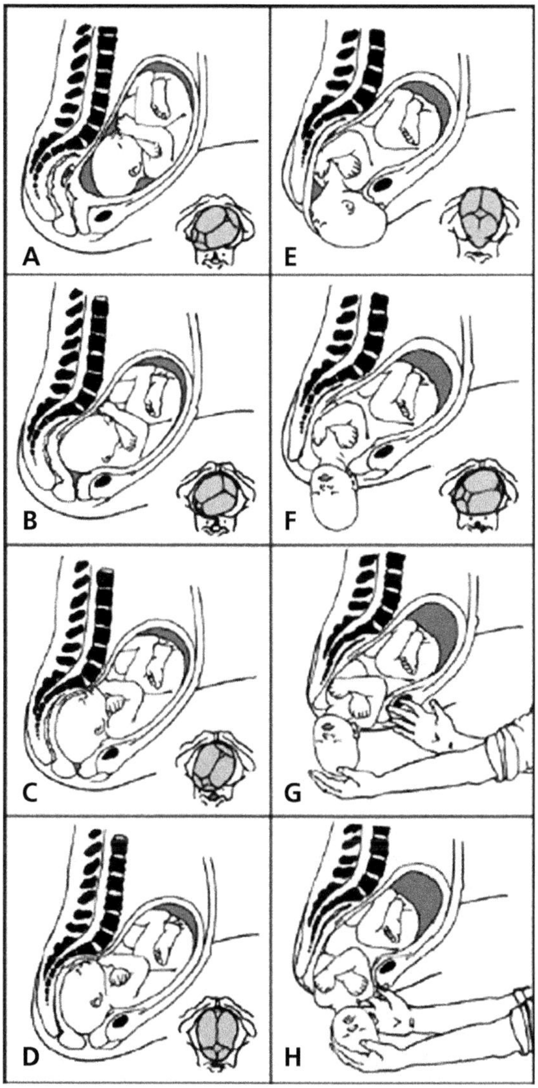

Figure 3.7 The emergence of the fetus through the bipedal human pelvis showing assistance at delivery Trevathan, (2010).

shoulders) and the dimensions of the mother's bony birth canal. Unlike birth canals in apes, which are anterior-posteriorly oval in cross section from inlet to outlet, human birth canals change in cross-sectional shape from a transverse oval at the inlet to an anterior-posteriorly elongated oval at the mid-plane and generally an approximate circle at the outlet. This means that the baby typically passes through the birth canal following a twisty path (referred to by obstetricians as the cardinal movements of labor) and emerges facing in the opposite direction as the mother (in an occiput-anterior position) (Figure 3.7). This somewhat complicated emergence pattern and the fact that our babies are relatively large compared to the babies of other apes (but also somewhat less mature developmentally) is the basis of the argument that human mothers benefit from the presence of helpers who can assist them and the neonate in a number of ways, increasing the likelihood of a successful outcome for both the mother and the baby. Trevathan (1988, 2017) and Rosenberg & Trevathan, (2002) have argued that birth assistance arose as an adaptation to this way of giving birth. Indeed, although there is a great deal of cross-cultural variation in the details of who attends birth, where it takes place, what rituals are associated with it etc., virtually all societies have attended birth as a norm. In contrast to nonhuman primates (and other mammals) who typically seek privacy when they give birth, humans give birth in a social setting with family members and perhaps others present to provide both physical assistance to the mother and baby (removing an umbilical cord from around a baby's neck, receiving the baby in their hands as it emerges facing away from the mother and putting it on the mother's breast, clearing its breathing passage, etc.) and, probably, equally importantly, emotional assistance in the form of encouragement and social support. When we say that assisted childbirth is a human adaptation, we refer to these behaviors and not to the highly medicalized (and arguably overmedicalized) way in which birth often occurs in clinical settings in many developed countries.

The value of social and emotional support is shown by the reduced rates of cesarean sections and other interventions for women who are accompanied by a doula during their labor. We have suggested (Rosenberg & Trevathan, 2002, 2018) that the fear and anxiety that many women experience giving birth in unfamiliar clinical settings may reflect a mismatch between the evolutionary legacy for a need for physical as well as emotional support and our modern clinical environment. Given the risks associated with cesarean section for both mothers and their babies (Rosenberg & Trevathan, 2018) and the rates of cesarean section in many developed countries well above the WHO recommendation, an evolutionary medicine perspective encourages us to examine the cultural, social, and psychological circumstances in which women labor and to provide social and emotional support before rushing to intervene surgically. This is becoming increasingly important as our understanding of the relationship between the birth process and the developing infant microbiome proceeds (Dominguez-Bello et al., 2010).

3.12 Menopause, the later life course, and grandmotherhood

An enduring question in evolutionary medicine studies of reproduction is why women cease to ovulate at about age 50, thus terminating their ability to reproduce at a point in their lives when many are still otherwise very healthy and active. If reproducing is the currency of the evolutionary process, why hasn't natural selection favored the ability to continue to reproduce as long as other aspects of health allow it? As we have noted, the age of menarche has decreased with improving health standards, so why hasn't the age of menopause been affected as well? Age 50 was once believed to be the maximum life expectancy of our ancestors so it was thought that very few women actually experienced the termination of reproduction until modern medicine and hygiene lengthened lifespans. Based on skeletal remains and studies of contemporary foragers, however, it appears that significant numbers of our ancestors lived well into their 60s, a decade or more beyond the cessation of reproduction (Blurton-Jones et al., 2002). Certainly, life expectancy has increased dramatically in most parts of the world in the last 100 years (rising from a high of 50 in 1900 to a high of 77 in 2001, (Riley, 2005), due primarily to improved living conditions, but women who lived years beyond their last births have likely made up a significant portion of group populations for several thousand years.

For several decades, menopause, the cessation of reproduction, was thought to be unique to humans, but now we know that all long-lived species stop reproducing several years before reaching maximum lifespan (Finch, 2015). Recalling that before birth a mammalian female has all of the eggs she will ever have and that they decrease in number from then on, it is clear that menopause occurs when all eggs have been depleted or greatly reduced in number, at approximately age 50 for all mammals. In other words, the life expectancy of the mammalian egg may be about 50 years, regardless of the life expectancy of the female mammal in whose body the egg is carried. It appears that the question of "why stop reproducing?" should be turned around to "why live so long after stopping reproducing?" In other words, what is unusual about humans and a few other mammals (e.g., whales and elephants) is that they live so many years beyond their last births and it may be post-reproductive longevity that natural selection has operated upon rather than cessation of reproducing (Sievert, 2006).

Why would selection favor extension of the human life for 12–15 years past the last birth? One explanation may be that those years are necessary to raise the highly dependent human offspring to the point of reproductive maturity and independence. Women who lived until they were in their mid-60 were able to invest in their last infants, boosting their survival rates and increasing their reproductive fitness. Furthermore, older women who were not producing their own infants would be freed to provide care for their grandchildren and other related young, further increasing their own reproductive success. This is the well-known "grandmother hypothesis" proposed by Hawkes (Hawkes et al., 1997; Hawkes et al., 1998; Hawkes, 2004) who argues that women who ceased reproducing and invested in their relatives would have more surviving descendants than those who continued to reproduce until their deaths. Although the evidence for this is inconclusive, there are several populations (e.g., Hadza, the Gambia) that show support for the economic and child-care contributions of grandmothers (Hawkes et al., 1997; Gibson & Mace, 2005; Sear & Mace, 2008). A recent study of historical life tables from four Western nations showed that populations with higher numbers of post-reproductive females also had lower rates of maternal mortality (Gemmill & Catalano, 2017).

3.13 Conclusions

We have seen in this review that one of the most important aspects of contemporary human female reproductive lives is how different they are from the reproductive lives of ancestral females. Today, in many populations, reproduction begins earlier with menarche as young as age 11 or 12 while childbearing begins at much earlier or much later ages (depending on the population) than it did for women as recently as 100 years ago. Contemporary women in resource-rich populations have far fewer children (or even no children) than their ancestors did. Breastfeeding lasts fewer months or none at all. All of these behaviors lead to far greater frequency of menstrual cycling than women in the past experienced. Major factors affecting these differences are found in the environments in which women today grow, develop, and reproduce. Most especially, environments of high resource availability associated with low levels of energy expenditure affect not only reproductive function, but also decisions that women make about whether or when to have children, whether to give birth "naturally" or surgically, and whether or not to breastfeed and for how long. The potential incompatibilities between modern lives and the environmental circumstances under which female reproductive biology evolved have resulted in a number of health challenges ranging from menstrual disorders to cancers of the reproductive organs. In many cases, clinical medicine is required to solve these challenges, but resolving the incompatibilities may help prevent some health problems without medical intervention.

What can the perspective of evolutionary medicine and an understanding of human evolutionary history do to mitigate some of the reproductive health challenges that women face today? Advice that includes maintaining high activity levels, especially during the growing years, and avoiding excessive calorie intake are somewhat simple lifestyle choices that have potential benefits for not only reproductive health, but also for overall health for women, men, and children. Medical education and public health campaigns could benefit from increased awareness of the variability in reproductive health parameters (e.g., hormone levels) and expansion of definitions of normal healthy functioning (Nunn et al., 2015; Palanza & Parmigiani, 2016). Reducing medical intervention in childbirth, including reducing the rates of elective cesarean section, would go a long way toward improving lifetime health of mothers and infants.

We have argued that the view from evolutionary medicine enhances understanding of all aspects of the human female reproductive life course, from menarche to menopause to grandmotherhood. Clinical medicine and reproductive decisions made by women and their families can be enhanced by this understanding. Most especially, one of the most significant aspects of human reproductive health in an evolutionary perspective is that women who are surrounded by supportive social networks from the time of menarche, through the childbearing and childrearing years and into the time of grandmotherhood have higher reproductive success than others. Like so many other aspects of our biology, our reproductive health must be understood in the context of our adaptation as a culture-bearing species. Understanding the evolutionary basis of many of the challenges to female reproductive health over the course of our life cycle makes a valuable contribution to medical school curricula (Abrahamson et al., 1999; Grunspan et al., 2018), and can inform how we understand and ultimately, how we choose to provide health care.

Acknowledgments

We thank Jay Schulkin for the invitation to contribute a chapter to this volume and Rob Schultz for drawing Figure 3.3.

References

Abrahamson, S., Baron, J., Elstein, A. S., Hammond, W. P., Holzman, G. B., Marlow, B., Taggart, M. S., & Schulkin, J. (1999). Continuing medical education for life: Eight principles. *Academic Medicine: Journal of the Association of American Medical Colleges*, 74(12): 1288–94.

Aktipis, C., Ellis, B. J., Nishimura, K. K., & Hiatt, R. A., (2015). Modern reproductive patterns associated with estrogen receptor positive but not negative breast cancer susceptibility. *Evolution, Medicine, and Public Health*, 2015(1): 52–74.

Barnett, D. K. & Abbott. D. H. (2003). Reproductive adaptations to a large-brained fetus open a vulnerability to anovulation similar to polycystic ovary syndrome. *American Journal of Human Biology*, 15(3): 296–319.

Bell, A. F. Erickson. E. N., & Carter. C .S. (2014). Beyond labor: The role of natural and synthetic oxytocin in the transition to motherhood. *Journal of Midwifery and Women's Health*, 59(1): 35–42.

Bentley, G. R. (1999). Aping our ancestors: Comparative aspects of reproductive ecology. *Evolutionary Anthropology: Issues, News, and Reviews*, 7(5): 175–85: 18.

Blurton Jones, N. G., Hawkes, Kristen, & O'Connell, James F. (2002). Antiquity of postreproductive life: Are there modern impacts on hunter-gatherer postreproductive life spans? *American Journal of Human Biology*, 14: 184–205.

Cunningham, F., Leveno, K., Bloom, S. Spong, C. Y., & Dashe, J. (2014). *Williams Obstetrics*. 24e. New York: Mcgraw-Hill.

Davis, E. P., Hobel, C. J., Sandman, C. A., Glynn, L., & Wadhwa, P. D. (2005). Prenatal stress and stress physiology influences human fetal and infant development. In M. Power & J. Schulkin, (eds.) *Birth. Distress and Disease: Placental-Brain Interactions* (pp. 183–201). Cambridge: Cambridge University Press.

DeSilva, J. M. & Lesnik, J. J. (2008). Brain size at birth throughout human evolution: A new method for estimating neonatal brain size in hominins. *Journal of Human Evolution*, 55(6): 1064–74.

Dieterich, C. M., Felice, J .P., O'Sullivan, E., & Rasmussen, K. M. (2013). Breastfeeding and health outcomes for the mother-infant dyad. *Pediatric Clinics of North America*, 60(1): 31–48.

Dominguez-Bello, M. G., Costello, E. K., Contreras, M., Magris, M., Hidalgo, G., Fierer, N., & Knight, R. (2010). Delivery mode shapes the acquisition and structure of the initial microbiota across multiple body habitats in newborns. *Proceedings of the National Academy of Sciences*, 107(26): 11971–5.

Doyle, Caroline, Swain Ewald, Holly A., & Ewald, Paul W. (2008). An evolutionary perspective on premenstrual syndrome: Implications for investigating infectious causes of chronic disease. In W. R. Trevathan, E. O. Smith, & J. J. McKenna, (eds). *Evolutionary Medicine and Health: New Perspectives* (pp. 196–215). New York: Oxford University Press.

Dunsworth, H. & Eccleston, L. (2015). The evolution of difficult childbirth and helpless hominin infants. *Annual Review of Anthropology*, 44: 55–69.

Dunsworth, H. M., Warrener, A. G., Deacon, T., Ellison, P. T., & Pontzer, H. (2012). Metabolic hypothesis for human altriciality. *Proceedings of the National Academy of Sciences*, 109(38):15212–6.

Eaton, S. B. & Eaton, S. B. III. (1999). Breast cancer in evolutionary perspective. In *Evolutionary Medicine*, (ed.) W. R. Trevathan, E. O. Smith, & J. J. McKenna, (pp. 429–42). New York: Oxford University Press.

Ellison, P. T. (1988). Human salivary steroids: Methodological considerations and applications in physical anthropology. *American Journal of Physical Anthropology*, 31(S9): 115–42.

Ellison, P. T. (2001). *On Fertile Ground: A Natural History of Human Reproduction*. Cambridge. MA: Harvard University Press.

Epstein, H. T. (1973). Possible metabolic constraints on human brain weight at birth. *American Journal of Physical Anthropology*, 39(1):135–6.

Finch. C. E (2014). The menopause and aging. A comparative perspective. *The Journal of Steroid Biochemistry and Molecular Biology*, 142: 132–41.

Frisch, R. E. & McArthur, J. W. (1974). Menstrual cycles: Fatness as a determinant of minimum weight for height necessary for their maintenance or onset. *Science*, 185(4155): 949–51.

Frisch, R. E. & Revelle, R. (1970). Height and weight at menarche and a hypothesis of critical body weights and adolescent events. *Science*, 169(3943): 397–9.

Gemmill, A. & Catalano, R. (2017). Do post-reproductive aged females promote maternal health? Preliminary evidence from historical populations. *Evolution. Medicine and Public Health*, 1: 136–43. https://doi.org/10.1093/emph/eox012

Gibson, Mhairi A. & Mace, Ruth. (2005). Helpful grandmothers in rural Ethiopia: A study of the effect of kin on child survival and growth. *Evolution and Human Behavior*, 26: 469–82.

Giesbrecht, G. F., Poole, J. C., Letourneau, N., Campbell, T., Kaplan, B. J., & APrON Study Team. (2013). The buffering effect of social support on hypothalamic-pituitary-adrenal axis function during pregnancy. *Psychosomatic Medicine*, 75(9): 856–62.

Gleicher, N. & Barad, D. (2006). An evolutionary concept of polycystic ovarian disease: Does evolution favour reproductive success over survival? *Reproductive Biomedicine Online*, 12(5): 587–9.

Gluckman, Chapter 1, this volume

Goddijn, M. & Leschot, N. J. (2000). Genetic aspects of miscarriage. *Best Practice & Research Clinical Obstetrics & Gynaecology*, 14(5): 855–65.

Grunspan, D. Z., Nesse, R. M., Barnes, M. E., & Brownell, S. E. (2018). Core principles of evolutionary medicine: A Delphi study. *Evolution, Medicine, and Public Health*, 2018(1): 13–23.

Guerneri, S., Bettio, D., Simoni, G., Brambati, B., Lanzani, A., & Fraccaro. M. (1987). Prevalence and distribution of chromosome abnormalities in a sample of first trimester internal abortions. *Human Reproduction*, 2(8): 735–9.

Hawkes, K., O'Connell, J. F., & Blurton Jones, N. G. (1997). Hadza women's time allocation, offspring provisioning, and the evolution of long postmenopausal life spans. *Current Anthropology*, 38: 551–77.

Hawkes, K., O'Connell, J. F., Jones, N. B., Alvarez, H., & Charnov, E. L. (1998). Grandmothering, menopause, and the evolution of human life histories. *Proceedings of the National Academy of Sciences*, 95(3); 1336–9.

Hawkes, K. (2004). Human longevity: The grandmother effect. *Nature*, 428(6979): 128.

Hidaka, B. H. & Boddy, A. M. (2016). Is estrogen receptor negative breast cancer risk associated with a fast life history strategy? *Evolution, Medicine, and Public Health*, 2016(1): 17–20.

Hollegaard, Birgitte, Lykke, J. A., & Boomsma, J. J. (2017). Time from pre-eclampsia diagnosis to delivery affects future health prospects of children. *Evolution. Medicine and Public Health*, 53–66. https://doi.org/10.1093/emph/eox004

Hrdy, S. B. (1999). *Mother Nature: A History of Mothers, Infants, and Natural Selection*. New York: Ballantine.

Hrdy, S. B. (2009). *Mothers and Others*. Cambridge: Harvard University Press.

Jasieńska, G. & Thune, I. (2001). Lifestyle, hormones, and risk of breast cancer. BMJ, 322:586–587.

Johnson, T .M. (1987). Premenstrual syndrome as a western culture-specific disorder. *Culture, Medicine and Psychiatry*, 11(3): 337–56.

Kaplan, H. (1994). Evolutionary and wealth flows theories of fertility: Empirical tests and new models. *Population and Development Review*, 20(4): 753–91.

Kunz, G., Beil, D., Huppert, P., & Leyendecker, G. (2007). Oxytocin–a stimulator of directed sperm transport in humans. *Reproductive Biomedicine Online*, 14(1): 32–9.

Kuzawa, C. W. (1998). Adipose tissue in human infancy and childhood: An evolutionary perspective. *American Journal of Physical Anthropology*, 107(S27):177–209.

Larsen, C. S. (2006). The agricultural revolution as environmental catastrophe: Implications for health and lifestyle in the Holocene. *Quaternary International*, 150(1): 12–20.

Leslie, Paul (2018). Fertility and fecundity. In W. Trevathan, ed. *The International Encyclopedia of Biological Anthropology*, New York: John Wiley and Sons. Inc.

Liao, J. B., Buhimschi, C. S., & Norwitz, E. R. (2005). Normal labor: Mechanism and duration. *Obstetrics and Gynecology Clinics*, 32(2): 145–64.

Liu, K., Case, A., Cheung, A. P., Sierra, S., Alasiri, S., Carranza-Mamane, B., Dwyer, C., Graham, J., Havelock, J., Hemmings, R., & Lee, F. (2011). Advanced reproductive age and fertility. *Journal of Obstetrics and Gynaecology Canada*, 33(11) 1165–75.

Lovett, J. L., Chima, M. A., Wexler, J. K., Arslanian, K. J., Friedman, A. B., Yousif, C. B., & Strassmann, B. I. (2017). Oral contraceptives cause evolutionarily novel increases in hormone exposure: A risk factor for breast cancer. *Evolution, Medicine, and Public Health*, 2017(1): 97–108.

Macaluso, M., Wright-Schnapp, T. J., Chandra, A., Johnson, R., Satterwhite, C. L., Pulver, A., Berman, S. M., Wang, R. Y., Farr, S. L., & Pollack, L. A. (2010). A public health focus on infertility prevention, detection, and management. *Fertility and Sterility*, 93(1): 16-e1.

Macklon, N. S., Geraedts, J. P., & Fauser, B. C. (2002). Conception to ongoing pregnancy: The "black box" of early pregnancy loss. *Human Reproduction Update*, 8(4): 333–43.

Martin, Emily. (1988). Premenstrual syndrome: Discipline, work and anger in late industrial societies. In Thomas Buckley & Alma Gottlieb, (eds.) *Blood Magic: The Anthropology of Menstruation* (pp. 161–81). Berkeley: University of California Press.

Martin, R. D. (1983). *Human Brain Evolution in an Ecological Context*, New York: American Museum of Natural History.

Martin, R. D. (2007). The evolution of human reproduction: A primatological perspective. *American Journal of Physical Anthropology*, 134(S45): 59–84.

Nunn, C. L., Wallace, I., & Beall, C. M., (2015). Connecting evolution, medicine, and public health. *Evolutionary Anthropology: Issues, News, and Reviews*, 24(4): 127–9.

Palanza, P. & Parmigiani, S., (2016). Why human evolution should be a basic science for medicine and psychology students. *Journal of Anthropological Sciences*, 94: 1–10.

Phillips, J. B., Abbot, P., & Rokas, A. (2015). Is preterm birth a human-specific syndrome? *Evolution, Medicine, and Public Health*, 2015(1): 136–48.

Pike, I. L. (2000). The nutritional consequences of pregnancy sickness: A critique of a hypothesis. *Human Nature*, 11: 207–32.

Pollard, T. M. & Unwin, N. (2008). Impaired reproductive function in women in Western and "Westernizing" populations: An evolutionary approach. In W. R. Trevathan, E. O. Smith, & J. J. McKenna, *Evolutionary Medicine and Health: New Perspectives* (pp. 169–80). New York: Oxford University Press.

Redman, C. W., & Sargent, I. L. (2005). Latest advances in understanding preeclampsia. Science, 308(5728), 1592–1594.

Reiber, Chris. (2008). An evolutionary model of premenstrual syndrome. *Medical Hypotheses*, 70: 1058–65.

Riley, J. C. (2005). Estimates of regional and global life expectancy. 1800–2001. *Population and Development Review*, 31(3): 537–43.

Rosenberg, K. & Trevathan, W. (2002). Birth, obstetrics and human evolution. *BJOG: An International Journal of Obstetrics & Gynaecology*, 109(11): 1199–206.

Rosenberg, K. R. & Trevathan, W. R. (2018). Evolutionary perspectives on cesarean section. *Evolution, Medicine, and Public Health*, 2018(1): 67–81.

Schiefenhövel, Wulf & Wenda Trevathan. (2019). Sexuality, reproduction and birth. In Brüne, Martin & Wulf Schiefenhövel (eds.). *Oxford Handbook of Evolutionary Medicine*. Oxford: Oxford University Press.

Sear, Rebecca & Mace, Ruth. (2008). Who keeps children alive? A review of the effects of kin on child survival. *Evolution and Human Behavior*, 29: 1–18.

Seckl, J. R., Drake, A. J.,& Holmes, M. C. (2005). Prenatal glucocorticoids and the programming of adult diseases. In M. L. Power & J. Schulkin, (eds.) *Birth, Distress, and Disease*. Cambridge, UK: Cambridge University Press.

Sievert, Lynette Leidy. (2006). *Menopause: A Biocultural Perspective*. New Brunswick. NJ: Rutgers University Press.

Sievert, L. L. (2008). Should women menstruate? An evolutionary perspective on menstrual-suppressing oral contraceptives. In W. R. Trevathan, E. O. Smith, & J., J. McKenna, *Evolutionary Medicine and Health: New Perspectives* (pp. 181–95). New York: Oxford University Press.

Stuebe, A. M. & Tully, K. P. (2010). Optimizing Maternal-Infant Health: Ameliorating the mismatch between evolved dyadic needs and the current culture of health around the 4th trimester. In: Jay Schulkin & Michael L. Power, (eds.) *Integrating Evolution into Medical Education—for Maternal and Child Healthcare Providers*. Oxford: Oxford University Press.

Tanner, J. M. (1987). Growth as a mirror of the condition of society: Secular trends and class distinctions. *Pediatrics International*, 29(1): 96–103.

Taylor, Diana. (2006). From "It's all in your head" to "Taking back the month": Premenstrual syndrome (PMS) research and the contributions of the Society for Menstrual Cycle Research. *Sex Roles*, 54: 377–91.

Trevathan, W. R. (2017). *Human Birth: An Evolutionary Perspective*. New York: Routledge.

Trevathan, W. R. (1988). Fetal emergence patterns in evolutionary perspective. *American Anthropologist*, 90(3): 674–81.

Trevathan, W. (2010). *Ancient Bodies. Modern Lives: How Evolution has Shaped Women's Health*, New York: Oxford University Press.

Trevathan, W. R. & Rosenberg. K. R. (eds.) (2016). *Costly and Cute: Helpless Infants and Human Evolution* Albuquerque, US: University of New Mexico Press.

Valeggia. Claudia & Ellison, Peter T. (2003). Lactational amenorrhoea in well-nourished Toba women of Formosa. Argentina. *Journal of Biosocial Sciences*, 36: 573–95.

Vitzthum, V. J., Spielvogel, H., Caceres, E., & Miller, A. (2001). Vaginal bleeding patterns among rural highland Bolivian women: Relationship to fecundity and fetal loss. *Contraception*, 64(5): 319–25.

Vitzthum, Virginia, Spielvogel, H., & Thornburg, J. (2004). Interpopulational differences in progesterone levels during conception and implantation in humans. *Proceedings of the National Academy of Sciences of the United States of America*, 101: 1443–8.

Vitzthum, Virginia, Spielvogel, Hilde, Caceres, Esperanza, & Gaines, Julia. (2000). Menstrual patterns and fecundity among non-lactating and lactating cycling women in rural highland Bolivia: Implications for contraceptive choice. *Contraception*, 62: 181–7.

Vitzthum, Virginia. (2001). Why not so great is still good enough. In P. T Ellison, (ed.) *Reproductive Ecology and Human Evolution* (pp. 179–202). New York: Aldine de Gruyter.

Whitcome, Katherine K., Shapiro, Liza J., & Leiberman, Daniel E. (2007). Fetal load and the evolution of lumbar lordosis in bipedal hominins. *Nature*, 450: 1075–8.

White, Y. A., Woods, D. C., Takai, Y., Ishihara, O., Seki, H., & Tilly, J. L. (2012). Oocyte formation by mitotically active germ cells purified from ovaries of reproductive-age women. *Nature Medicine*, 18(3): 413–21.

WHO Facts Sheet: Maternal Mortality. [https://www.who.int/news-room/fact-sheets/detail/maternal-mortality]

Williams, G. C. & Nesse, R. M. (1991). The dawn of Darwinian medicine. *The Quarterly Review of Biology*, 66(1): 1–22.

Worthman, Carol M. & Stallings. J. F. (1994). Measurement of gonadotropins in dried blood spots. *Clinical Chemistry*, 40: 448–53.

Worthman, Carol M. & Stallings, J. F. (1997). Hormone measures in finger-prick blood spot samples: New field methods for reproductive endocrinology. *American Journal of Physical Anthropology*, 104: 1–21.

Wu, C. S., Nohr, E. A., Bech, B. H., Vestergaard, M., Catov, J. M., & Olsen. J. (2009). Health of children born to mothers who had preeclampsia: A population-based cohort study. *American Journal of Obstetrics and Gynecology*, 201(3): 269-e1.

CHAPTER 4

Optimizing Maternal-Infant Health: Ameliorating the Mismatch between Evolved Dyadic Needs and the Current Culture of Health around the 4th Trimester

Alison M. Stuebe and Kristin P. Tully

4.1 Introduction

In the months following birth, a mother must navigate numerous challenges: she invests significant time and energy to protect and nurture her infant while at the same time recovering from childbirth. In the early postpartum period, mothers negotiate both physical sequelae of birth, such as lochia discharge, potential pelvic floor trauma and the onset of milk production, and psychological changes entrained by huge shifts in reproductive hormone levels following delivery of the placenta. These physiological changes occur amid adaptations to new roles and relationships (Tully, Stuebe, & Verbiest, 2017). The first 12-weeks postpartum is a significant transition period that has only recently begun to receive attention for most families in the U.S. and beyond (Stuebe et al., 2018). In this chapter, we describe the importance of the concept of the "4th trimester." Framing the period between birth and the first 3 months of motherhood as a "trimester" (Kitzinger, 1975) draws attention to the fetal-to-infant and woman-to-mother continuum. This outlook enables a deeper, more holistic understanding of perinatal health so that clinical, research, and lay fields can provide effective support that addresses not only proximate but also ultimate reasons for perinatal experiences.

Given the critical importance of reproduction and infant survival to humanity and the myriad approaches to reaching family and parenting goals, it is not surprising that an evolutionary lens on mother-infant dyads provides insights applicable to modern medical education, public health, and individual health guidance (Jasienska et al., 2017; Wells et al., 2017). We are not the way we are by happenchance. Our life history contributes to features of the 4th trimester experience, which in turn present as clinical and psychological challenges facing mothers and infants. Central to this process is negotiation of trade-offs between parents and offspring. Although caregivers and infants are deeply intertwined, physiologically and emotionally as dyads, they are also individuals with differing optimal circumstances. There is not one "right way" to parent,

Stuebe, A. M. and Tully, K. P., *Optimizing Maternal-Infant Health: Ameliorating the Mismatch between Evolved Dyadic Needs and the Current Culture of Health around the 4th Trimester* In: *Integrating Evolutionary Biology into Medical Education—for maternal and child healthcare students, clinicians, and scientists*. Edited by: Jay Schulkin and Michael L. Power, Oxford University Press (2020). © Oxford University Press. DOI: 10.1093/oso/9780198814153.003.0005

but instead a series of negotiations to manage from pregnancy through infancy and beyond.

Moreover, the demands of post-industrial, Western society differ from the environments that shaped human birth and childrearing outcomes for our ancestors. Genetic change occurs over many generations, and the human life course is relatively long, so we are said to "expect" to encounter conditions that have prevailed for millions of years. However, these deeply engrained expectations differ from current lived experiences, resulting in mismatches that impact health and wellbeing. Novel circumstances—such as cesarean section childbirth—can be life-saving or life-enhancing, but can also heighten risk of poor outcomes.

In this chapter, we explore ways in which life history theory and parent-offspring conflict contribute to maternal and infant health in the months following birth. This context can enhance current medical training and inform provision of more empathetic and effective clinical care. We begin by outlining life history theory as a tool for understanding that infant and maternal traits co-evolved and are therefore more appropriately treated in relation to each other.

4.2 Life history theory

Life history theory is a set of principles for how natural selection operates across the life cycle. This framework draws attention to coevolution of traits that improve the number of off-spring that survive to reproduce, which is referred to as an organism's "fitness." Organisms optimize fitness through trade-offs that impact growth, health maintenance, immune function, and reproduction. Examples include an "age and size at maturation, number and size of offspring, number of reproductive events, and ageing and lifespan" (Wells et al., 2017). All species are responsive or "plastic" to environmental conditions, to varying degrees. This flexibility has promoted survival through variable ancestral conditions. For humans, this process of being "biologically susceptible to context" seems to begin in utero and continue throughout development (Belsky & Pluess, 2013; Dipietro, 2012; Ellis, Boyce, Belsky, Bakermans-Kranenburg, & van Ijzendoorn, 2011; Enlow et al., 2017; Pluess & Belsky, 2011). If bio-psycho-social responses did not differ based on environmental cues and constraints, people would not have been able to thrive in such a wide range of settings.

For example, lactation is more than a means for infant nutrition—it is a survival strategy. Power and Schulkin describe lactation as a mammalian, primate, and human adaptation in detail (Power & Schulkin, 2016). Among mammals, humans in particular benefit from being "lactivores" in the early part of our lives because food sources were unreliable during the most recent formative "environments of evolutionary adaptedness" (Bowlby, 1969). Our ancestors relied on foraging and hunting. The Pleistocene environments in southern and eastern Africa were open and xeric with dispersed resources (Aiello & Key, 2002; Benyshek & Watson, 2006). Through lactation, a mother could buffer her breastfed infant from periodic shortfalls in food supply because milk production can utilize maternal adipose tissue. In contrast, avian species forage for food and regurgitate it for their young, and thus require a reliable food supply (Dall & Boyd, 2004). In addition to the ability of the mother to mobilize stored fat reserves, lactation would have co-evolved with gestational length, developmental status of newborns, the rate of infant growth including brain development that require mother-infant proximity (Ball , 2008). The care of human infants is costly, not only through transfer of physical stores from the mother to the infant but also in relation to the time. Our life course is long, and conditions vary over time; thus an evolutionary lens predicts variation in the extent to which caregivers "invest" in offspring. Our young are totally dependent on caregivers in the 4th trimester and then gradually become independent over years.

There are two general life history approaches, both of which are documented among humans in present times (Bereczkei, 2001; Scheper-Hughes, 1992). First, organisms facing high risk for mortality "favor" a faster life history, with earlier reproduction, more frequent pregnancies and greater numbers of smaller offspring, who receive relatively minimal care during infancy and beyond. On the other end of the spectrum, in settings of low mortality risk, organisms tend to have greater fitness in the evolutionary sense when they exhibit a "slow" life history: Infants are larger at birth, individuals have a more delayed onset of reproduction, mothers have fewer pregnancies and produce fewer offspring, and caregivers invest more heavily in their young. Life history strategies vary among species. Additionally, within a given species, such as among human populations, ecological conditions shape the approaches. Context matters. Thus, in low-resource settings or where infant mortality risk is high, early childbearing with shorter intervals between pregnancies may be adaptive in the ultimate sense that more offspring live to pass on their genes; in contrast, in high-resource settings, later childbearing and longer inter-pregnancy intervals may be the biologically optimal approach.

Awareness of life history theory is important for healthcare in the United States and elsewhere because it situates biological processes, such as increasingly earlier onset of menarche, and emotional health, such as the high prevalence of postpartum anxiety, on a deeper level than commonly considered. It addresses the "why." Ultimate levels of explanation for phenomena are observed and as expected are complementary to proximate discussion of mechanisms. Tinbergen put forth these interrelated questions to address not only how traits develop and operate (proximate levels of explanation), but to also draw attention to the evolutionary history (phylogeny) and forces that shaped the advantage that may be conferred (ultimate levels of explanation) (Tinbergen, 1963). Lewis (Lewis, 2008) suggested medical education is limited in that takeaways are "catalogs of clinically relevant information rather than explications of what that information reveals about the organism" (p. 400). Thinking through the combination of "how" and "why" questions are essential for fuller understanding and support of human health. Without this holistic view on the purpose of evolved suites of traits, the potential to classify phenomenon like infant night waking as pathologic is greater. That mindset of burden without acknowledgement of the positive function to survival can lead to frustration with our bodies and minds. We hope that consideration of how the current culture of health around the 4th trimester and other phases of human life are at odds with the ways in which humans have adapted to survive will lead to more appreciation and perhaps acceptance and support. We are primates and our heritage shapes our needs. The "lactation strategy" of frequent suckling, dilute milk, and a long period of nursing is consistent with the primate norm (Power & Schulkin, 2016), yet we often expect families to thrive with routine mother-infant separations and then place blame on women when they do not achieve their breastfeeding goals.

4.2.1 Lactation

After birth, all mammals, by definition, are primed to provide immunological protection and nutrition to their young through lactation. The thousands of extant mammals are testimony to the effectiveness of the class's unique adaptation. Milk secretion is thought to have first evolved about 310 million years ago (MYA) among synapsids (Oftedal, 2012). Synapsids laid parchment-like eggs that required a source of moisture; lactation emerged as a strategy to provide moisture and bathe the egg in an antibacterial secretion. Over the period from 310 to 160 MYA milk delivery evolved into mammary glands that transfer nutrients and protection directly to infants, and milk composition became tailored to fulfill the needs of a wide range of species. Milk is an evolved substance, as described in detail by Power and Schulkin (Power & Schulkin, 2016).

Among modern mammals, features of lactation vary by the life history of the species, including milk composition, frequency of feeding, and duration of lactation. Lactation and breastfeeding are biologically normative for humans and therefore the model against which alternatives are measured with regard to short- and long-term health outcomes (American Academy of Pediatrics, 2012). Among mothers, never breastfeeding or early weaning is associated with increased risks of breast cancer, ovarian cancer, diabetes, hypertension, and myocardial infarction (Stuebe & Schwarz, 2010). For infants, never breastfeeding or early weaning is associated with increased risks of otitis media, diarrhea, lower respiratory tract infection, sudden infant death syndrome, leukemia, and type 1 diabetes (Ip et al., 2007). Based on these associations, all major medical organizations recommend 6 months of exclusive breastfeeding, with continued breastfeeding through the infant's first year and beyond (American Academy of Family Physicians, 2014; American Academy of Pediatrics, 2012; American College of Obstetricians and Gynecologists, 2016). When, for example, vitamin D supplementation is advised, it is not due to an inherent deficiency in human milk—the need for the mother-infant dyad to take in extra levels reflects the significantly less exposure to sunlight that most people obtain relative to our ancestors. Importantly, human milk is evolved in relation to other parameters of our lives—lactation is not "perfect" because that is not how natural selection operates (Power & Schulkin, 2016).

Lactation is a dynamic process that provides individually tailored nutrients, immune factors, and physical contact to support normal growth and development. For most mammalian mothers and their young, there is a period of exclusive suckling after birth, followed by a transitional period during which both maternal milk and other foods are provided. The duration of transitional feeding varies among mammals, with longer periods being more common among mammals with a single, "precocial" offspring (Sellen, 2009) with slower life histories, such as primates. Communicating our infants' capabilities to patients may be helpful in encouraging cue-based parenting and minimizing the tendency for adults to attempt to foster independence in their children when it is developmentally inappropriate. Parenthood too is a developmental process. Clinical counseling can support families as they negotiate the intensity and rewards of childrearing over time.

4.2.2 Precocial vs. altricial offspring

The maturity of the neonate varies greatly across mammalian taxa and can be characterized along a continuum of "altricial" to "precocial." Altricial offspring have limited neurological development at birth and are typically born in litters with fused eyes and ears and minimal body hair; precocial offspring are more developed, typically born as singletons, and have open eyes and ears and a more extensive hair coat. The precocial offspring can locomote soon after delivery or cling to the caregiver. Martin and MacLarnon found more rapid and extensive maternal metabolic investment among organisms with altricial offspring, compared with lower growth and reduced metabolic demand on mothers in precocial offspring (Martin, 1985). Insectivores, rodents, and carnivores typically have altricial offspring, reflecting, in part, the need for mothers to hunt or forage for food while leaving immature litters in the nest between feedings; other mammalian orders, including primates, have relatively precocial young.

Human infants can see, hear, and have hair at birth—but we are referred to as being "secondarily altricial" (Portmann, 1990) because of the extensive neurological development that occurs following birth. Our infants are immobile as newborns and unable to cling to their mothers or others. Hrdy summarizes the extreme dependency of *Homo* infants on their caregivers in stating, "to lose touch was death" because "essentially the mother was the baby's niche" (Hrdy, 2000). The biology of our infants operates on a notion of close-contract; we are not a "nesting" species. The ways that our infants function and develop are interrelated with maternal physiology and behavior, which can be referred to as the "maternal ecology." The mother (or other caregivers) is the infant's environment of evolutionary adaptedness. For example, infant vision is developed at term birth to see the breast, which is the default center of their interactions in the neonatal period—and to help the infant see the breast, the maternal areola darkens during pregnancy. Similarly, skin-to-skin care is now understood in the medical literature to regulate infant temperature, breathing, and heart rate as well as promote breastfeeding (Moore, Bergman, Anderson, & Medley, 2016). Without considering the "need" that infants have for sustained physical contact, the impact of skin-to-skin may be viewed as transient. In reality, this caregiving approach is not only valuable in the immediate postpartum period as one of the *Ten Steps to Successful Breastfeeding*, but whenever it is practiced throughout the life course.

Continuous close contact with a caregiver was required for survival before the advent of bassinet warmers or home heating—so the practice can be expected to promote health for mothers and their newborns. Our infants could not, and do not, survive in isolation. Winnicott wrote, "There is no such thing as a baby … if you set out to describe a baby, you will find you are describing a baby and someone" (Winnicott, 1945). These expected patterns of behavior can be impeded by modern technology, thus creating iatrogenic obstacles for mothers and their infants. For example, standard hospital bassinets have four high sides that impede both visual and physical contact for newly postpartum mothers, inhibiting mothers from learning and responding to infant cues (Tully & Ball, 2012). An evolutionary perspective draws attention to these needs and informs human-centered design. To address this gap in patient-centered care, our team has recently developed an innovative medical bassinet to transform the experience of healthy mothers and their infants on the postnatal unit (see, Couplet Care Bassinet™ at sph.unc.edu/cgbi/hospital_infant_bassinet/).

Other great apes manage infant care in part through newborn clinging to caregivers with prehensile hands and feet. This ability to grasp the mother's coat permits the infant to remain at or near the breast continuously without disrupting the mother's ability to forage. However, while human infants manifest a grasping reflex, they are unable to support their weight, and human mothers lack graspable chest hair. Martin proposes that human ancestors solved this need for close infant contact with the creation of the sling. Further, Trevathan and Rosenberg (2016) propose that one of the factors contributing to newborn humans being born with a relatively high proportion of body fat is because appearing "cute" serves to engaging non-related adults as "allo-mothers" that assist with caregiving. Many people have strong opinions on infant care. This focus of attention on babies has evolutionary roots, in that investment from the wider community has been a core part of successfully raising children throughout our evolutionary history. It may be helpful to counsel new mothers that parenting advice—often unsolicited—reflects our heritage. We are not "neutral" to infants, and by extension—their caregivers. However, such opinions are not always accompanied by material support for the new mother, leaving mothers feeling judged and isolated.

4.2.3 Human life history compared with non-human primates

Human infants are more dependent on their caregivers than infants of other primates, so a community of childrearing support might be expected to be even more critical for us. The human neonatal brain comprises only 25% of its adult volume, in contrast to the > 45% present in non-human primate infants (Martin, 2007). In fact, human infants are the most immature of all mammalian newborns other than marsupials (Lawrence, 2005). Other mammals are born with 50% of their adult brain volume (Jurmain., 2008), and most singleton offspring have more than 75% (Lancaster, 1993). Evolution is based on bodily system constraints that enable individuals to live long enough to produce and raise surviving young, not refinement towards an ideal set of characteristics to make life easier, more healthful, or more prolonged. The 4th trimester is inherently challenging for caregivers, and it is a byproduct of the remarkable growth and development that infants undergo over the first postpartum year and beyond.

When we compare human life history with our nearest non-human primate relatives, we see several differences. Human reach sexual maturity considerably later; however, as Sellen (Sellen, 2007) notes, interbirth intervals are typically shorter in humans, compared with wild gorillas, chimpanzees, and orangutans. A pattern of shorter intervals between pregnancies allows humans to produce more offspring during their reproductive career. Sellen argues that this strategy succeeds in the setting of multiple adaptations related to feeding and caring for young that directly impact the 4th trimester (Sellen, 2009). Among large non-human primates, lactation lasts as long as 6 to 7 years, and juveniles are not fully weaned until they are able to forage and provision themselves with adult food; in contrast, human children are generally weaned well-before they are or would be able to forage for themselves.

Infant feeding is a prime example of how disjointed the current culture of health in the U.S. and many other settings is for families, and how an evolutionary perspective can serve to strengthen understanding of maternal-infants needs and inform strategies for more effective support. Breastfeeding is implicitly expected to be a part of the many daily activities women accomplish on their own, without accommodations to enable the bio-psychologically intensive activity. With regards to the biological level, the energy cost of human lactation is about 26% above the female non-reproductive (maintenance) state (Dufour, Reina, & Spurr, 2002). Direct physiological investment by the mother to her offspring continues until complete weaning (Valeggia, 2001). Lactation is the final stage of labor (Labbok, 2001), with its termination as the physiological completion of the reproductive cycle, but women are confronted with a myriad of challenges in realizing their infant feeding goals.

In addition to human metabolic costs of lactation, extensive time is needed for infant caretaking and childcare. Human milk composition, like other large primates, is relatively low in fat, necessitating frequent feeding. Viewing humans as a biologically "continuous contact" species (Shaul, 1963) draws attention to the sustained intensity of childrearing. The limited neurological development at birth was possible in our evolution only because parental and allo-parental care adapted to meet the high infant needs. However, a potential drawback to this system is that human maternal commitment to offspring has been documented as being more contingent on environmental circumstances in comparison with other primates (Hrdy, 2000). We have the capacity to exhibit "facultative" responses to young—humans can heavily invest in infants, not do so at all, or anywhere in between. Whether consciously or not, prioritization occurs based on the predicted costs and benefits of the competing demands of growth, maintenance and reproductive effort. We provide overview of the concept of "parent-offspring conflict" to ground parenting strategies in relation to trade-offs. These trade-offs reflect the constraints that caregivers must negotiate to promote overall family well-being. Despite the ubiquity of this process, it is infrequently addressed clinically. Framing family life in relation to efforts to optimize outcomes for each family member may be a helpful tool in elevating parental strengths and then discussing areas of care that the mothers and others might like to alter.

4.2.4 Parent-offspring conflict

As Hrdy (Hrdy, 2016) has noted, humans stand apart from great apes as "cooperative breeders," who benefit from community support in nurturing our infants and raising young. It is often said that "it takes a village" to raise a child. Non-human primates such as marmosets use a similar strategy—marmosets birth twins, and a single mother cannot carry and care for two infants at the same time. As such, she breastfeeds, but delegates carrying and other caregiving to other community members. Notably, if such support is not available, marmosets will abandon or kill their infants.

Haig (Haig, 1993, 2008) builds on Trivers's (Trivers, 1974) view that discordant interests are inherent between parents and offspring, with offspring selected to acquire more parental resources than is in the parents' best interest to provide. This physiological and behavioral "tug of war" is due to the infant striving to be as healthy as possible without draining the caregiver to a degree that she/he can no longer invest. In contrast, the parental strategy is to produce healthy offspring (that lives to reproduce) at the lowest possible cost. Vitzthum (Vitzthum, 2008) explains that it is in a human mother's best interest to balance resources among previous, current, and future offspring to ensure their survival to reproductive maturity. The high-level, subconscious conflict is theorized to be ultimately due to sexual reproduction because a child's genes are 50% different than those of either parent.

The observation that genetic relatedness does not guarantee parental investment has led to a mathematical formula to predict investment outcomes. Hamilton's Rule (Hamilton, 1964) comprises three key variables: degree of relatedness between the donor and recipient (r); benefit of investment to the recipient (B); and cost of the resources to the donor (C). With his model, investment is evolutionarily "worthwhile" if $rB > C$. The prediction therefore is that animals, including humans, will preferentially invest in close relatives, individuals with perceived high potential for future reproduction, and those who incur relatively low costs (Strassman, 2008). Evolutionary theorists have framed aspects of women's reproductive experience previously regarded as pathological, such as nausea during pregnancy (Flaxman & Sherman, 2008; Profet, 1992) and miscarriage (Haig, 1993; Vitzthum, 2008) as energetically protective for the mother. Morning sickness is considered a defensive mechanism against maternal ingestion of teratogens that could harm the developing embryo, and early pregnancy loss can be a screening tool for conceptions that are not likely to survive long term. Viewed within this broader life-history context, lactation too can be understood as a balance between maternal self-investment and care for her infant.

Even in conditions of plenty, humans face opportunity costs. For example, like many aspects of parenting, women determine the duration and intensity of breastfeeding based on their circumstances, experiences, and preferences. Instead of carrying out a "fixed action" behavior to breastfeed for a set period of time, parturient women are physiologically primed to breastfeed and then invest at different levels. As outlined previously, an evolutionary lens predicts less maternal investment in the form of breastfeeding when costs of infant care are high or the benefit of the input is low. Mothers may decide to expend a portion of their finite time and effort towards breastfeeding, or they can draw upon alternative pathways for infant care such as human milk substitutes, donor human milk, or even lack of engagement.

The maternal "instinct" to raise few children and invest heavily in each is a specific historical construction. In the face of high infant mortality or other critical variables, infant attachment may be initially withheld. This clinically concerning and seemingly contradictory maternal behavior towards her offspring may serve to protect women's emotional and physical engagement with their infants. Reiches and colleagues (Reiches et al., 2009) discuss that among human females, it is adaptive to differentially care for offspring because the level of investment required to gestate, lactate, and otherwise raise them is considerable. They summarize the "energy budget" and human life history theoretical perspective, "It is adaptive to commit to these expenditures only when prospects for success are reasonable and only to a degree that optimizes lifetime reproductive success" (Reiches et al., 2009).

This balance is influenced by factors that are intrinsic and extrinsic to the mother-infant dyad. Tully and Ball (Tully & Ball, 2013; Tully, 2018) use the concept of marginal returns to consider the degree to which breastfeeding or other parenting investment is "worth it" in a given context. For each woman, there is theoretically a maximum "profit" where the difference between the benefit to the infant and the cost to herself is greatest. The hypotheses generated from this model are: 1) reduction in the maternal cost of breastfeeding promotes greater breastfeeding outcomes; 2) increase in the infant benefit of breastfeeding promotes greater breastfeeding outcomes; and 3) reduction in the maternal cost and increase in infant benefit (or perceptions thereof) will be more effective in promoting greater breastfeeding outcomes than addressing either maternal cost or infant benefit in isolation. These predictions can be used to inform counseling through validating the "work" of breastfeeding alongside the joy and self-fulfillment women may experience meeting their infant feeding goals. Every effort should be taken both to educate on health impacts of parenting practices and to establish environments in which women are enabled to realize their goals.

4.2.5 Integrated systems

Care of human offspring is so intensive and so prolonged that it is theorized that the importance of "allo-mothers" may have contributed to human ancestors living beyond their reproductive phase: grandmothers provided advantage in the raising of multiple dependent children (Sellen, 2007). Despite the prevailing notion in modern Western society that mothers can "parent alone," extended family, peer support, postpartum doulas, and other approaches have the potential to recapitulate the cooperative parenting "village" that seems to have been a foundational part of human reproductive success. Just as human mothers do not generally birth alone (Trevathan & Rosenberg, Chapter 3, this volume), optimal childrearing from both the child and maternal perspectives does not occur in the absence of sustained practical and emotional support. This shared responsibility for parenting continues as children mature: Van Noordwijk and colleagues note that human children are weaned before they would be able to keep up with their foraging mothers, and thus often stayed behind in a community under the care of allo-mothers (Van

Noordwijk, Kuzawa, & Van Schaik, 2013). "Nursery school" is part of our evolutionary heritage.

In settings of limited resources, frequent, full breastfeeding suppresses ovulation, preventing pregnancy until the mother regains sufficient energy stores to support a pregnancy. Lactational amennorhea is neither a pathology nor coincidental process. On a proximate level, an infant breastfeeds for years for physical and emotional growth and development; frequent lactation further benefits the infant by suppressing ovulation and delaying the arrival of another sibling. To the extent that local conditions "favor" a slower life history, this delay in ovulation benefits the mother as well. Prolonged lactation can lengthen the time between pregnancies; conversely, earlier weaning allows shorter inter-birth intervals. Ultimately, however, the mother's interest in producing additional offspring will conflict with the child's "interest" in high levels of sustained care. Trivers (Trivers, 1974) proposed that this trade-off could be quantified by considering the benefits to the offspring and the cost to the mother of the continued parental investment, as well as the cost to the offspring of deferring the arrival of another sibling with shared genetic inheritance. These trade-offs manifest today in tensions not only between siblings but also between the postpartum demands of infant caregiving and maternal self-care. Mothers must care for themselves both to be able to fully support their infants and also because, fundamentally, "mamas matter."

4.3 Applying an evolutionary lens to health needs in the 4th trimester

To optimally support women becoming mothers, health care providers can acknowledge the inherently challenging process of becoming a new family, educate on the physical and emotional trade-offs associated with health behaviors, and provide realistic, individually-tailored guidance for parents to realize their health goals. The evolutionarily-derived features of human reproduction contribute to common clinical concerns in the 4th trimester. The Listening to Mothers III survey queried American women regarding major and minor challenges in the 2 months following birth (Figure 4.1) and found considerable levels of morbidity (Declercq, Sakala, Corry, Applebaum, & Herrlich, 2013). These issues reflect a legacy of human physiology in the context of a post-industrial society that does not align with the needs of the mother-infant dyad. Many postpartum health challenges are more pronounced than the support that is provided.

Our heritage of socially-supported childbirth and postpartum care is at odds with contemporary norms in many groups, particularly in the United

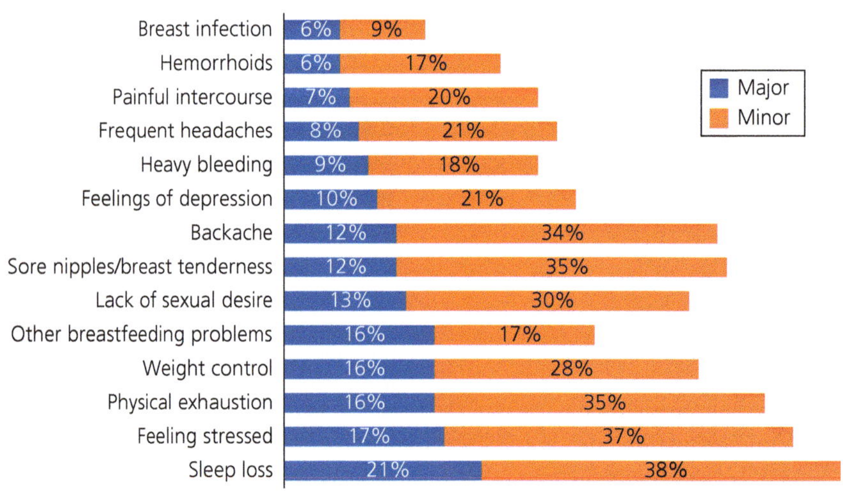

Figure 4.1 Maternal health problems in the first 2 postpartum months
From Declerq et al. (2013).

States. As of this writing, we are one of four nations without any national policy for paid family leave. A recent study found that 23% of employed women in the U.S. return to work within 10 days after birth (Klerman, Daley, & Pozniak, 2014), unable to access, or able to access but unable to afford, the unpaid job protected leave provided by the Family and Medical Leave Act (FMLA). Moreover, economic constraints have required a growing proportion of men and women to work full time, such that the traditional "village" of community members that might contribute meaningful care for the mother and newborn no longer exists. Furthermore, women who have completed childbearing are more likely to be engaged in paid work than in prior generations (U.S. Department of Labor Women's Bureau), further limiting the population of allo-mothers available. Alongside these employment constraints, American culture lacks a tradition of "lying in" following birth that is found in many traditional cultures (Dennis et al., 2007). These traditions date back to Biblical times—it is written that Mary brought Jesus to the temple at 40 days, after a period of rest and recovery (Luke 2:22). However, in the U.S. in the early twenty-first century, the only vestige of a 40-day period of recovery was the "six week postpartum check" (Stuebe et al., 2018). Instead of an evidence-based arrangement for health care service delivery, this culturally-constructed approach means many mothers are devoid of personal and professional support when they need it most. Instead of being woman-centered, the standard postnatal care that many mothers experience can be "inhumane" in that it does not reflect our real needs. Our evolutionary history means we function in ways at odds with current medical, economic, and social systems.

4.3.1 Mismatches

Based on non-human primates, the average time for complete weaning of human infants would be expected to be 4 to 7 years (Dettwyler, 2004). However, it appears that many human populations have long weaned earlier than this cross-species prediction (Humphrey, 2010). As discussed, infant and young child feeding practices are highly sensitive to ecological context, such that women consider (consciously or not) the trade-offs of any and continued breastfeeding (Tully & Ball, 2013). Such constraints play out in contemporary societies, where a woman's ability to provision herself and her child is influenced by the availability (or lack thereof) of paid parental leave and otherwise accommodating workplaces, perceptions of optimal parenting, and social support.

In traditional agricultural societies, women's activities can be shaped around the need for frequent infant contact for feeding, but industrial workplaces do not accommodate a baby at breast, such as while mothers work along an assembly line. As Wolf (Wolf, 1998, 2003) has detailed, industrial norms have been orthogonal to infant feeding behavior; nineteenth-century doctors advised women to nurse on a schedule, to prepare their infants for the rigors of modern life. This scheduling precipitated an epidemic of insufficient milk supply because milk production relies on frequent suckling. Medical publications at the time decried the modern woman's inexplicable inability to produce sufficient quantity of milk and drew attention to the attendant rise in infant mortality due to feeding with unpasteurized cow's milk. The solution adopted was a public health campaign to improve the sanitation of the supply chain for cow's milk. By the 1920s in the US, feeding of pasteurized cow's milk had largely replaced breastfeeding.

Conflicts between societal expectations and feeding physiology continue a century after the infant feeding contributors that Wolf chronicled. Individuals regarded as parenting experts throughout the twentieth century recommended scheduled feedings at intervals incompatible with exclusive breastfeeding, which is summarized by Ball (2007). Baby supply stores now teem with swaddle wraps, swings, shushing machines, pacifiers, and other devices that claim to obviate the need for infants to be held and fed at physiologically normal intervals. These products are a direct response to the substantial loss of allomothers. Most American women do not have access to any paid leave after birth, especially when considering beyond accrued sick and vacation time, and among those who do, leave is often limited to 6 to 8 weeks after birth. This duration is 20 weeks shy of the recommended 6 months of exclusive breastfeeding. Further, breastfeeding conveys more than nutrition, including child comfort for emotional development and hormones tailored to the day or nighttime (Cubero et al., 2005; Nozhenko, Asnani-Kishnani, Rodriguez, & Palou, 2015).

In the context of global comparisons of maternity and paternity leave, provisions in the U.S. stand out as inappropriate. In an attempt to mitigate unmet health needs, breast pump manufacturers offer a variety of devices that women can transport to and from work, along with hands-free bras and other supplies to facilitate milk expression and storage. High-quality childcare where professional allomothers feed and care for infants while their mothers attend to paid work is prohibitively expensive for many. As a result, few women meet their breastfeeding goals, yet little is done about the systematic factors that create such high opportunity costs of parenting. Little attention is devoted to offering flexible work arrangements and valuing the effort that goes into sustaining offspring life in addition to one's own. Subsequently, women whose families suffer the most are those marginalized by poor quality education, language barriers, systemic racism, and limited job.

More evolutionary-appropriate solutions for promoting human health exist; in the Babies at Work model (babiesatwork.org), businesses and parents arrange for infants to be worn and carried for during the working day, so that mothers can attend to both their paid work and their reproductive effort. Similarly, on-site childcare can allow mothers to feed their infants directly during the working day. However, such options are more common among highly-paid professionals, for whom innovative work options are an effective incentive to recruit and retain the most sought-after employees. Such options are less common among low-wage, hourly workers, contributing to health disparities. Indeed, as Burtle and Bezruchka (Burtle & Bezruchka, 2016) have written, "The lack of policies substantially benefitting early life in the United States constitutes a grave social injustice: those who are already most disadvantaged in our society bear the greatest burden." Indeed, among oppressed populations, social constraints coupled with production and marketing of infant formula have pushed the plasticity of human age at weaning to the limit: mothers essentially are forced to forgo breastfeeding and much of their time infant nurturing in order to return to paid work.

Instead of centering the unique, irreplaceable, and critical role of mothers so that they have the opportunity to realize their parenting and other goals, the dominant discourse of Western feminism in the 1970s focused on freeing women from caregiving responsibilities. As Galtry writes (Galtry, 2015):

> Women's rights with regard to infant feeding were often construed as the right NOT to breastfeed. It did not always occur to policymakers and legislators (as evidenced, for example, by the United States' "equality versus difference" debates) that, the lack of structural provisions such as breastfeeding breaks in the workplace meant that many women did not actually have the right to breastfeed (p. 6).

Within this context, it is not surprising that the Convention on the Elimination of Discrimination Against Women (CEDAW) had very little content related to breastfeeding, limited only to a parenthetical comment that women had a right to adequate nutrition during pregnancy and lactation.

In the US, recent breastfeeding policy has incorporated structural determinants, reducing emphasis from an individual mother's responsibility to sustain breastfeeding to her wider context (US Department of Health and Human Services, 2011). Healthy People 2020 goals for breastfeeding include maternity practices that support breastfeeding and workplace lactation support. However, national legislation regarding paid, job protected family leave has not been adopted. Although breastfeeding initiation rates are relatively high in the U.S., rates of exclusivity and continued duration are not—which also reflect significant disparities.

The mismatch between evolutionary imperatives and modern realities can frame instructive conversations in medical education, particularly given that consensus medical recommendations for 6 months of exclusive breastfeeding do not conform to many women's lived experiences in contemporary modern life. These conflicts manifest in the lived experience of medical students and physicians in training as well: a recent qualitative analysis describing the experience of pediatric residents with breastfeeding was titled, '"Frustrated,' 'Depressed,' and 'Devastated' Pediatric Trainees" (Dixit, Feldman-Winter, & Szucs, 2015). These adverse experiences may stimulate physician-parents to become more knowledgeable about lactation and breastfeeding support for their own patients, or they may manifest as dismissal of breastfeeding recommendations as impractical or unrealistic. Such tensions may explain why, comparing surveys in 1995 and 2014 (Feldman-Winter et al., 2017), pediatricians were less likely to report that the benefits of breastfeeding outweigh the difficulties (1995: 70%, 2014: 50%, $p<.05$).

Medical educators should also be cognizant of the extent to which policies and facilities accommodate (or fail to accommodate) the needs of lactating students. Are policies in place making infants welcome for preclinical lectures and small group discussions? What policies are in place for leave of absence for birth and postpartum recovery? Are accommodations proactively communicated or available only through parental inquiry or advocacy? How do student insurance coverage policies and tuition payment requirements intersect to support or penalize women who bear children during medical training? And how do board certification requirements for residency impact childbearing and childrearing for physician-parents?

4.4 Perinatal depression and anxiety

Parenting requires sustained attention of caregivers to ensure the health and safety of offspring (Hahn,-Holbrook, Holbrook, & Haselton, 2011). This diligence manifests as increased sensitivity to threats during pregnancy and the postpartum period, which ultimately serves to protect the mother and young. On a proximal level, this sensitivity may be mediated, in part, by falling levels of progesterone after delivery (Nappi et al., 2001). Progesterone is produce by the placenta, and levels are elevated markedly in pregnancy. This reproductive steroid is metabolized to allopregnanolone, a neurosteroid that is a potent agonist of the gamma amino butyric acid (GABA) receptor, conferring an anxiolytic affect. After delivery, allopregnanolone levels fall, resulting in increased irritability and anxious behavior, particularly in the days following birth. The loss of progesterone may underlie the "Baby Blues," a period of emotional lability that affects about 80% of mothers and lasts about 10 to 14 days. Researchers have hypothesized that typical maternal physiology includes upregulation of expression of the GABA receptor to compensate for falling allopregnanolone levels, and more severe depressive symptoms may occur in women with less upregulation of GABA receptor expression. Indeed, in an animal model, knock-out of a subunit of the GABA receptor resulted in markedly impaired mothering behavior (Maguire & Mody, 2008). More recently, clinical trials of an allopregnanolone analogue have shown promise as a treatment for severe postpartum depression (Kanes et al., 2017).

Molecular mechanisms notwithstanding, there is a sound evolutionary basis for anxious parental behavior. As Nesse (Nesse, 2001) has articulated, it is advantageous for defenses against potential threats to be more sensitive than specific. Faced with a sound behind a bush that may be a predator, an animal is likely to flee, as the cost of fleeing is considerably smaller than potentially being attacked: "A relatively inexpensive defense, like flight, should be expressed if there is even a small chance of a catastrophic harm, like attack by a predator." Consistent with this "smoke detector principle," a perceived threat to a mother's offspring prompts a protective response. Being "on alert" to a certain level serves to protect the health and well-being of families; however, it can also cause undue emotional and physiological strain.

A corollary to the smoke detector principle is that, when there is no fire, there needs to be a mechanism to shut off the alarm. Hahn-Holbrook and colleagues (Hahn-Holbrook et al., 2011) elaborate on this concept in their review of evolution and maternal anxiety, noting that motivation systems involve appraisal, response selection, and satiety:

> Parenthood is stressful, in part, because parents' precaution systems naturally generate worry over potential future harm coming to their children. The reward response derived from affiliative interactions with offspring buffers otherwise deleterious levels of heightened anxiety related to parenting, helping to satiate precautionary parental drives.

In this framework, close contact between mother and infant stimulates reward networks, via dopamine and oxytocin, resolving episodes of anxiety. A failure of this "reset" mechanism may manifest as perinatal depression and anxiety, which affects 15% of mothers (Gavin et al., 2005). Of note, postpartum anxiety and depression are associated with dysregulation of the hypothalamic-pituitary-adrenal axis, with symptomatic mothers experiencing increases in pituitary adrenocorticotropic hormone (ACTH) without the expected risk in adrenal secretion of cortisol, impeding negative feedback (Jolley, Elmore, Barnard, & Carr, 2007). We further found that levels of oxytocin were lower during breastfeeding among women with depression/anxiety symptoms. These results suggest that the calming effect of breastfeeding and close infant contact is disrupted among women with mood symptoms (Stuebe, Grewen, & Meltzer-Brody, 2013). Moreover, to the extent that close infant contact mitigates maternal anxiety, anxiety may be worsened by routine separation such as hospitalization in neonatal intensive care, routine removal to the newborn nursery, or constraints that require women to return to paid work.

In other cases, adverse life experiences may undermine the reward circuitry that counterbalances the maternal "smoke alarm." Evidence from animal models suggests that poor care in infancy changes the epigenetics of the oxytocin system, compromising maternal behavior in adult animals (Beery, McEwen, MacIsaac, Francis, & Kobor, 2016). Similar disruption of oxytocin circuitry may mediate associations between adverse childhood experiences and postpartum mood disorders. Dysregulation in childhood may lead to pathological responses during parenthood.

Insufficient social support may also manifest as postpartum depression. Nettle argues that low mood states may be adaptive in the setting of restricted resources (Nettle, 2009). Discussing foraging behavior in the setting of poor food supply, he argues that a hungry, but not starving, individual may opt not to forage, in order to avoid expending energy, rather than expending effort and using up stored energy with the possibility of no return on the investment. Faced with overwhelming odds, doing nothing may be wiser than doing something. This line of thought is of particular interest in the setting of human cooperative childrearing. Our reproductive strategy relies on the support of allomothers to sustain mother and child in the early months after birth. This perspective suggests that if such support is not present, a mother may withdraw from her infant and from her caregiving responsibilities, just as a marmoset mother rejects her infants if there is not sufficient social support on hand (Hrdy, 2016). Citing Hrdy, Crouch proposes that some forms of postpartum depression reflect isolation encountered in modern parenting (Crouch, 2002).

Applying an evolutionary lens to the interplay of molecular, neurologic and social influences on maternal mood underscore the need for a multifaceted approach for management of perinatal mood disorders. Treatment strategies must alleviate maternal symptoms, repair mother-infant interactions, and ensure sufficient social support. The are many, interrelated ways to recapitulate the "village" of allo-mothers that is our evolutionary legacy.

4.5 Sleep and fatigue

In the Listening to Mothers III survey, sleep disruption was the most prevalent concern of mothers in the first 2 months postpartum (Declercq et al., 2013). Whereas infants usually wake to feed every 1 to 2 hours, adults generally benefit from more prolonged bouts of sleep. This mismatch in health needs between parents and their young is aggravated by Western expectations of independent sleep and settling for infants, with focus on "sleeping through the night" in the first few months after birth. Adding further complexity and discord is the positive association of exclusive and continued breastfeeding with parent-infant bedsharing when sharing the same sleep surface is currently understood to be a risk factor for Sudden Infant Death Syndrome (SIDS). The way we understand the 4th trimester shapes public health messaging around maternal and child health. Educational campaigns in the United States have deployed billboards with headstones as a part of adult bed headboards with messages such as, "Don't let this be your baby's final resting place" (city.milwaukee.gov/health/

Safe-Sleep-Campaign). Parent-infant room sharing and the particular sleep arrangements families experience at home are complex, reflecting family preferences, needs, and constraints (Tully & Sullivan, 2018).

As a result of shame-based approaches to nighttime parenting, mothers have been fearful of breastfeeding while in bed, instead using couches or in rocking chairs for infant feeding (Kendall-Tackett, 2010). In addition to disrupting parental sleep and instilling a "battle" mentality around bedtime, this messaging may actually increase infant risk of infant falls and suffocation, in that an infant whose parent falls asleep on a couch or in a rocking chair is more likely to be dropped and become lodged in a crevice than an infant whose parent falls asleep on a bed. To this end, the American Academy of Pediatrics (Moon & Task Force On Sudden Infant Death, 2016) currently recommends room-sharing without bedsharing, but suggests nighttime feedings occur in an adult bed. This strategy is in recognition of fatigue and of oxytocin released during milk let-down, which often leads to unintentional shared sleep. Parent-infant sleep on an adult bed is safer than elsewhere. Other sleep environment guidance holds true in the variety of arrangements infants experience—the sleep surface should be firm, free from soft bedding, pillows, and toys, and responsive parenting should not be undermined by pharmacological substances or alcohol. Nighttime infant sleep locations are multi-faceted too because of the inherent parent-infant conflict involved with the developmentally-based sleep patterns, moral judgments (Tomori, Palmquist, & Dowling, 2016), and limited dialogue in clinical settings, especially in the U.S. (Ball, 2017). Bias about what constitutes healthful and appropriate infant sleep is also evident in scientific studies (Tully & Sullivan, 2018).

Here, again, an evolutionary perspective can be integrated into medicine to illuminate mismatches, which in this case contribute to parental exhaustion. McKenna has proposed that the biological norm for nighttime mother-infant interaction is "breastsleeping" (McKenna & Gettler, 2016), a state of close physical contact that supports synchronized dyadic breathing, feeding, and thermoregulation. Ball et al. have studied maternal nighttime position, finding that a breastfeeding mother characteristically turns toward her infant, forming a protective "C" shape, and provides access to the breast throughout the night (Ball, 2006). Review of SIDS data further suggest that, in the exclusively breastfeeding, non-smoking mother, such shared sleep does not increase mortality risk (MV-adjusted OR 1.1, 95% CI 0.6-2.0 for infants less than 3 months old) (Blair, Sidebotham, Pease, & Fleming, 2014). Risk reduction efforts often focus on education and message exposure, but families live in diverse conditions with varying infant needs and responses to particular parenting approaches. Caregivers adopt infant sleep practices for a variety of reasons, which may differ from original plans and reflect considerations more central to their daily lives then other factors (Tully, Holditch-Davis, & Brandon, 2015). Volpe and colleagues (Volpe, Ball, & McKenna, 2013) apply life history theory to clinical guidance around safe sleep environments:

> Clinicians may use this awareness of maternal trade-offs to inform conversations with mothers, such as by asking them if there are specific barriers that prevent them from implementing safe sleep guidelines or by exploring how they prioritise their own needs with different aspects of infant care. Furthermore, understanding and acknowledging the complexity of maternal strategies and the reality of negotiating trade-offs suggests that if clinicians want mothers to adopt specific risk-reduction behaviours, they must assist mothers in figuring out how to tolerate the potential "costs" of these behaviours and help them implement strategies to ameliorate such costs.

However, mismatches exist between nighttime parenting behavior and modern constraints. Shared sleep contrasts with Western values related to individualism, self-sufficiency, and understanding of infant development (Ball & Volpe, 2013); moreover, parents may express concerns about privacy and resumption of sexual activity when the infant is sleeping in the same room. And to the extent that frequent night feeding, even as breastsleeping, prevents consolidated parental sleep, it can be unsustainable when parents are expected to participate fully in daytime paid work.

In the face of this tension, an extensive industry has emerged, including infant "sleep training" manuals, apps, and consultants who advise on strategies for reducing night waking and parental responses to infants. A Google search for "my newborn doesn't sleep at night" yields 58.8 million results; an Amazon search for "baby sleep" yields more than 100,000 results. Infant sleep engenders considerably controversy, with some arguing that efforts to limit nighttime parenting adversely affects child stress reactivity and that deep, prolonged infant sleep may actually contribute to SIDS risk (McKenna, Ball, & Gettler, 2007; McKenna & McDade, 2005), and others that responding to night waking exhausts parents with limited benefit to the child. From an evolutionary standpoint, these debates reflect an ancient conflict between parent and offspring, in which trade-offs depend on relative risks within the current ecological context. Sleep is a developmental process which is subject to bidirectional influences, but our understandings of "appropriate" infant and parental behavior is not often recognized as being influenced by both biology and culture.

4.6 Contraception and birth spacing

According to contemporary U.S. guidelines, there are no data to support common recommendations that women delay resuming intercourse until 6 weeks postpartum; the risk for infection is minimal at 2 weeks after birth. This clinical guidance differs from taboos in many cultures that prohibit intercourse before postpartum bleeding has ceased (Dennis et al., 2007). Longer periods of abstinence are recommended in some traditional cultures: a 1972 analysis of ethnographic data found that postpartum taboos longer than one year were more common in polygynous cultures where power was consolidated among older males (Saucier, 1972). Clinicians should be aware that there is wide variation in cultural proscriptions against resuming intercourse after childbirth and that the 6 week recommendation is derived from the traditional timing of the standard obstetrical postnatal visit, not evidence.

Provision of modern contraception is a primary focus of postpartum health care services, appearing on the top of the checklist for medical students and residents rounding on postpartum patients. Recent policy efforts have focused on providing long-acting reversible contraception (LARC) prior to hospital discharge. These methods are highly effective, with failure rates of less than 1%, and last for 3 to 10 years (Curtis et al., 2016); once removed, women can become pregnant almost immediately.

This laser-like attention to preventing another pregnancy is sometimes not aligned with the priorities of new mothers, particularly in the first few days after birth. A focus on preventing the next pregnancy can be of concern for marginalized populations. Counseling to avoid an unintended pregnancy can be perceived as an effort to limit reproduction, especially among groups for whom births are less valued. One Latina focus group participant commented:

> Birth control in general makes me really wary because it was meant to keep people like me from procreating and having more of us, right? .. I don't really trust doctors because I don't know what subconscious things are going on when certain methods are being recommended to me. (Higgins, Kramer, & Ryder, 2016, p. 1936).

Recognizing the complex history of birth control provision among women of color, the SisterSong Reproductive Justice Collective has drafted guidelines for counseling on these long-acting methods (SisterSong, 2017).

Tensions also exist regarding the potential impact of hormonal contraception on breastfeeding. Progesterone withdrawal after birth is thought to trigger onset of milk secretion, and exogenous progesterone might inhibit milk production; however, randomized controlled trials have not found an adverse effect of hormonal contraception on breastfeeding outcomes (Lopez et al., 2015). Limitations of published studies include relatively small sample size and inclusion of healthy mothers with term infants; these studies may not be applicable to mothers at high risk of breastfeeding difficulties, such as those who are expressing milk for hospitalized infants.

Across diverse human populations, frequent suckling of the infant at the maternal breast suppresses fecundity, which is the ability to conceive and develop offspring. Ovulation resumes as breastfeeding becomes less frequent, depending, in part, on when maternal energy stores are adequate to sustain another pregnancy. Under natural fertility conditions—meaning no commercial pharmacological or barrier birth control methods—in preindustrial cultures, births occurred at intervals of about 3 years (Sellen, 2007).

Modern conditions have substantially altered the relationship between lactation and fertility. First, whereas modern hunter-gatherer populations such as the !Kung wear their infants in slings and nurse multiple times an hour, women in modern cultures typically breastfeed at intervals of at least 2 to 3 hours. Lactation can be an effective contraceptive for women who are fully breastfeeding at intervals of less than every 4 to 6 hours, have not resumed menses, and are less than 6 months postpartum (World Health Organization, 2015). Once any one of these conditions is no longer met, however, another form of contraception is needed to prevent pregnancy. Women who are expressing milk and bottle-feeding may have longer intervals between breast stimulation, and it is not known whether a breast pump is as effective as an infant in suppressing ovulation.

Changes in maternal energy balance also influence the return of fecundity. Gibson and Mace (Gibson & Mace, 2006) studied interbirth interval in rural Ethiopia in villages with and without access to water taps. In villages without taps, women carry water for an average of 3 hours per day during the dry season; in village with taps, water carrying was reduced to 15 minutes per day. The authors found that this decreased energy expenditure burden for mothers was associated with a shorter interbirth interval. Ultimately, human fertility is responsive to such conditions because pregnancy and lactation are substantial time and energy investments, and it would not be advantageous to develop offspring if resources were so limited that survival was jeopardized. For most U.S. women however, plentiful nutrition and reduced physical activity from our ancestral past means that access to energy often does not limit return of ovulation.

The purpose of offering postpartum contraception options to women is to help them achieve their birth spacing goals. Evidence from high-income countries suggests that inter pregnancy intervals (IPI) of greater than 18 months improve maternal and infant outcomes for the subsequent pregnancy; however, such studies are confounded by differences in socioeconomic status. This confounding is not surprising, when considered through the lens of life history theory. As noted previously, in poor ecological conditions, a "fast" life history has advantages: early menarche, rapidly followed by many short-interval pregnancies that result in numerous small offspring maximizes genetic "throughput" into the next generation. Poverty and oppression can be contributing factors of short IPI, for which the solution is not only communication of health trade-offs and access to acceptable, safe, and affordable contraception but also improvement in material circumstances.

4.7 Physical recovery from birth

The human form reflects function. The development of bipedalism in our ancestors corresponds with pelvic anatomy, which impacts the birth process because of the encephalization that occurred more recently in our evolutionary history. Wells et al. propose that mismatches in human evolutionary history and current conditions may be exacerbating the "obstetrical dilemma" (Wells, DeSilva, & Stock, 2012). As humans shifted from foraging to agricultural diets, lower protein intake and higher glycemic load resulted in shorter maternal stature and greater fetal growth, increasing the risk of obstructed labor. Additionally, with the transition to upright posture, the pelvic floor assumed responsibility for supporting the abdominal viscera (Schimpf & Tulikangas, 2005). The soft tissue of the pelvic floor must be strong enough to keep the uterus, bladder, and bowel situated inside a woman's body, and yet flexible enough to allow passage of the neonate.

Interestingly, recent work suggests that the dimensions of the female pelvis change across the lifespan to accommodate the changing priorities of birth and prevention of pelvic organ prolapse. Huseynova and colleagues (Huseynov et al., 2016) analyzed CT scans in a cross-sectional sample of males and females across the lifespan and found that from the age of 15 to 25, developmental changes unique to women create a "wide, obstetrically favorable birth canal." Later, at ages 40 to 45, further developmental changes transform the pelvis to a shape more similar to 20 to 25-year-old males. The authors speculate that this narrowing improves abdominopelvic stability during post-menopausal life. The same constraints might contribute to the higher cesarean section rate observed among mothers in their forties (Vaughan, Cleary, & Murphy, 2014).

Tradeoffs between pelvic floor support and sufficient space for passage of the neonate contribute to maternal morbidity and mortality. In the US, nearly one third of postpartum women report urinary and bowel problems in the first 2 months after birth. In settings without access to cesarean delivery, obstructed labor can be fatal; it is cited as accounting for 12% of maternal deaths worldwide (Wells et al., 2012).

Bipedalism is also hypothesized to have contributed to the evolution of the human placenta and to postpartum bleeding. In humans, vaginal bleeding persists for weeks following birth—a recent systematic review found that blood loss persists for a mean of 24 to 36 days after birth, with durations as long as 90 days in some studies (Fletcher, Grotegut, & James, 2012). Abrams and Rutherford (Abrams & Rutherford, 2011) link postpartum bleeding to bipedalism. They note that the hominoid placenta is considerably more invasive than other mammalian species. During early placental development, fetal cells induce changes in the morphology of uterine blood vessels to increase fetal perfusion and nutrient transfer; placental tissue penetrates the uterine epithelium, the endometrial stroma, and the upper third of the myometrium. Humans and great apes are thought to share similar placental morphology, although data on great ape placentae are limited. Abrams hypothesizes that this highly invasive morphology arose because walking upright reduced perfusion of uterus through compression of the aorta and the inferior vena cava. This deep invasion into the uterine wall is thought to contribute to considerable blood loss after birth, which has only been described in humans. This advantage for infant growth and development can come at a price of the mother, in that postpartum hemorrhage is a leading cause of maternal mortality.

Abrams et al. further propose that selective pressures to prevent maternal hemorrhage may explain the persistence of genetic thrombophilias, such as Factor V Leiden (FVL). This genetic polymorphism increases risk for venous thromboembolism, but it is also associated with a reduced risk of postpartum hemorrhage and perioperative bleeding (Lindqvist, 2015; Lindqvist, Svensson, Dahlback, & Marsal, 1998). Lindqvist (Lindqvist et al., 1998) has proposed that protection from maternal mortality due to hemorrhage conferred a protective advantage for FVL carriers, explaining the prevalence of this variant, which ranges from 3% in Southern Europe to as high as 15% in Sweden. However, the variant is rare in Asia, Africa and South America.

Increasing a woman's risk of thromboembolism in order to reduce her risk of hemorrhage was a favorable tradeoff in our evolutionary past. In the developing world, embolism remains a far less common cause of maternal death than hemorrhage (3.1%, 95% CI 1.7-5.4 vs. 27.1, 95% CI 19.9-36.4) (Say et al., 2014). However, in the setting of modern obstetrics, where uterotonics, surgery, and blood transfusion can mitigate the risks of blood loss, venous thromboembolism is now nearly as common a cause of maternal death as hemorrhage (13.8%, 95% CI 10.1-22.0 vs. 16.3%, 95% CI 11.1-24.6). In a sense, selective pressure to reduce blood loss at the cost of venous thromboembolism may be mismatched with obstetric care in high income countries.

4.8 Conclusion

This chapter has outlined ways in which components of the postpartum period—infant feeding decisions, maternal sensitivity and emotional wellbeing, nighttime parenting practices, parental leave policies, employment arrangements, gender equity, fecundity, and physical recovery from childbirth—are interrelated. Differences between evolved strategies for childrearing and modern life illustrate how the evolutionary concept of "mismatch" impacts human health.

This illustration is highly relevant to medical students and residents, who are navigating tensions between reproductive and productive work in their own lives as they contemplate whether and when to fit childbearing into medical training. When one author (A.S.) was in medical school, she organized a highly-attended, "Pregnancy Paranoia Panel," inviting female faculty with children to discuss how they juggled call schedules, board exams, and exposure to teratogenic infections such as parvovirus during their pregnancies. Frank discussions regarding allo-mothers and the "village" required for child-rearing may help trainees to anticipate needs and craft strategies to garner support.

New curricula for medical education provides additional opportunities to consider the health of mothers and infants through an evolutionary lens. At the University of North Carolina, the third-year clerkships have been organized into larger blocks, such that students complete obstetric and mother-baby rotations in tandem, following families from onset of labor through newborn care. Framing this transition as the start of the 4th trimester would build context for physical exam findings such as the newborn grasp and rooting reflexes, as well as facilitate understanding of causes of severe maternal morbidity, including hemorrhage, thrombo-embolism, and postpartum preeclampsia.

More broadly, integration of evolutionary perspectives across the continuum of medical education provides an opportunity to strengthens clinical counseling through conceptualization of humans as a species, with both proximate- and ultimate-level explanations for the ways we function. An evolutionary perspective provides a starting point for considering how contemporary environments can be made more accommodating to dyads and ways in which practical and social support can better "mother the mother." And in one-on-one conversations with new families, sharing an evolutionary perspective can demystify the demands of child-rearing and encourage mothers to engage a broader village of support, rather than struggling to navigate through the postpartum period in isolation. Understanding our evolutionary history as collective breeders informs a growing appreciation of the importance of social determinants of health and wellbeing. Just as marmosets who lack sufficient support abandon or kill their infants, people who are isolated and marginalized suffer profound physiologic effects.

The effects of stress on physiology are illustrated in Niles Newton's classic study on the effects of pain and stress on milk let down (Newton & Newton, 1948). Newton was working on her PhD dissertation and had learned that when a cow is stressed, she produces less milk. Newton hypothesized that stress might similarly impact lactation in humans, and she and her husband designed an experiment. Newton was nursing her 7-month-old daughter, and each morning, she nursed her baby while enduring a variety of stressful interventions. Newton's feet were put in ice water; her husband asked her difficult math questions, and shocked her if she answered incorrectly, and he wrapped a piece of gauze around her great toe, yanking until it was painful. During these manipulations, she also received a dose of either oxytocin or placebo. Weights of her infant before and after feeding showed that pain and stress prevented milk transfer—but that milk transfer was rescued with an injection of oxytocin. When we teach medical students, we point out that pain and stress can disrupt a mother's ability to provide milk for her infant. Conversely, a few minutes of conversation at the bedside can ease a mother's anxiety, resulting in fast-flowing milk and a gulping baby. Human physiology is tightly linked to our surroundings, and "taking feet out of the ice water" is a central element of the art of medicine.

The early postpartum period and beyond are critical for both child and maternal health. Women's health needs are currently inadequately supported in the U.S. The concept of the "4th trimester" draws attention to the health and social characteristics of this important transition period. Medical education that accounts for evolved traits can enable delivery of healthcare that effectively addresses complex bio-psycho-social processes by anticipating patients' subconscious weighting of costs-benefits, not only in relation to infant outcomes but also for women's health. We may better serve women and their families by explicitly addressing the trade-offs that women negotiate in striving to optimize the health of their infants and themselves.

References

Abrams, E. T. & Rutherford, J. N. (2011). Framing postpartum hemorrhage as a consequence of human placental biology: An evolutionary and comparative perspective. *Am Anthropol*, 113(3): 417–30. doi:10.1111/j.1548–1433.2011.01351.x

Ahrens, K. A., Hutcheon, J. A., Ananth, C. V., Basso, O., Briss, P. A., Ferre, C. D., Frederiksen, B. N., Harper, S., Hernandez-Diaz, S., Hirai, A. H., Kirby, R. S., Klebanoff,

M. A, Lindberg L, Mumford, S. L., Nelson, H. D., Platt, R. W., Rossen, L. M., Stuebe A. M., Thoma M. E., Vladutiu C. J., & Moskosky S. (2018). Report of the Office of Population Affairs' expert work group meeting on short birth spacing and adverse pregnancy outcomes: Methodological quality of existing studies and future directions for research. *Paediatr Perinat Epidemiol*, 9(10).

Aiello, L. C. & Key, C. (2002). Energetic consequences of being a Homo erectus female. *Am J Hum Biol*, 14(5): 551–65. doi:10.1002/ajhb.10069

American Academy of Family Physicians. (2014). Breastfeeding, Family Physicians Supporting (Position Paper). Retrieved from https://www.aafp.org/about/policies/all/breastfeeding-support.html

American Academy of Pediatrics, & Section on Breastfeeding. (2012). Breastfeeding and the use of human milk. *Pediatrics*, 129(3): e827–841. doi:10.1542/peds.2011–3552

American College of Obstetricians and Gynecologists. (2016). Optimizing support for breastfeeding as part of obstetric practice. Committee Opinion No. 658. *Obstet Gynecol*, 127, e86–e92.

Ball, H. (2006). Parent-infant bed-sharing behavior: Effects of feeding type and presence of father. *Hum Nat*, 17(3): 301–18. doi:10.1007/s12110-006-1011-1

Ball, H. L. (2017). The Atlantic divide: Contrasting U.K. and U.S. recommendations on cosleeping and bed-sharing. *J Hum Lact*, 33(4): 765–9. doi:10.1177/0890334417713943

Ball, H. L. & Volpe, L. E. (2013). Sudden Infant Death Syndrome (SIDS) risk reduction and infant sleep location—moving the discussion forward. *Soc Sci Med*, 79: 84–91. doi:10.1016/j.socscimed.2012.03.025

Ball, H. L. & Klingaman, K. P. (2008). Breastfeeding and mother-infant sleep proximity: Implications for infant care. In W. R. Trevathan, E. O. Smith, & J. J. McKenna (eds.), *Evolutionary Medicine and Health: New Perspectives* (pp. 226–41). Oxford: Oxford University Press.

Beery, A. K., McEwen, L. M., MacIsaac, J. L., Francis, D. D., & Kobor, M. S. (2016). Natural variation in maternal care and cross-tissue patterns of oxytocin receptor gene methylation in rats. *Hormones and Behavior*, 77: 42–52. doi:https://doi.org/10.1016/j.yhbeh.2015.05.022

Belsky, J. & Pluess, M. (2013). Beyond risk, resilience, and dysregulation: Phenotypic plasticity and human development. *Dev Psychopathol*, 25(4 Pt 2): 1243–61. doi:10.1017/S095457941300059X

Benyshek, D. C. & Watson, J. T. (2006). Exploring the thrifty genotype's food-shortage assumptions: A cross-cultural comparison of ethnographic accounts of food security among foraging and agricultural societies. *Am J Phys Anthropol*, 131(1): 120–6. doi:10.1002/ajpa.20334

Bereczkei, T. (2001). Maternal trade-off in treating high-risk children. *Evol Hum Behav*, 22(3): 197–212.

Blair, P. S., Sidebotham, P., Pease, A., & Fleming, P. J. (2014). Bed-sharing in the absence of hazardous circumstances: Is there a risk of sudden infant death syndrome? An analysis from two case-control studies conducted in the UK. *PLoS One*, 9(9): e107799. doi:10.1371/journal.pone.0107799

Bowlby, J. (1969). *Attachment and Loss*. New York: Basic Books.

Burtle, A. & Bezruchka, S. (2016). Population health and paid parental leave: What the United States can learn from two decades of research. *Healthcare (Basel)*, 4(2): E30. doi:10.3390/healthcare4020030

Trevathan, W. R. & Rosenberg, K.R. (2016). Human evolution and the helpless infant. In W. R. Trevathan & K. R. Rosenberg (eds.) *Costly and Cute: Helpless Infants and Human Evolution* (pp. 1–28). Santa Fe: School for Advanced Research: Albequerque: University of New Mexico Press.

Crouch, M. (2002). Bonding, postpartum dysphoria, and social ties. *Human Nature*, 13(3): 363–82. doi:10.1007/s12110-002-1020-7

Cubero, J., Valero, V., Sanchez, J., Rivero, M., Parvez, H., Rodriguez, A. B., & Barriga, C. (2005). The circadian rhythm of tryptophan in breast milk affects the rhythms of 6-sulfatoxymelatonin and sleep in newborn. *Neuro Endocrinol Lett*, 26(6): 657–61.

Curtis, K. M., Tepper, N. K., Jatlaoui, T. C., Berry-Bibee, E., Horton, L. G., Zapata, L. B., Whiteman, M. K. (2016). U.S. medical eligibility criteria for contraceptive use, 2016. *MMWR Recomm Rep*, 65(3): 1–103. doi:10.15585/mmwr.rr6503a1

Dall, S. R. & Boyd, I. L. (2004). Evolution of mammals: Lactation helps mothers to cope with unreliable food supplies. *Proc Biol Sci*, 271(1552): 2049–57. doi:10.1098/rspb.2004.2830

Declercq, E. R., Sakala, C., Corry, M. P., Applebaum, S., & Herrlich, A. (2013). *Listening to Mothers III: New Mothers Speak Out*. Retrieved from New York: Childbirth Connection. https://transform.childbirthconnection.org/reports/listeningtomothers/

Dennis, C. L., Fung, K., Grigoriadis, S., Robinson, G. E., Romans, S., & Ross, L. (2007). Traditional postpartum practices and rituals: A qualitative systematic review. *Womens Health (Lond)*, 3(4): 487–502. doi:10.2217/17455057.3.4.487

Dettwyler, K. A. (2004). When to wean: Biological versus cultural perspectives. *Clin Obstet Gynecol*, 47(3): 712–23.

Dipietro, J. A. (2012). Maternal stress in pregnancy: Considerations for fetal development. *J Adolesc Health*, 51(2 Suppl): S3-8. doi:10.1016/j.jadohealth.2012.04.008

Dixit, A., Feldman-Winter, L., & Szucs, K. A. (2015). "Frustrated," "Depressed," and "Devastated" pediatric trainees: US academic medical centers fail to provide adequate workplace breastfeeding support. *J Hum Lact*, 31(2): 240–8. doi:10.1177/0890334414568119

Dufour, D. L., Reina, J. C., & Spurr, G. B. (2002). Energy intake and expenditure of free-living, lactating Colombian women in an urban setting. *Eur J Clin Nutr*, 56(3): 205–13. doi:10.1038/sj.ejcn.1601302

Ellis, B. J., Boyce, W. T., Belsky, J., Bakermans-Kranenburg, M. J., & van Ijzendoorn, M. H. (2011). Differential susceptibility to the environment: An evolutionary—neurodevelopmental theory. *Dev Psychopathol*, 23(1): 7–28. doi:10.1017/S0954579410000611

Enlow, M. B., Devick, K. L., Brunst, K. J., Lipton, L. R., Coull, B. A., & Wright, R. J. (2017). Maternal lifetime trauma exposure, prenatal cortisol, and infant negative affectivity. *Infancy*, 22(4): 492–513. doi:10.1111/infa.12176

Feldman-Winter, L., Szucs, K., Milano, A., Gottschlich, E., Sisk, B., & Schanler, R. J. (2017). National trends in pediatricians' practices and attitudes about breastfeeding: 1995 to 2014. *Pediatrics*, 140(4). doi:10.1542/peds.2017-1229

Flaxman, S. M. & Sherman, P. W. (2008). Morning sickness: Adaptive cause or nonadaptive consequence of embryo viability? *Am Nat*, 172(1): 54–62. doi:10.1086/588081

Fletcher, S., Grotegut, C. A., & James, A. H. (2012). Lochia patterns among normal women: A systematic review. *J Womens Health (Larchmt)*, 21(12): 1290–4. doi:10.1089/jwh.2012.3668

Galtry, J. (2015). Strengthening the human rights framework to protect breastfeeding: A focus on CEDAW. *Int Breastfeed J*, 10: 29. doi:10.1186/s13006-015-0054-5

Gavin, N. I., Gaynes, B. N., Lohr, K. N., Meltzer-Brody, S., Gartlehner, G., & Swinson, T. (2005). Perinatal depression: A systematic review of prevalence and incidence. *Obstet Gynecol*, 106(5 Pt 1): 1071–83. doi:10.1097/01.AOG.0000183597.31630.db

Gibson, M. A. & Mace, R. (2006). An energy-saving development initiative increases birth rate and childhood malnutrition in rural Ethiopia. *PLoS Med*, 3(4): e87. doi:10.1371/journal.pmed.0030087

Hahn-Holbrook, J., Holbrook, C., & Haselton, M. G. (2011). Parental precaution: Neurobiological means and adaptive ends. *Neurosci Biobehav Rev*, 35(4): 1052–66. doi:10.1016/j.neubiorev.2010.09.015

Haig, D. (1993). Genetic conflicts in human pregnancy. *Q Rev Biol*, 68(4): 495–532.

Haig, D. (2008). Intimate relations: Evolutionary conflicts of pregnancy and childhood. In S. C. K. Stearns, & J. C. Koella (eds.), *Evolution in Health and Disease* (pp. 65–76). 2 ed. Oxford: Oxford University Press.

Hahn-Holbrook, J. et al. (2011). *Neuroscience and Biobehavioral Reviews*, 35: 1052–66.

https://www.ncbi.nlm.nih.gov/pmc/articles/PMC3345294/pdf/nihms369367.pdf

Hamilton, W. D. (1964). The genetical evolution of social behaviour. *J Theor Biol*, 7(1), 1–16.

Higgins, J. A., Kramer, R. D., & Ryder, K. M. (2016). Provider bias in Long-Acting Reversible Contraception (LARC) Promotion and removal: Perceptions of young adult women. *Am J Public Health*, 106(11): 1932–7. doi:10.2105/AJPH.2016.303393

Hrdy, S. B. (2000). *Mother Nature: Maternal Instincts and how they Shape the Human Species.* 1st Ballantine Books ed. New York: Ballantine Books.

Hrdy, S. B. (2016). Variable postpartum responsiveness among humans and other primates with "cooperative breeding": A comparative and evolutionary perspective. *Horm Behav*, 77: 272–83. doi:10.1016/j.yhbeh.2015.10.016

Humphrey, L. T. (2010). Weaning behaviour in human evolution. *Semin Cell Dev Biol*, *21*(4): 453–61. doi:10.1016/j.semcdb.2009.11.003

Huseynov, A., Zollikofer, C. P., Coudyzer, W., Gascho, D., Kellenberger, C., Hinzpeter, R., & Ponce de Leon, M. S. (2016). Developmental evidence for obstetric adaptation of the human female pelvis. *Proc Natl Acad Sci U S A*, 113(19): 5227–32. doi:10.1073/pnas.1517085113

Ip, S., Chung, M., Raman, G., Chew, P., Magula, N., DeVine, D., & Lau, J. (2007). Breastfeeding and maternal and infant health outcomes in developed countries *Evid Rep Technol Assess (Full Rep)* (153): 1–186.

Jasienska, G., Bribiescas, R. G., Furberg, A. S., Helle, S., & Nunez-de la Mora, A. (2017). Human reproduction and health: An evolutionary perspective. *Lancet*, 390(10093): 510–20. doi:10.1016/S0140-6736(17)30573–1

Jolley, S. N., Elmore, S., Barnard, K. E., & Carr, D. B. (2007). Dysregulation of the hypothalamic-pituitary-adrenal axis in postpartum depression. *Biological research for nursing*, 8(3): 210–22. doi:10.1177/1099800406294598

Jurmain., R. N., H.; Kigore, L.; & Trevathan, W. R. (2008). *Introduction to Physical Anthropology.* Belmont, CA.: Thomson Wadsworth.

Kanes, S., Colquhoun, H., Gunduz-Bruce, H., Raines, S., Arnold, R., Schacterle, A., & Meltzer-Brody, S. (2017). Brexanolone (SAGE-547 injection) in post-partum depression: A randomised controlled trial. *Lancet*, 390(10093): 480–9. doi:10.1016/S0140-6736(17), 31264–3

Kendall-Tackett, K., Cong, Z. & Hale, T.W. (2010). Mother-infant sleep locations and nighttime feeding behavior. *Clinical Lactation*, 1(1): 27–31.

Kitzinger, S. (1975). The fourth trimester? *Midwife Health Visit Community Nurse*, 11(4): 118–21.

Klerman, J. A., Daley, K., & Pozniak, A. (2014). *Family and Medical Leave in 2012: Technical Report*. Retrieved from Cambridge, MA: http://www.dol.gov/asp/evaluation/fmla/fmla2012.htm

Labbok, M. H. (2001). Effects of breastfeeding on the mother. *Pediatr Clin North Am*, 48(1): 143–58.

Lancaster, J. B. (1993). The evolutionary biology of women. In A. J. Almquist, A. Manyak (eds.), *Milestones in Human Evolution* (pp. 21–37). Prospect Heights, IL: Waveland Press.

Lawrence, R. A. & R. Lawrence (2005). *Breastfeeding: A guide for the medical profession* (6th ed.). Philadelphia: Elsevier Mosby.

Lewis, S. (2008). Evolution at the intersection of biology and medicine. In W. R. Trevathan, E. O. Smith, & J. J. McKenna (eds.), *Evolutionary Medicine and Health: New Perspectives* (pp. 399–415). Oxford: Oxford University Press.

Lindqvist, P. G. (2015). On the evolutionary advantage of coagulation factor V Leiden (FVL). *Curr Med Chem*, 22(32): 3676–81.

Lindqvist, P. G., Svensson, P. J., Dahlback, B., & Marsal, K. (1998). Factor V Q506 mutation (activated protein C resistance) associated with reduced intrapartum blood loss—a possible evolutionary selection mechanism. *Thromb Haemost*, 79(1): 69–73.

Lopez, L. M., Grey, T. W., Stuebe, A. M., Chen, M., Truitt, S. T., & Gallo, M. F. (2015). Combined hormonal versus nonhormonal versus progestin-only contraception in lactation. *Cochrane Database Syst Rev*, 3, Cd003988. doi:10.1002/14651858.CD003988.pub2

Maguire, J. & Mody, I. (2008). GABA(A)R plasticity during pregnancy: Relevance to postpartum depression. *Neuron*, 59(2): 207–13. doi:10.1016/j.neuron.2008.06.019

Martin, R. D. (2007). The evolution of human reproduction: A primatological perspective. *Am J Phys Anthropol Suppl* 45: 59–84. doi:10.1002/ajpa.20734

Martin, R. D. & Maclarnon, A. M. (1985). Gestation period, neonatal size and maternal investment in placental mammals. *Nature*, 313(5999): 220–3.

McKenna, J. J., Ball, H. L., & Gettler, L. T. (2007). Mother-infant cosleeping, breastfeeding and sudden infant death syndrome: What biological anthropology has discovered about normal infant sleep and pediatric sleep medicine. *Am J Phys Anthropol*, Suppl 45: 133–61. doi:10.1002/ajpa.20736

McKenna, J. J. & Gettler, L. T. (2016). There is no such thing as infant sleep, there is no such thing as breastfeeding, there is only breastsleeping. *Acta Paediatr*, 105(1): 17–21. doi:10.1111/apa.13161

McKenna, J. J. & McDade, T. (2005). Why babies should never sleep alone: A review of the co-sleeping controversy in relation to SIDS, bedsharing and breast feeding. *Paediatr Respir Rev*, 6(2): 134–52. doi:10.1016/j.prrv.2005.03.006

Moon, R. Y. & Task Force on Sudden Infant Death Syndrome. (2016). SIDS and other sleep-related infant deaths: Evidence base for 2016 updated recommendations for a safe infant sleeping environment. *Pediatrics*, 138(5): pii: e20162940. doi:10.1542/peds.2016–2940

Moore, E. R., Bergman, N., Anderson, G. C., & Medley, N. (2016). Early skin-to-skin contact for mothers and their healthy newborn infants. *Cochrane Database Syst Rev*, 11: CD003519. doi:10.1002/14651858.CD003519.pub4

Nappi, R. E., Petraglia, F., Luisi, S., Polatti, F., Farina, C., & Genazzani, A. R. (2001). Serum allopregnanolone in women with postpartum "blues". *Obstet Gynecol*, 97(1): 77–80.

Nesse, R. M. (2001). The smoke detector principle. Natural selection and the regulation of defensive responses. *Ann N Y Acad Sci*, 935: 75–85.

Nettle, D. (2009). An evolutionary model of low mood states. *J Theor Biol*, 257(1): 100–3. doi:10.1016/j.jtbi.2008.10.033

Newton, M. & Newton, N. R. (1948). The let-down reflex in human lactation. *The Journal of Pediatrics*, 33(6): 698–704.

Nozhenko, Y., Asnani-Kishnani, M., Rodriguez, A. M., & Palou, A. (2015). Milk leptin surge and biological rhythms of leptin and other regulatory proteins in breastmilk. *PLoS One*, 10(12): e0145376. doi:10.1371/journal.pone.0145376

Oftedal, O. T. (2012). The evolution of milk secretion and its ancient origins. *Animal*, 6(3): 355–68. doi:10.1017/s1751731111001935

Pluess, M. & Belsky, J. (2011). Prenatal programming of postnatal plasticity? *Dev Psychopathol*, 23(1): 29–38. doi:S0954579410000623

Portmann, A. (1990). On the uniqueness of biological research. 1949. *J Med Philos*, 15(5): 457–72.

Power, M. L. & Schulkin, J. (2016). *Milk: The Biology of Lactation*. Baltimore, Maryland: Johns Hopkins University Press, p. 234.

Profet, M. (1992). Pregnancy sickness as adaptation: A deterrent to maternal ingestion of teratogens. In J. E. C. Barkow, & Tooby, J. (eds.), *The Adapted Mind: Evolutionary Psychology and the Generation of Culture* (pp. 327–65). Oxford: Oxford University Press.

Reiches, M. W., Ellison, P. T., Lipson, S. F., Sharrock, K. C., Gardiner, E., & Duncan, L. G. (2009). Pooled energy budget and human life history. *Am J Hum Biol*, 21(4): 421–9, 442. doi:10.1002/ajhb.20906

Saucier, J.-F. (1972). Correlates of the long postpartum taboo: A cross-cultural study. *Current Anthropology*, 13(2): 238–58.

Say, L., Chou, D., Gemmill, A., Tunçalp, Ö., Moller, A.-B., Daniels, J., & Alkema, L. (2014). Global causes of maternal death: A WHO systematic analysis. *The Lancet Global Health*, 2(6): e323–e333. doi:10.1016/S2214-109X(14)70227-X

Scheper-Hughes, N. (1992). *Death Without Weeping: The Violence of Everyday Life in Brazil*. Berkeley: University of California Press.

Schimpf, M. & Tulikangas, P. (2005). Evolution of the female pelvis and relationships to pelvic organ prolapse. *Int Urogynecol J Pelvic Floor Dysfunct*, 16(4): 315–20. doi:10.1007/s00192-004-1258-1

Sellen, D. W. (2007). Evolution of infant and young child feeding: Implications for contemporary public health. *Annu Rev Nutr*, 27: 123–48. doi:10.1146/annurev.nutr.25.050304.092557

Sellen, D. W. (2009). Evolution of human lactation and complementary feeding: Implications for understanding contemporary cross-cultural variation. *Adv Exp Med Biol*, 639: 253–82. doi:10.1007/978-1-4020-8749-3_18

Shaul, D. M. B. (1963). The composition of the milk of wild animals. *International Zoo Yearbook*, 4(1): 333–3. doi:10.1111/j.1748–1090.1963.tb03691.x

SisterSong Women of Color Reproductive Justice Collective, & National Women's Health Network. (2017). Long-acting reversible contraception statement of principles. Retrieved from https://docs.google.com/document/d/1ID4cEuaV1oSAXSWdJmSi4YMs5TLCGhnomjOX0In5odU/edit

Strassman, B. I. & Mace, R. (2008). Perspectives on human health and disease from evolutionary and behavioural ecology. In S. C. K. Stearns & J. C. Koella (eds.), *Evolution in Health and Disease* (pp. 109–22). 2nd ed. Oxford: Oxford University Press.

Stuebe, A., Auguste, T., & Gulati, M. (2018). American College of Obstetricians and Gynecologists' Presidential Task Force on Redefining the Postpartum Visit, & Committee on Obstetric Practice. Optimizing postpartum care. Retrieved from https://www.acog.org/Clinical-Guidance-and-Publications/Committee-Opinions/Committee-on-Obstetric-Practice/Optimizing-Postpartum-Care.

Stuebe, A. M., Grewen, K., & Meltzer-Brody, S. (2013). Association between maternal mood and oxytocin response to breastfeeding. *Journal of Women's Health*, 22(4): 352–61. doi:10.1089/jwh.2012.3768

Stuebe, A. M. & Schwarz, E. B. (2010). The risks and benefits of infant feeding practices for women and their children. *J Perinatol*, 30(3): 155–162. doi:jp2009107 [pii]

Tinbergen, N. (1963). On aims and methods of ethology. *Zeitschrift für Tierpsychologie*, 20: 410–33.

Tomori, C., Palmquist, A. E., & Dowling, S. (2016). Contested moral landscapes: Negotiating breastfeeding stigma in breastmilk sharing, nighttime breastfeeding, and long-term breastfeeding in the U.S. and the U.K. *Soc Sci Med*, 168: 178–85. doi:10.1016/j.socscimed.2016.09.014

Trivers, R. L. (1974). Parent-offspring conflict. *American Zoologist*, 14(1): 249–64.

Tully, K. P. & Ball, H. L. (2012). Postnatal unit bassinet types when rooming-in after cesarean birth: Implications for breastfeeding and infant safety. *J Hum Lact*, 28(4): 495–505. doi:10.1177/0890334412452932

Tully, K. P. & Ball, H. L. (2013). Trade-offs underlying maternal breastfeeding decisions: a conceptual model. *Matern Child Nutr*, 9(1): 90–8. doi:10.1111/j.1740–8709.2011.00378.x

Tully, K. P., Holditch-Davis, D., & Brandon, D. (2015). The relationship between planned and reported home infant sleep locations among mothers of late preterm and term infants. *Matern Child Health J*, 19(7): 1616–23. doi:10.1007/s10995-015-1672-7

Tully, K. P., Stuebe, A. M., & Verbiest, S. B. (2017). The 4th trimester: A critical transition period with unmet maternal health needs. *Am J Obstet Gynecol*. 27(1): 37–41. doi:10.1016/j.ajog.2017.03.032

Tully, K. P. & Sullivan, C. S. (2018). Parent-infant room-sharing is complex and important for breastfeeding. *Evid Based Nurs*, 21(1): 18. doi:10.1136/eb-2017–102801

Tully, K. P. & Ball, H. L. (2018). Understanding and enabling breastfeeding in the context of maternal-infant needs. In C. E. P. Tomori, A. E. L.Palmquist, & E. A. Quinn (eds.), *Breastfeeding: New Anthropological Approaches* (pp. 199–211). Oxford: Routledge Books.

U.S. Department of Health and Human Services. (2011). *The Surgeon General's Call to Action to Support Breastfeeding*. Washington, DC: U.S. Department of Health and Human Services, Office of the Surgeon General.

Valeggia, C. R. & Ellison, P. T. (2001). Lactation, energetics and postpartum fecundity. In P. T. Ellison (ed.), *Reproductive Ecology and Human Evolution* (pp. 85–105). New York: Aldine de Gruyter.

Van Noordwijk, M. A., Kuzawa, C. W., & Van Schaik, C. P. (2013). The evolution of the patterning of human lactation: A comparative perspective. *Evol Anthropol*, 22(5): 202–12. doi:10.1002/evan.21368

Vaughan, D. A., Cleary, B. J., & Murphy, D. J. (2014). Delivery outcomes for nulliparous women at the extremes of maternal age—a cohort study. *Bjog*, 121(3): 261–8. doi:10.1111/1471–0528.12311

Vitzthum, V. J. (2008). Evolution and endocrinology: The regulation of pregnancy outcomes. In S. O. H. Elton, P. (ed.), *Medicine and Evolution: Current Applications, Future Prospects* (pp. 99–126). London: CRC Press.

Volpe, L. E., Ball, H. L., & McKenna, J. J. (2013). Nighttime parenting strategies and sleep-related risks to infants. *Soc Sci Med*, 79: 92–100. doi:10.1016/j.socscimed.2012.05.043

Wells, J. C., DeSilva, J. M., & Stock, J. T. (2012). The obstetric dilemma: An ancient game of Russian roulette, or a variable dilemma sensitive to ecology? *Am J Phys Anthropol, 149 Suppl*, 55: 40–71. doi:10.1002/ajpa.22160

Wells, J. C. K., Nesse, R. M., Sear, R., Johnstone, R. A., & Stearns, S. C. (2017). Evolutionary public health: Introducing the concept. *Lancet*, 390(10093), 500–9. doi:10.1016/s0140-6736(17)30572-x

Winnicott, D. W. (1945). Primitive emotional development. *Int J Psychoanal*, 26(3–4), 137–43.

Wolf, J. H. (1998). "Don't kill your baby," Feeding infants in Chicago, 1903–1924. *Journal of the History of Medicine and Allied Sciences*, 53(3): 219–53. doi:10.1093/jhmas/53.3.219

Wolf, J. H. (2003). Low breastfeeding rates and public health in the United States. *Am J Public Health*, 93(12): 2000–10.

World Health Organization. (2015). *Medical Eligibility Criteria for Contraceptive Use*. Retrieved from Geneva: https://www.who.int/reproductivehealth/publications/family_planning/Ex-Summ-MEC-5/en/

SECTION II

BIOLOGICAL REGULATION

CHAPTER 5

Can Evolution Inform Insights into Normal and Adverse Pregnancy Outcomes and Ultimately Improve Maternal and Fetal Care?

Heide Aungst, Robert Rossi, Heather Brockway, Sam Mesiano, and Louis J. Muglia

Applying an evolutionary approach to pregnancy is not only a logical endeavor, but also one that should be considered essential. If we broadly consider human health and disease, three general influences shape outcomes: an individual's current environmental exposures, individual behaviors, and innate physiology, or overall genetic-biological make-up. Past environments, with adaptations and selection to these, increase fitness in those environments and shape an individual's physiology and overall genetic makeup. As such, current genetic make-up reflects the influence of past evolution, but may not be optimal for current environmental conditions. This mismatch may predispose to disease via mismatches that differ from past selective pressures (Godfrey et al., 2007). Evolution does not provide selection for optimizing long-term health or longevity—maximal transmission of genetic information to the next generation is the priority, even at the expense of compromising other life-history traits and phenotypes. This concept, from an evolutionary perspective, is that selection records on the genotype, but acts on the phenotype (Stearns et al., 2010).

A fundamental characteristic of evolution is that there is not a priori design to generate the optimal outcome for pregnancy, or any aspect of health or disease risk. Rather, evolution must work through constraints and trade-offs (for example a shared tool-box across phenotypes where a change in one trait leads to a change in another—e.g., pleiotropy). In other words, a traditional medical view would see organisms as machines, engineered for efficiency and health, but an evolutionary perspective views organisms as "bundles of compromise shaped by natural selection to maximize reproduction, not health" (Stearns et al., 2010). Therefore, the study of reproduction must look at how each part evolved and what influence that evolution has today on human conception, pregnancy, and birth timing.

The traditional medical framework seeks to define mechanisms as *proximate* causes of normal physiology and disease that can serve as the basis for potential preventive or therapeutic interven-

Aungst, H., Rossi, R., Brockway, H., Mesiano, S., and Muglia, L., *Can Evolution Inform Insights into Normal and Adverse Pregnancy Outcomes and Ultimately Improve Maternal and Fetal Care?* In: *Integrating Evolutionary Biology into Medical Education—for maternal and child healthcare students, clinicians, and scientists*. Edited by: Jay Schulkin and Michael L. Power, Oxford University Press (2020). © Oxford University Press. DOI: 10.1093/oso/9780198814153.003.0006

tions. For example, preeclampsia is associated with inadequate maternal-fetal interface remodeling of the spiral arteries and elevated soluble FLT-1 arising from the fetoplacental unit. Evolutionary thought seeks to elucidate *ultimate* causes—why does shallow invasion and inadequate spiral artery invasion result in preeclampsia? In this circumstance, the fetal drive for maximal nutrient extraction from the mother is compromised by the shallow invasion and lack of spiral artery remodeling to maintain a low resistance placental vascular circuit. To adapt to this lower flow state, the fetus attempts to compensate by initiating humoral signals to increase maternal vascular resistance and provide a greater fraction of blood flow to the fetal interface (Haig, 2007). Another critical example of looking for ultimate causes relates to pregnancy, as a whole: Why doesn't the mother's immune system attack the fetus? Inflammation is important to embryo implantation, but it also can threaten the pregnancy. It has been argued that this "inflammation paradox" may be the transformation of a process needed for implantation evolving to be able to control any negative effects on the fetus (Chavan et al., 2017). Evolutionary thinking has even impacted pharmaceutical companies that market birth control pills that allow menstruation only four times a year, rather than monthly, simulating that ancestral women spent more time pregnant or lactating, having as few as 160 cycles in a lifetime, compared to modern women who have as many as 450 cycles during their reproductive years (Trevathan, 2007).

In 1973, geneticist and evolutionary biologist Theodosius Dobzhansky stated, "Nothing in biology makes sense except in light of evolution" (Dobzhansky, 1973). And, this is especially true when we look at reproduction. Evolution acts to optimize reproductive fitness and maximize transmission of genes to the next generation. Mammalian pregnancy must balance optimizing gene transmission of maternal genes by the mother over successive pregnancies by limiting provisioning to any one pregnancy and fetus, while the fetus (and paternal genome) seeks to maximize outcomes for the one pregnancy it is part of, and extract maximal resources. Successful transmission of genes for both the mother and fetus thus requires successful pregnancy outcomes. These outcomes include growth of the fetus, duration of the gestation, and regulation of maternal and fetal metabolism to avoid complications, such as preeclampsia and gestational diabetes.

Indeed, while the maternal-fetal crosstalk, or "tug of war," in healthy pregnancy reaches a stable equilibrium for maternal and fetal wellbeing throughout the pregnancy, David Haig and others have convincingly argued that achieving this equilibrium depends on balancing conflicting fetal goals of optimizing nutrition partitioning for its pregnancy, while the maternal response is to provide adequate but limited nutrients over the course of the current pregnancy to maximize gene transmission over all of her pregnancies (Haig, 1993). Recent research, however, shows that the balance needed to maintain pregnancy may be more of a modulation of gene regulatory or cell communication networks that contributed to the evolution of pregnancy (Lynch et al., 2011). Many of the adverse outcomes of pregnancy—preeclampsia, gestational diabetes, and preterm birth—may represent maladaptive complications in attempting to achieve balance in a "tug of war" scenario or disruptions in signaling during development in a regulatory or modulation construct.

Elucidating the drivers of conception, maintenance, and termination of human pregnancy have proven challenging, likely as a result of the strong selective pressures that shape optimal outcomes for every organism. Animal model systems used to investigate the regulation of gestational duration and fetal growth often reveal mechanisms that clearly do not apply to humans. For example, the critical role of the fetal hypothalamic pituitary adrenal axis in activating labor in ovine pregnancy is not reflected in the biology of human pregnancy; nor is the abrupt fall in circulating maternal serum progesterone concentration that prompts initiation of labor and delivery in most non-primate mammalian species seen in human parturition. As we begin to catalogue the differences that have been optimized across different species, such as the number of offspring per pregnancy, duration of gestation due to metabolic or structural constraints (such as pelvic remodeling with bipedalism in humans), and uterine shape, these differences should perhaps be expected rather than be a conundrum.

5.1 Why care about pregnancy outcomes?

We can broadly define pregnancy outcomes and complications from both the maternal and neonatal perspective, as well as from the public health/societal aspect, as well. Poor perinatal outcomes are still commonplace despite the advances of modern medicine including obstetric anesthesia, blood transfusion, aseptic techniques, and antibiotics.

One of the first and important studies of pregnancy "outcomes" is the story of Ignaz Semmelweis who is known as one of the earliest pioneers of antiseptic procedures. In one of the premier European teaching maternity wards of the mid-nineteenth century, the maternal mortality rate was approximately 10 percent due to puerperal fever. The rate in the adjacent midwife clinic was less than 4 percent. Semmelweis meticulously studied the differences between clinics and found that the only difference between them was that the medical students would work with cadavers and then attend deliveries. He correctly hypothesized that contamination from the cadavers was likely the cause of childbed fever and the high mortality rates. By simply instituting chlorine handwashing after cadaver lab, the rate of maternal death decreased to below 2 percent within the entire hospital over the course of one month. Despite Semmelweis dying well before the discovery of Streptococcus pyogenes or even before the theory of "germs," he had saved the lives of millions of childbearing women (National Research Council. 2004). His story illustrates the power and simplicity to which epidemiology can identify major health burdens and lead to changes that improve survival and quality of life (Semmelweis & Carter, 1983).

Jean Baptiste Lamarck once theorized that a species will strive or achieve greater complexity and perfection over time. However, the imperfection of human pregnancy is a prime illustration of the various conflicting evolutionary influences between the mother and fetus even in the setting of optimized environmental conditions (adequate nutrition and prenatal care) (Lamarck & Elliott, 2011/orig.1914). And so, human pregnancy, while complex, is a far from perfect process with high pregnancy loss rates (30%) early in gestation (Wilcox et al., 1988; Wang et al., 2003), high prevalence of birth defects (3 percent), and high rates of pregnancy related complications, such as pregnancy induced hypertensive disorders (6–17 percent) (Yoder et al., 2009; Hauth et al., 2000; Buchbinder et al., 2002), gestational diabetes (10%) (Hartling et al., 2012; Moyer, 2014), placental abruption (1 percent) (Ananth et al., 2015; Ruiter et al., 2015), pathological fetal growth restriction (2–3 percent) (Barker et al., 2013), and preterm birth (10 percent) (http://www.who.int/maternal_child_adolescent/documents/born_too_soon/en/), even within developed nations.

In fact, preterm birth is the leading cause of infant mortality throughout the world, with some urban areas in the United States having higher rates than several developing countries. Infants who survive prematurity not only suffer the immediate consequences of prematurity, such as cerebral palsy and breathing difficulties, but their prematurity also has long-term health consequences throughout the lifespan and a deep societal impact for the child's ongoing needs.

Every pregnancy is not only a personal event between fetus and the mother physiologically, but it is also a sociological and public health issue that can arise with pregnancy complications and childbirth. Infant mortality is not the only possibly preventable death resulting from a pregnancy. Every day more than 800 women die worldwide and approximately 300,000 women die annually from preventable causes related to pregnancy and childbirth (World Health Organization. 2016).

Therefore, pregnancy and reproductive outcomes are perhaps the most important measures of a population's overall health. Birth represents a pivotal and major life event with tremendous emotional and economical import and lifelong health consequences. A healthy pregnancy ensures not just the wellbeing of the mother, but also the wellbeing of the fetus, the newborn, the family, and society at large. The concept of "Evolutionary Public Health" introduced in 2017 by leaders in the field—Jonathan

Wells, Randolph Nesse, Rebecca Sear, Rufus Johnstone, and Stephen Stearns—captures the importance of environmental impact on mortality risk, nutrition, and disease that affect reproduction—and ultimately, health throughout the life course (Wells et al., 2017). Therefore, if a species is to survive, the population as a whole must be successful in not only reproducing but also maintain enough genetic variation as to adapt to changing conditions (Grunspan et al., 2017). As such, pregnancy outcomes are considered health indicators, or quantifiable characteristics of a population, used by researchers as parameters of population health.

Before taking the wide view of population health, it is critical to examine what we know about reproduction. A mystery remains that is critically important to the health and wellbeing of both women and infants: what is the mechanism that triggers birth? If we knew that, we could determine what "goes wrong" in premature births, and, ultimately, discover therapies to prevent it. This could relieve many health burdens on those who live with the lifelong health consequences of prematurity. And, perhaps along the way, we also could learn what triggers the common causes of death in childbirth and find potential ways to stop that, as well.

5.2 Normal pregnancy: What must happen

A successful pregnancy is the result of an investment of both the fetus and mother to ensure reproductive success. This investment, at times, acts more like a molecular "tug of war" between the resource needs of each or a disruption in regulatory or modulating mechanisms—think of how quickly even a small power surge can disrupt or burn out electronics. Complications can occur when the relationship is out of balance. That's why it is critical to understand the physiology of a "normal" pregnancy.

We define "normal" term birth, or parturition, as occurring at about 40 weeks' gestation and preterm birth as at less than 37 weeks of gestation. Viability is considered at about 22–23 weeks. In recent years, two babies between 21 and 22 weeks are known to have survived (Howard, 2017). The rate of survival of extremely premature babies has improved slightly in recent years, with one study showing an increase from 16–20% of 22–24-week gestation babies surviving without neurodevelopmental impairment (Younge et al., 2017)—a vital measurement of success.

A myriad of physiological changes must occur in both maternal and fetal tissues over the course of the pregnancy. The 40 weeks of pregnancy are divided into three trimesters of about 14 weeks each. The first trimester begins the last day of the normal menstrual cycle through the first 14 weeks of gestation. The second trimester encompasses weeks 15 through 28, with the final trimester beginning at week 29 and continuing through 40–42 weeks of gestation. The developmental processes occurring in the fetus, the placenta, and the maternal tissues in each trimester are essential for a successful pregnancy. Any aberrant process during this time puts the pregnancy at risk for preterm or post-term delivery.

However, studying these developmental processes in human pregnancy is difficult and constrained by ethical limitations. Thus, we use animal models to study the morphological changes and molecular mechanisms underpinning the processes of pregnancy and parturition in vivo. Unfortunately, there is not one type of model that replicates reproduction in humans, but different animals help us understand components of the process, although none, not even the monkey or ape, is a completely representative model.

In addition to commonplace mouse models—easy to use because of short gestation, but with extremely different reproductive physiology to a human—sheep are often used to study human pregnancy for several reasons. Pregnancy in women and ewes is more similar in gestation length, number of offspring, and the shifting of progesterone production in the corpus luteum of the ovary to the placenta. However, they have a bicornate uterus, allowing them to carry two fetuses and a cotyledonary placenta, rather than a discoid placenta of humans and primates. Guinea pigs have a placenta much closer to humans, and also do not require systemic progesterone withdrawal to initiate labor, like humans. Non-human primates, with a five-month gestation period, cause similar problems, in addition to the expense of maintaining the animals and the difficulty of genetic manipulation—although their reproductive physiology is much closer to a human (Bezold et al., 2013).

Although different animal models can help us study different aspects of human reproduction—the uterus, the placenta, hormonal influence, parturition—much of the physiology of normal pregnancy remains poorly understood, in large part because of the lack of a singular identical model. Genetic studies, however, hold great promise in disentangling the basis of human birth timing and its failings. Multiple types of studies have looked at genetic influences on preterm birth. For example, some genetic diseases in babies also are associated with preterm birth, such as myotonic dystrophy, Smith–Lemli–Opitz syndrome, and neurofibromatosis (Lengyel et al., 2014). However, these are rare presentations that have not led to new knowledge about the mechanism of birth timing. Other types of genetic studies have revealed some clues, although not a one-size-fits-all answer. These include family-based linkage studies, genome-wide association studies (GWAS), rare variant and candidate gene analysis, mitochondrial genetics, and microRNA-based pathways. Twin studies have allowed researchers to look at the influence of maternal and fetal genetic factors. These studies have attributed a 21% variation in birth timing to maternal influence and 13% to fetal genetic factors. Other studies have put the mother's genomic contribution closer to 40% (Swaggert et al., 2014).

5.3 Establishing pregnancy

The first trimester encompasses several essential processes for the establishment of the maternal-fetal interface that precedes the actual placental development. Prior to the actual establishment of the placental interface, eggs are released and fertilized; there is development of the blastocyst, receptivity and deciduation of uterine tissues, implantation, and, then finally, the development of the placenta. Each of these steps has a unique set of developmental processes that are not yet fully understood. The placenta is a two-sided organ with the maternal contribution to the placenta originating in the uterus and its tissues: the endometrium and myometrium with fetal tissues originating from the blastocyst. The placenta develops very early in the pregnancy: as the blastocyst develops, its cells differentiate into an outer shell of trophoectoderm—destined to become the placenta—and the inner cell mass which will become the fetus. As the pregnancy proceeds, it is the developing placenta that maintains the fetal growth trajectory in response to maternally supplied nutrients, and oxygen (Dimasuay et al., 2016). In addition to adapting to the ever-changing maternal environment, the placenta must not only maintain maternal immunotolerance established in the first trimester, but also act as a barrier to environmental hazards, such as xenobiologics or maternal sex hormones that could harm the developing fetus (Burton & Fowden, 2015; Maltepe & Fisher, 2015). The role of the placenta is increasingly critical as gestation progresses. As the maternal environment shifts, it must adapt to the growing needs of the fetus. The placenta is key in modulating the allocation of nutrients to the growing fetus and "senses" the maternal environment through the mechanistic target of rapamycin (mTOR) signaling pathway, adapting to ensure that the increasing metabolic fetal needs are met (Diaz & Jansson, 2014; Fowden & Moore, 2012). These signals are critical to the ultimate health of the fetus and the newborn, since poor fetal growth—possibly determined by the signals in the placenta—is associated with higher rates of chronic disease in adult life (Lewis et al., 2012). Therefore, the placenta's function is absolutely essential during pregnancy, yet it is the least understood of any organ (Burton & Fowden, 2013).

Central to the development of the placenta—and subsequently the initiation of pregnancy—is the receptivity of the fetal blastocyst for implantation into the endometrium. The endometrium consists of surface epithelial cells, stromal cells, and glands. If the endometrial tissues are not receptive to the implanting blastocyst, placentation cannot take place, leading to a spontaneous miscarriage or another adverse pregnancy outcome (Cha et al., 2012). Examination of the genes and pathways in the endometrium through the evolutionary history of mammalian gene expression in endometrial tissues may be one way to understand—and perhaps eventually treat—many reproductive disorders.

Gulum Kosova and colleagues used this method to identify dysregulated genes in women with recurrent early pregnancy loss and determined, in fact, that these women have out-of-phase histology and elevated cyclin E levels in the endometrial glands (Kosova et al., 2015).

Mouse models of receptivity provide a clearer understanding of the complex cellular signaling, differentiation, and proliferation that takes place in the endometrial tissues (James et al., 2012; Cha et al., 2012). In humans, receptivity takes place regardless of mating status, occurring in a cyclic process known as the menstrual cycle. The ovary is the primary regulator of the menstrual cycle, producing increasing levels of estrogen from day zero to day 14 when ovulation occurs. Under the influence of high estrogen levels, the endometrium grows through cellular proliferation of the various cell types. Once ovulation occurs, the endometrium thickens in response to the progesterone the corpus luteum (CL) produces. The window of receptivity is between days 20 to 24 in the cycle. If a blastocyst is present producing chorionic gonadotropin (CG) as an endocrine signal to prevent regression of the CL—leading to sustained progesterone production—the endometrium will not be shed the uterus and implantation will take place. However, if there is no blastocyst present or if it is not able to properly implant (for example a lack of hCG), then there is an immediate reduction in the levels of estrogen and progesterone due to CL regression. Then, the thickened endometrial layer sheds (menses) and the cycle resets (Cha et al., 2012; Gellersen & Brosens, 2014).

The dynamics of implantation are quite complicated and still not completely understood. Implantation of the blastocyst in the great apes, including humans, is interstitial in nature, meaning that the blastocyst is fully encapsulated in the uterine wall, whereas in mice, the blastocyst attaches more superficially covered by only a small pocket of maternal tissue (James et al., 2012; Cha et al., 2012). In humans, molecules drive blastocyst attachment to the surface epithelial cells, including transcription factors, cytokines and their receptors, enzymes, morphogens, and hormones (Cha et al., 2012; Gellersen & Brosens, 2014; Mori et al., 2016). The initiation of implantation of the blastocyst not only facilitates the growth of the fetal tissues, but also the differentiation of the endometrial stromal cells through the process of decidualization to form decidual cells. This process is dependent on the presence of progesterone. Human decidualization takes place in response to numerous endocrine signals from the implanting blastocyst such as cyclic adenosine monophosphate (cAMP) and CG. Furthermore, human decidual cells are not polyploid in nature, yet secrete many of the same proteins as mouse decidual cells to facilitate a more invasive implantation.

The role of the decidual cells is highly complex. In the earliest stages of pregnancy, they provide nutritional and oxygen support to the developing fetus prior to the establishment of the placenta (Mori et al., 2016; Oreshkova et al., 2012). During placentation, they help modulate trophoblast invasion through alteration to the extracellular matrix that subsequently leads to vascular remodeling of the maternal spiral arteries, as well as angiogenesis (Cha et al., 2012; Gellersen & Brosens, 2014; Mori et al., 2016). Furthermore, decidual cells interact with various maternal immune cells to establish an immuno-tolerance to the invading trophoblasts (Gellersen & Brosens, 2014; Mori et al., 2016; Oreshkova et al., 2012). Decidualization is the result of a complex web of molecular signaling to alter the maternal endometrium in response to the presence of a blastocyst. However, decidualization is also at the root of the molecular "tug of war" between the fetus and mother that ultimately governs the success of the pregnancy. Absolute failure of the processes of receptivity, implantation, and decidualization will lead to either infertility or immediate termination of the pregnancy (Cha et al., 2012).

Once the blastocyst is encapsulated in the decidual tissue, the trophoectoderm continues to proliferate becoming the placental cone and differentiates into the trophoblast lineages. In humans, these are the cytotrophoblast, the syncytiotrophoblast, and the extravillous trophoblast (Maltepe & Fisher, 2015; Burton & Jauniaux, 2017). As implantation and placentation continues, the layer of cells just beneath the placental cone becomes the extra-embryonic ectoderm, which then further develops into the chorionic epithelium. The fetal allantosis from the posterior of the embryo aligns with the chorion, allowing for the growth and development of the fetal vasculature on the fetal side of the placenta.

As gestation continues, the human placental cone differentiates into villous cytotrophpblasts and the syncytiotrophoblasts which form large columnar branched villus structures in the intervillous space. These villi are structured such that the fetal endothelial cells are incased in stromal cells, then cytotrophoblast, with the syncytiotrophoblast forming the outer layer of the villi. Microvilli make up the apical membrane of the syncytiotrophoblast (Maltepe & Fisher, 2015). At the tips of these columnar villi, interstitial cytotrophoblasts differentiate into extravillous trophoblast, invading the decidua and maternal spiral arteries, facilitating maternal immune tolerance and vascular remodeling (James et al., 2012; Cha et al., 2012; Maltepe & Fisher, 2015). When the vascular remodeling is complete, the highly branched villous structures are bathed in maternal blood, allowing for maternal-fetal exchange (James et al., 2012; Cha et al., 2012; Maltepe & Fisher, 2015).

5.4 Successful pregnancy

Another crux of the molecular "tug of war" is the establishment of maternal immuno-tolerance to the presence and invasion of fetal tissue to facilitate maternal vascular remodeling. To establish maternal blood flow into the placenta, invading trophoblasts remodel the maternal spiral arteries in the endometrium to a high flow capacity into the intervillous space in the human placenta. However, the invasion of fetal trophoblasts into the maternal endometrium creates an immunological paradox in that they should be rejected by provoking a strong immune response from the mother. Yet, in a successful pregnancy, this does not occur—although there are numerous immune cells found throughout the decidua (Erlebacher, 2013). Extensive research has uncovered that instead of rejecting the fetal trophoblasts, the maternal immune system modifies during pregnancy to accept the fetal "graft" for the length of gestation and, furthermore, works with the trophoblasts to modulate maternal vascular remodeling and angiogenesis (Erlebacher, 2013; Hammer, 2011). It is in these processes that mouse and human are similar—although research shows that human invasion is much deeper and extends into the myometrium.

There are several different immune cells found in the decidual stromal cell population: dendritic cells, T-cells, and uterine natural killer cells (uNK). Dendritic cells are the "first responders" of the immune system, detecting the exposure of pathogens and, in the case of pregnancy, invading fetal tissue, and thus directly modulating the inflammatory response upon invasion. Their numbers in the decidua are few after implantation, which suggests a suppression of the inflammatory response that is typically seen in organ transplants (Oreshkova et al., 2012). T-cells are present in the decidua, yet their exact role in establishing maternal immunotolerance is not clearly understood (Oreshkova et al., 2012). Uterine natural killer (uNK) cells—the most numerous of all the types throughout gestation—secrete various chemokines, cytokines, and angiogenic factors to modulate immune response and facilitate vascular remodeling, working with the invading trophoblast cells rather than against them (Oreshkova et al., 2012). Furthermore, these cells are essentially non-cytotoxic, displaying a unique receptor phenotype that is distinct from their more cytotoxic counterparts found elsewhere in the human body.

Among those receptors are the killer cell inhibitory receptors (KIRs) and a select set of CD receptors/ligands (Erlebacher, 2013; Hammer, 2011; Bouteiller & Bensussan, 2017). Invading trophoblasts display a select range of human leukocyte antigen (HLA) class I molecules on cell surface—HLA-G, HLA-C, HLA-E, and HLA-F—all which act to further inhibit any remaining cytotoxicity of the uNK and T-cells thus suppressing the maternal immune response (Bouteiller & Bensussan, 2017). Interestingly, there has been a dynamic co-evolution of the KIR and HLA class I molecules throughout primate evolution that likely corresponds to the differences in invasiveness in the various primate species, as well as the deep invasiveness of the human trophoblasts and the extensive vascular remodeling of the human maternal arteries (Parham & Moffett, 2013). However, maternal immune suppression is not limited to the decidua. The syncytiotrophoblast do not display any HLA molecules on the apical surface, which is in direct contact with the maternal blood, and thus cannot trigger any immune response from the maternal blood immune cells (Bouteiller & Bensussan, 2017).

The remodeling of the maternal spiral arteries is a culmination of crosstalk between the invading trophoblast, uNKs, and maternal vascular cells which occurs early in the first few weeks of pregnancy (Cartwright et al., 2010). Until the remodeling finishes, the growing embryo is reliant of the decidual cells for nutrition and oxygen for survival. One evolutionary perspective on decidual cells is that they evolved to control the invasion of trophoblast cells, as well as to suppress the mother's immune response to fight off the invading fetus (Wagner et al., 2014). Invading trophoblasts perform several particular tasks to facilitate invasion: attachment to and breakdown of the extracellular matrix via secretion of a mix of proteases, and then migration into the space previously occupied by the matrix (O'Tierney-Ginn & Lash, 2014). A multitude of factors regulate these functions including: endocrine factors, immune cell secretions, oxygen levels, and maternal blood flow (Green et al., 2016; O'Tierney-Ginn & Lash, 2014; Chen & Zheng, 2013).

5.5 Maintaining pregnancy and growth

From the earliest stages of development, the placenta has an endocrine role, acting on the maternal metabolism to increase blood glucose concentrations so there are sufficient amounts for fetal growth. The placental hormones drive increased food intake and the resulting energy storage, not necessarily for maternal use, but for fetal use throughout gestation (Burton & Fowden, 2016). The invasiveness of the fetal tissue has been described as an investment by the fetus to ensure the maximum access to maternal nutrients, yet it has to be contained by the mother to facilitate her survival, thus creating in a sense a tug-of-war between the fetus and mother as described by Haig (Haig, 1997; 1996; 1996; 1993). Therefore, molecular compromise between the fetus and mother, as the invading fetal tissue is contained, allows for the development and growth of the fetus and the survival of the mother. It is only when that compromise is disrupted that we observe adverse pregnancy outcomes, such as preeclampsia, placenta accreta, and gestational trophoblastic disease.

It is also worth noting that placental metabolism increases throughout gestation, and its own metabolic needs must also be met if the pregnancy is to continue to term. Any alteration to placental sufficiency at this point will not only compromise fetal growth and placental function, but the pregnancy itself. Thus, optimal maternal nutrition is essential for a successful term pregnancy. Moreover, the placental function not only modulates fetal growth, but also the early fetal programing of health. If there is any constraint on the maternal environment during pregnancy, it can directly affect the growing fetus and its lifetime health outcome. This has been observed in both mouse models and human populations (Sferruzzi-Perri & Camm, 2016). This is absolutely essential to maintain the optimal maternal environment for not only a successful pregnancy outcome, but also for the post-natal health of the fetus.

5.6 Gestational length and parturition

What triggers the birth process? We know there are molecular mechanisms that regulate gestational length and pregnancy maintenance, along with the initiation of labor and birth, yet we know little about what determines birth timing. Much of the mechanisms that trigger parturition remains a mystery.

The steroid hormone progesterone (P4)—via its interaction with intracellular nuclear P4 receptors (PRs)—is essential for the establishment and maintenance of pregnancy. It is likely a disruption in P4/PR signaling triggers parturition. This occurs in all viviparous species. Therefore, evolutionary selection for an optimal gestational length allows for fetal growth and development. Humans have a unique "obstetrical dilemma" when it comes to reproduction and birth: a pelvis remodeled for obligate bipedal locomotion that must also be able to accommodate the passage of a large fetal cranium; the product of selection for encephalization—the amount of brain mass relative to body mass—(Dunsworth et al., 2012; Rosenberg & Trevathan, 2002; Smith, 2007). This imposes a growth restriction on the fetus as it must be able to pass through the pelvic opening at birth, otherwise both mother and fetus risk death. Encephalization is unique to the human ancestral lineage; other primates do not have such an issue at birth. Their pelvic openings are quite large and can accommodate not only the neonatal head but the rest of the body easily (Rosenberg & Trevathan, 2002). Thus, on a morphological level, there appears to have been evolutionary selection along the human ancestral lineages for a gestational length that allows for optimal fetal development and growth coupled with the ability to pass a large-headed fetus through a remodeled pelvic opening.

However, advances in molecular studies have suggested there is a biological clock that measures the length of gestation and initiates a trigger to parturition (Philippe & Philippe, 2017). One of the first molecules to be associated with birth timing was corticotropin-releasing hormone (CRH) which the placenta produces and secretes into the maternal serum and suggests the existence of a placental clock that determines birth timing (Smith, 2007). More recent evidence from fetal villous tissues and membranes—along with placentally secreted proteins measured in maternal serum levels—provides additional support for this placental clock.

Given the central role of the placenta to fetal growth, the transient nature of the organ, and the endocrine regulation of pregnancy physiology, a biological clock hypothesis is highly plausible. Studies of gene expression and methylation patterns in normal pregnancy suggest there are quite a few changes between first trimester and term (Lim et al., 2017; Mayne et al., 2016; Roberts, 2016). Yet, a precise understanding of these changes in relation to normal pregnancy progression is not entirely clear. Furthermore, additional studies in various adverse pregnancy outcomes such as preterm birth, preeclampsia, fetal growth restriction, and even stillbirth, all implicate a prematurely aging placenta in their respective etiologies (Mayne et al., 2017; Leavey et al., 2017; Maiti et al., 2017; Morgan, 2016). These studies suggest that premature aging of the placenta impacts the pathophysiology related to these outcomes via changes in methylation patterns, gene expression, villous morphology, and maternal serum markers. Therefore, understanding the normal molecular mechanisms regulating gestational length and initiation of parturition is essential to understanding how these mechanisms are perturbed in adverse pregnancy outcomes.

5.7 The role of progesterone in the maintenance and termination of pregnancy

Previous understanding of gestational length has come from evolutionary studies examining the impact of body size, brain development, and fitness of both mother and fetus (Phillips et al., 2015). Parturition, in all viviparous species, likely requires the withdrawal of P4/PR signaling. P4/PR promotes pregnancy by: 1) Inducing the secretory endometrium to produce an intrauterine environment conducive to embryo survival and implantation; 2) modulating the maternal immune system to allow tolerance of the implanted allogenic embryo, and 3) promoting uterine quiescence in the form of myometrial relaxation, cervical closure, and decreased responsiveness of uterine cells to pro-labor/pro-inflammatory stimuli. The removal of the physiologic and pathologic signals of the P4/PR block trigger parturition (Csapo, 1956). Natural selection has favored parturition processes and timing mechanisms that maximize reproductive efficiency. The timing of parturition would have adapted in response to specific selective pressures to maximize reproductive efficiency within the context of a species' overall reproductive strategy and environmental niche.

At parturition, the myometrium contracts synchronously and rhythmically in response to uterotonic hormone such as prostaglandin (PG) $F_{2\alpha}$ ($PGF_{2\alpha}$) and oxytocin, to become the engine for birth. The uterine cervix softens and dilates to open the gateway for birth; and the amnion membrane weakens and ruptures to release the fetus from the amniotic sac. Parturition usually occurs at a time in gestation, referred to as term (37 to 42 weeks in a human), when the fetus is sufficiently mature to survive outside of the uterus and the mother is competent to nurture her neonate(s). Term varies between species due to strong Darwinian selection to optimize reproductive efficiency. The probability for successful reproduction is maximal at term and decreases exponentially with the extent of pre- and

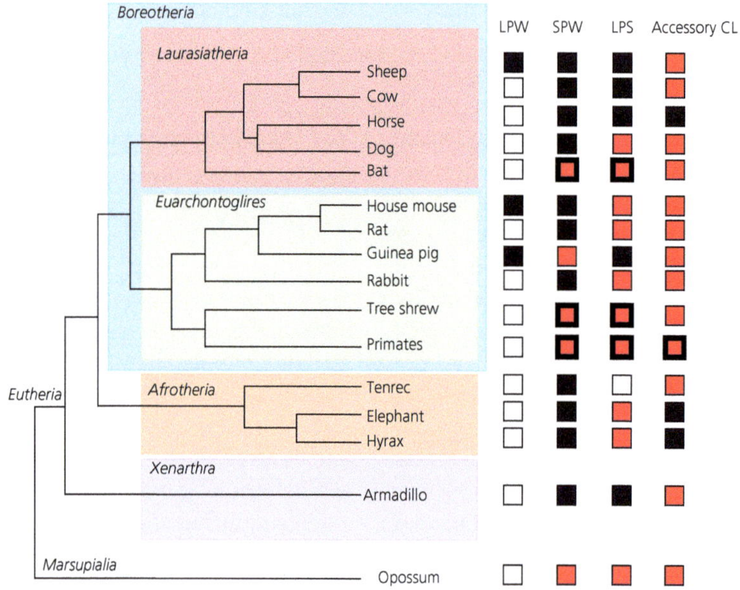

Figure 5.1 Progesterone withdrawal
Data obtained from: Hoyert D. L. (2007). Maternal mortality and related concepts.

post-term parturition. Extant pathways that trigger term parturition define a species' "normal" parturition trigger mechanism. However, multiple pathways for parturition likely evolved and may coexist with secondary mechanisms in place in response to specific physiologic signals. This redundancy and variability is in part due to species' differences in the mechanism and control of P4/PR withdrawal (Figure 5.1).

Natural selection would favor a mechanism that links birth timing with fetal maturation and competence to survive outside the uterus to ensure that birth does not occur before the fetus is sufficiently mature to survive as a neonate. Thus, fetal signals would be favored that maintain pregnancy until its critical organ systems have matured. However, post-term parturition increases morbidity and mortality risk for the mother and fetus due to difficulties associated with birthing an overgrown baby. In this case, natural selection would favor a mechanism for the mother to trigger parturition in response to pregnancy-related stressors.

In most species, a decrease in systemic maternal P4 levels trigger parturition. The mechanism that causes, this, however, varies according to the source of P4 and how its synthesis/secretion decreases. The maternal corpora lutea (CL) is the primary source of P4 that promotes the establishment of pregnancy. In some species (e.g., mouse, rat, rabbit, goat) the CL persists as the sole source of P4 throughout pregnancy. In these animals, CL regression, leading to systemic P4 withdrawal, triggers parturition at term.

In other species (e.g., human, non-human primates, sheep), the placenta, in addition to the maternal CL, produces P4 and becomes the principal source of P4 in the latter stages of pregnancy. In some of these species, such as sheep, the induction of the P450c17 steroidogenic enzyme in placental cells converts P4 to androstenedione leading to a fall in maternal P4 levels that triggers parturition (Flint et al., 1975). In this system, a prepartum surge of cortisol, produced by the fetal hypothalamic-pituitary-adrenal axis, induces placental P450c17 expression. Importantly, cortisol also promotes the functional maturation of organ systems that the fetus will need to survive as a neonate. This mechanism links fetal maturation with the timing of parturition.

Local P4 withdrawal also may occur at a uterine tissue-specific level (e.g., in the cervical stroma). For this mechanism, P4 metabolizes to a PR-inactive form. Studies in transgenic mice show that at the time of labor conversion of P4 to the inactive 20α-dihydroprogesterone increases in endocervical fibroblasts (Mahendroo et al., 1999; Andersson et al., 2008). Studies in human myometrial cells suggest that at parturition, P4 metabolizes to a form that does not bind the PR, and unliganded PR exerts transcriptional activity to increase expression of genes encoding contraction associated proteins (Nadeem et al., 2016). Whether this pathway for P4 withdrawal is linked to fetal maturation or maternal stress is not clear.

In some species, including humans, most non-human primates, and guinea pig, placental P4 production, and consequently the circulating maternal P4 level, does not decrease at term, and, consequently, parturition occurs without a systemic P4 withdrawal (Walsh et al., 1979). In these species, parturition is thought to be triggered by a functional P4/PR withdrawal whereby P4 target cells in the gravid uterus become refractory to P4/PR signaling. Various potential mechanism for functional P4/PR withdrawal have been identified, each involving modulation of PR transcriptional activity.

A single gene, progesterone receptor gene (*PGR*), encodes nuclear progesterone receptors (*PRs*). *PGR* arises from duplications of a primordial estrogen receptor gene early in vertebrate evolution (Thornton, 2001). In most species, two promotors control PGR expression. This produces two major PR isoforms: the full-length PR-B, and the truncated (by 164 amino N-terminal acids) PR-A (Kastner et al., 1990). The ligand binding domain on each PR is conserved across diverse species and has a high (Kd ~1nM) affinity for progesterone. Consistent with other steroid hormone receptors, the PR isoforms function primarily as ligand-activated transcription factors (Evans, 1988; O'Malley & Tsai, 1992; Tsai & O'Malley, 1994; Evans, 2005). The relative levels of the PR-A and PR-B affect responsiveness to P4, as does post-translational modifications, especially site-specific serine phosphorylation, in a cell type- and context-specific manner. At some gene promoters, PR-A represses the activity of PR-B and P4 responsiveness is inversely related to the relative amount of PR-A and PR-B. This concept led to the PR-A/PR-B hypothesis for functional progesterone withdrawal, which posits that PR-B mediates the pro-gestational actions of P4 and that the trans-repressive actions of PR-A causes functional P4/PR withdrawal.

Consistent with this hypothesis, labor in women at term is associated with increased PR-A abundance in myometrial cells leading to a rise in the PR-A/PR-B ratio (Mesiano, 2002; Merlino et al., 2007; Ke et al., 2016). And, in cultured human myometrial cells, increased abundance of PR-A represses the transcriptional activity of PR-B (Merlino et al., 2007; Hardy et al., 2006; Tan et al., 2012). More recent studies suggest that site-specific serine phosphorylation induced by pro-inflammatory/pro-labor stimuli controls the inhibitory effect of PR-A (Amini et al., 2016; Peters et al., 2016). The data suggest that PR-A—phosphorylated at serine residues -344 and -345 (pSer$^{344/345}$-PRA)—inhibit the PR-mediated, labor-blocking actions of P4. The generation of pSer$^{344/345}$-PRA in response to pro-inflammatory stimuli suggest that stressors impacting the gravid uterus trigger labor by inducing pSer$^{344/345}$-PRA-mediated functional P4/PR withdrawal. Other mechanisms for functional progesterone withdrawal include increased expression of PR-B co-repressors (Condon et al., 2003), inhibition of PR activity by other transcription factors (Kalkoven et al., 1996), and reduced binding of PRs to deoxyribonucleic acid (DNA) (Nadeem et al., 2016). In addition, specific microRNAs may inhibit PR signaling in uterine cells (Mahendroo et al., 1999; Andersson et al., 2008; Williams et al., 2012). Thus, multiple redundant processes may cause functional P4/PR withdrawal to trigger human parturition in response to maternal or fetal signals.

Viewing this from an evolutionary perspective, the shift in progesterone is highly species specific. One theory is that estrogen and progesterone signaling interacts to communicate at the cervical cell level—evolving after the most common ancestor of marsupials and eutherian mammals, although the role of prostaglandins and relaxin are even older (Wagner et al., 2017). Or, it may be that functional P4/PR withdrawal evolved from mutations in *PGR*, in conjunction with the hominid traits of encephalization and bipedalism. Comparative studies of human and non-human primate genomes suggest rapid adaptive evolution of the *PGR* gene in the hominid lineage after the divergence of hominids from the chimpanzee lineage around 10 million years ago (Moorjani et al., 2016).

Consequently, the human *PGR* is distinct from that of other primates due to an increased rate of non-synonymous mutations that produced distinct non-polymorphic amino acid substitutions. The changes are mainly in the N-terminal region that qualitatively and quantitatively affect PR transcriptional activity, and interestingly, includes multiple serine residues that are known phosphorylation sites (Clark et al., 2003; Chen et al., 2008). Thus, changes in *PGR* may have conferred a human-specific mechanism for the control of P4/PR transcriptional activity, possibly by serine phosphorylation, to provide mechanisms for functional P4/PR withdrawal. This may have been critical for altering the timing of human parturition in response to the selective pressures of bipedalism and encephalization such that the factors that trigger parturition, possibly maternal and/or fetal stressors, cause term birth to occur before the fetal head becomes larger than the pelvic outlet.

Studies in mice suggest that multiple mechanisms for parturition coexist and activate in response to specific physiologic signals. Although systemic P4 withdrawal from the regression of the maternal CL triggers term parturition in mice, mouse parturition also can be induced preterm, and prior to CL regression and systemic P4 withdrawal, by treating animals with an inflammatory stimulus such as E coli lipopolysaccharide (LPS) (Hirsch & Hirsch, 1995). In this experimental model, parturition occurs in response to E. coli or LPS and without systemic P4 withdrawal, which suggests that a mechanism for stress/inflammation-induced functional P4/PR exists in the mouse as a secondary parturition trigger mechanism.

This concept of coexisting parturition trigger pathways may be important for unraveling the pathophysiology of human preterm birth. It is possible that the response of specific maternal and fetal stressors induces distinct mechanisms for functional P4/PR withdrawal. Such redundancy would be expected given the critical role of parturition in reproduction. Further studies of the comparative evolutionary biology, and especially the molecular genetics of *PGR* and P4/PR signaling, as well as gene expression testing in the cervix that shows broad scale histone deacetylation in labor (Wagner et al., 2017) will provide a deeper understanding of the hormonal control of parturition and provide important insights into the pregnancy disorders, especially preterm birth.

5.8 Diseases of pregnancy

Hormonal interactions have evolved through natural selection to maximize the chance of successful pregnancy. Although natural selection operates at the species level by optimizing overall reproductive efficiency, the unintended consequences of traits derived from genetic conflict between maternal and fetal/paternal interests (Haig, 1993) may have increased the risk for certain diseases of pregnancy. The risk is tolerated at the individual level, as long it does not affect reproductive efficiency at the species level.

5.9 Infertility and recurrent miscarriage

The rate of infertility in human couples is remarkably high with an estimated 50% of zygotes failing to implant and around 15% of established pregnancies ending in the first trimester (Elish et al., 1996; Garcia-Enguidanos et al., 2002). This is counterintuitive if natural selection operates by optimizing reproductive efficiency. However, in this context, natural selection is strongly influenced by a battle between the sexes that is fought between the ovum and spermatozoa.

Natural selection would favor increased spermatic efficiency manifested as numerous, robust, and aggressive spermatozoa. The consequence of this evolutionary battle over millions of years is a fine tuning of the male and female systems to produce a compromise leading to successful reproduction, most of the time. However, a slight shift in the balance of power in either direction, may lead to instability and fertilization failure. The resultant pathology is infertility caused by polyspermy, due to weak maternal restriction to multiple sperm penetration, and infertility due to oligospermia (insufficient sperm to overcome female barriers). Such a system would be highly sensitive to environmental conditions that increase or decrease sperm number. Oligospermia is clearly a major cause of male infertility. However, more is not necessarily better since abnormally high sperm density can overcome ovum defenses and increase the incidence of polyspermy (Ven et al., 1985).

Most cases of first trimester miscarriage are due to embryo aneuploidy attributable to non-disjunction in the oocyte. Upwards of 30% of eggs are thought to aneuploid as extrapolated from IVF studies. From an evolutionary perspective this represents a logical form of natural selection to avoid the expenditure of maternal resources on a genetically abnormal conceptus. However, it also is plausible that maternal systems would evolve to monitor the fitness of a euploid embryo to further ensure her return on investment. Endocrine signals between the conceptus and the mother wage this battle—whereby the mother monitors fetal fitness by the strength and quality of its hormonal signals. In this scenario, the fetuses with a stronger signal would have an advantage. An example of such an interaction is the production of chorionic gonadotropin (CG) by the peri-implantation embryo that prevents regression of the maternal CL and maintains maternal P4 levels to prevent menstruation. Embryo fitness, in this case, would directly correlate with the strength of the CG signal. Similar interactions likely would occur through multiple endocrine and paracrine signals to form critical checkpoints on the establishment and progression of pregnancy.

The conflict hypothesis predicts that fetal traits, likely derived from paternal genes, would be selected to satisfy maternal checkpoints. However, early in pregnancy, stringency of maternal restriction may be higher than those of the fetus to ensure that the mother commits to a pregnancy that will provide return on investment. Consequently, the rate of spontaneous first trimester miscarriage in established—and presumably euploid—pregnancies is around 15%, which is relatively high compared with other viviparous species (Garcia-Enguidanos et al., 2002; Zinaman et al., 1996). A major factor in this selection process involves that maternal immune system. Spontaneous miscarriage of an otherwise normal conceptus may arise from the maternal immune system's rejection of the allogenic conceptus. Natural selection would favor a compromise whereby the maternal immune system evolves a certain level of tolerance for the fetal allograft, and as a counterbalance, the fetus evolves a mechanism to avoid maternal immune rejection.

In 1953 Sir Peter Medawar in his studies of graft rejection proposed that the fetus is protected from maternal immune rejection by: 1) anatomical separation between fetal cells and maternal immune cells; 2) camouflage of the fetal cells and local modulation of the maternal immune cells; and/or, 3) global modulation of the maternal immune system to produce refractoriness to allogenic fetal cells (Madawar. 1953). It is now clear that, although the maternal-fetal interface has a highly organized anatomy, a physical barrier between maternal immune cells and fetal cell expressing alloantigens does not exist, thus eliminating Medawar's first tenet. In the 1990s, studies using transgenic mice and high-fidelity flow cytometry added detail to Medawar's second tenet. For example, it was proposed that the fetal tissues inhibit the proliferation of maternal T cells at the maternal-fetal interface by catabolizing tryptophan, an essential amino acid needed for all rapidly dividing cells including T cells (Munn et al., 1998; Mellor et al., 2001). It also was proposed that trophoblast cells express Fas ligand that induces apoptosis of activated maternal T lymphocytes (Coumans et al., 1999; Makrigiannakis et al., 2001; Kalantaridou et al., 2007). At around that same time, studies of invading trophoblasts led to the hypothesis that fetal cells camouflage themselves from maternal immune attack by expressing human leukocyte antigen G (HLA-G), a unique non-polymorphic HLA that inhibits natural killer cell activation (Fisher & Damsky, 1993; Rouas-Freiss et al., 1997; van Beekhuizen et al., 2010; Liu et al., 2011). However, with HLA antigens, fetal trophoblast cells at the maternal-fetal interface have been shown by the development of antibodies that have increased specificity to groups to express classical HLA-A, HLA-B, and HLA-C antigens (46–50) (Sunderland et al., 1981; King et al., 1996; King et al., 2000; Blaschitz et al., 2001; Apps et al., 2008). Likewise, normal fertility and pregnancy outcomes have been shown for HLA-G-null individuals (Castro et al. 2000). Together, these results highlight the existence of additional, and likely more dominant, mechanisms for maintaining maternal immunological tolerance to genetically foreign fetal antigens. This likely involves the third tenet of Medawar's postulate.

Although maternal lymphocytes with the capacity to attack and eliminate fetal cells exist throughout gestation (Tafuri et al., 1995; Thellin et al., 2000), they are somehow globally suppressed during pregnancy such that the maternal immune system tolerates the fetal allograft (Jiang & Vacchio, 1998). This tolerogenic state is exemplified clinically with the decrease in severity of autoimmune disease during pregnancy, and its relapse after parturition (Beagley & Gockel, 2006) in most women. And in experiments, mice show increased tolerance for engraftment of paternally derived tumor cells in sites outside of the uterus during pregnancy (Tafuri et al., 1995). It is now clear that the tolerogenic state of pregnancy involves a subset of T cells known as regulatory T (Treg) cells, characterized as CD4+CD25+ and expression of forkhead box P3 (FOXP3), that promote tolerance to non-self-antigens and inhibit inappropriate immune responses against self-antigens (Sagakuchi et al., 1995). Studies have shown that the pool of maternal Treg cells increases during pregnancy and that Treg cell expansion is essential for establishing immune tolerance (Aluvihare et al., 2004). Interestingly, both exposure to fetal alloantigen (Rowe et al., 2011) and pregnancy hormones, especially progesterone, synergistically drive Treg cell expansion (Mao et al., 2010). Again, conflict or disruption between the mother and fetus produces a delicate immunologic equilibrium that easily could be disturbed to cause pregnancy failure.

One intriguing hypothesis is that the stringency with which the maternal immune system accepts or rejects the fetal allograft is preprogrammed by the frequency of exposure to sperm antigens before pregnancy. Davis and Gallup (Platek & Shackelford, 2006) proposed that frequent exposure to the same sperm/semen is a biologic measure of the likelihood of paternal investment in the mother during pregnancy and the mother and neonate after parturition. Their hypothesis posits that a pregnancy derived from a "familiar" sperm/semen antigen profile reflects commitment of the father to the pregnancy and offspring and, as such, pregnancies and children sired by that male have a higher probability of receiving paternal investment then pregnancies and offspring from unfamiliar sperm/semen. A conceptus derived from unfamiliar sperm may be subjected to increased immunologic scrutiny and risk for rejection that could increase the risk for miscarriage. Studies show that the incidence of miscarriage is higher in single polyandrous females compared with women in monogamous partnership (Osborn et al., 2000). This may be another check mechanism by which the mother decides on her investment to continue the pregnancy.

5.10 Preeclampsia

Preeclampsia is clinically defined as pregnancy-associated maternal hypertension with proteinurea (Sibai et al., 2005). It usually involves multiple maternal organ systems and, as its severity increases, can progress to eclampsia and organ failure (Turner, 2010). The only effective treatment for severe preeclampsia—and the prevention of eclampsia—is termination of the pregnancy, usually be Cesarean section. Preeclampsia and chronic hypertension (that can lead to preeclampsia) are among the main reasons for "medically indicated" preterm birth (Carreno et al., 2013). Maternal blood pressure returns to normal after removal of the placenta, suggesting that fetal/placental endocrine signals induce maternal hypertension.

From a Darwinian perspective, it is counterintuitive that the fetus would impose such a life-threatening risk on the mother on which it is completely dependent. However, the capacity for fetal/placental endocrine signals to control maternal physiology and, especially maternal blood supply to the placenta, is a distinct advantage. As pregnancy advances, the nutrient needs of the growing fetus (and placenta) increase and, as such, must be matched by maternal supply. The capacity for maternal blood flow to the placenta is, therefore, a major limit to fetal growth. In this environment, the fetus benefits from increased blood supply to the placenta so that it can freely extract nutrients needed for continued growth and maturation. It is plausible, therefore, that natural selection favored fetal genes that increase maternal blood pressure, whereas maternal genes were selected that counteract those effects to prevent the life-threatening risks of hypertension. Indeed, maternal hypertension is a normal adaptation to pregnancy and has benefits for the fetus. However, those benefits are realized only if maternal blood pressure is kept within safe limits. In this context, preeclampsia is another example of the unintended consequences from the maternal/fetal conflict, where a trait that benefits the fetus may impose a cost to the mother (and fetus) if it is not stringently controlled by a maternal counterbalance.

The pathogenesis of preeclampsia is thought to involve abnormal trophoblast invasion into the maternal uterine spiral arteries (Brown & Crombleholme, 1993). A common feature of eutherian mammals is that the embryo implants into the endometrium soon after fertilization to establish a placental connection to the maternal compartment. In humans this occurs during the second week after ovulation and implantation is well-established within several days. Humans are unique, however, in that a second phase of implantation occurs at around the twelfth week of pregnancy (Robillard et al., 2003). During this burst of implantation, trophoblasts invade deep into the endometrium/myometrium and enter the spiral arteries where they replace the endothelium and smooth muscle. This remodeling increases vessel diameter and prevents the constriction of blood flow through the key arteries supplying maternal blood to the placenta. Thus, the fetus subverts maternal control of placental blood flow and remodels the vessels to maximize flow. Limitations to the depth into the arterioles that trophoblasts invade counteract maternal restriction to this process.

Preeclampsia is thought to be caused by this inadequate depth of invasion. This is not likely to be a problem early in pregnancy when fetal resource needs are easily matched by unmodified uterine blood supply. However, with advancing gestation, fetal demands increase, and the fetus may not receive adequate nutrients due to suboptimal blood flow to the placenta in cases with shallow trophoblast invasion. This may induce a fetal/placental response to secrete endocrine factors into the maternal compartment that increase systemic blood pressure as a compensatory effort to increase blood flow to the placenta. It is thought that fetal genes evolved under positive selection in the placenta to produce a phenotype whereby the trophoblasts exhibit the secondary invasion into the uterine arterioles and the fetus can increase maternal blood pressure. Maternal genes that limit these effects would evolve in response. Under normal conditions, the expression of the fetal/paternal and maternal traits are symmetrically counterbalanced to produce a healthy pregnancy and newborn. However, in keeping with the theme of genetic conflict, the extant system is sensitive to perturbations leading to pathophysiology. This inadequate trophoblast invasion—due to overactive maternal traits or underperforming fetal traits—would increase the risk for subsequent compensatory increases in maternal blood pressure later in pregnancy leading to preeclampsia.

The evolutionary origin of the second phase of trophoblast invasion—and by implication preeclampsia—has been linked to the hominid trait of encephalization and the increased nutritional needs of the fetus to support rapid brain growth (Robillard et al., 2003). During the third trimester, 60% of nutrients the human fetus extracts are devoted to brain growth, compared with 20% for other primates (Cunnane et al., 1993). It is possible that the strong selective pressure for encephalization favored the evolution of backup systems to compensate for inadequate nutrients to support brain growth in the later stages of pregnancy. In this context, the benefits of encephalization outweigh the costs of maternal risk for preeclampsia due to hypertension, inducing an increased blood supply to the placenta. A similar dynamic of unintended consequences stemming from the drive to increase hominid brain size is also thought to be the evolutionary basis for preterm birth.

5.11 Gestational diabetes

Gestational diabetes mellitus (GDM) is glucose intolerance discovered during pregnancy. It is the most common disease of pregnancy. Pregnant women are screened for abnormal glucose tolerance between 24 and 28 weeks. With a global obesity epidemic, the incidence of type 2 diabetes has doubled worldwide in recent years (World Health Organization, 2016), and similarly, the rates of diabetes in pregnant women is skyrocketing, with WHO estimating that GDM impacted 21 million births in 2013—or about 16% of pregnant women worldwide.

GDM causes long-lasting health complications in both the mother and baby. Mothers are more likely to be diagnosed with type 2 diabetes, especially within the first 5 years following the birth (Kim et al., 2002). Babies of mothers with GDM tend to have a higher birth weight and face a variety of complications, including premature birth and, even death. These complications seem to continue throughout the lifespan, with the offspring of mothers with GDM more likely to be obese, and have hypertension, heart, and metabolic disease (Gluckman & Hanson 2010; Horikoshi et al., 2016; Martin & Sacks, 2018). In addition, the economic burden of caring for the complications resulting from GDM, in the United States alone, was estimated to be $1.3 billion in 2012, a 103% increase since 2017 (Dall et al., 2014).

From an evolutionary viewpoint, Sheau Yu Teddy Hsu of Stanford and colleagues found the *GIP* gene may have once protected pregnant women in an agriculturally based society during lean times by boosting insulin production. But through about eight thousand years, the evolution of energy-balance regulation and the populations' diets have changed, making women with copies of this gene more vulnerable to diabetes during pregnancy (Chang et al., 2011). An example is the Dutch Hunger Winter. Pregnant women who experienced starvation early in their pregnancy, gave birth to children who experienced poor health and heart disease as adults. For those mid-gestation, kidney disease with associated diabetes and lung disease occurred in the offspring as adults. And, for those already late in pregnancy, their children were more likely to have type 2 diabetes as adults (Stearns & Medzhitov, 2016).

In addition, a study of genome-wide associations of birth weight and the correlation with adult disease found 60 genetic loci associated with birth weight, including both maternal and fetal genetic effects connected to birth weight and the risk for type 2 diabetes in later life (Horikoshi et al., 2016). Therefore, the growing obesity epidemic throughout the world may arise from the prenatal environment and genes dating back through our ancestors, is another example of the importance of viewing reproduction through an evolutionary lens.

5.12 Preterm birth

Globally each year, about 15 million women deliver prematurely with 450,000 preterm births in the United States (World Health Organization, 2013). Causes include infection, poor nutrition, socioeconomic and environmental stress, as well as a personal or maternal family history of preterm birth, including whether or not the woman herself was born prematurely (Porter et al., 1997) and if her sisters have delivered prematurely (Winkvist, 1998). There does not seem to be an influence on the father's side, such as if his birth was premature or if his sisters delivered preterm (Boyd et al., 2009). Gene-environment interactions, such as smoking which can restrict intrauterine growth, also have been linked to preterm birth. Birth before term increases the chance of neonatal morbidity and mortality due to fetal organ system immaturity. Prolonged gestation and post-term birth also is problematic and increases risk for the mother and fetus due to the difficulties of birthing a fetus that has grown too large to exit the birth canal.

Genetics and ancestral background related to premature birth are foci of current research. Louis Muglia and colleagues discovered six genes associated with the duration of gestation and risk for preterm birth using data from 43,500 women of European descent. The discovery cohort came from the direct-to-consumer company, "23andMe" and the findings were replicated in three Northern European cohorts Zhang et al. 2017. But women with African ancestry have a much higher rate of preterm birth. In fact, black women are twice as likely to deliver prematurely than European ancestry women, regardless of socioeconomic status or education. African-American women typically deliver an average of about 1 week earlier than white women (ncbi.nlm.nih.gov/pubmed/2292982), bringing into question whether "normal" gestation length should be defined differently given a woman's ancestral or genetic heritage.

Mis-timed birth is relatively rare in most species consistent with the timing of this event being critical for a successful pregnancy. In humans, however, the incidence of preterm birth and the consequent neonatal death is unusually high, affecting 10–15% of pregnancies. This is counterintuitive to the principals of natural selection.

Since the trigger is not strictly linked to fetal maturation, parturition could be induced by any condition that increases the inflammatory stimulus at any time. This may explain the strong association of preterm birth with intrauterine infection/inflammation. Thus, the extant trigger mechanism for initiating and timing human parturition is not perfect, but evolution operates at the species rather than individual level. As such, the benefit of encephalization conferred by the extant parturition trigger mechanism outweighs the costs of neonatal loss in a proportion of pregnancies. This system also reflects another aspect of maternal/fetal tug of war since maternal survival is balanced against neonatal survival.

5.13 Causes of maternal morbidity and mortality

Maternal death is a catastrophic outcome for families, communities, and caregivers, and as such, receives tremendous public health and media attention. Maternal deaths and the more common "near miss" or severe morbidity associated with pregnancy also serve as surrogates for abnormal pregnancy outcomes from a maternal perspective. Most of these deaths (99%) occur in low resource regions including sub-Saharan Africa (50%) and South Asia (30%) (World Health Organization. 2016; Wilmoth, 2012).

Maternal death is defined by the World Health Organization (WHO) as demise of any woman from any pregnancy-related cause while pregnant or within 42 days of pregnancy termination. Maternal death can be considered direct (obstetrical complication), indirect (disease prior to pregnancy or developing during pregnancy and exacerbated by the physiology of pregnancy), or nonmaternal (obstetric death from accidental cause). The maternal mortality ratio (MMR) is presented as maternal deaths (direct, indirect, or nonmaternal) per 100,000 live births (Gabbe et al., 1991). Worldwide distribution of MMR in 2015 can be seen in Figure 5.2.

For developed countries, the MMR is comparably low at twelve per 100,000 live births, whereas the estimated lifetime risk of maternal death for an adolescent woman from a designated "fragile" country is as high as one in fifty-four (Figure 5.3).

National Center for Health Statistics. Vital Health Stat 3(33).

The Maternal Mortality Rate has not always been low, however, in resource rich countries. At the turn of the twentieth century, the mortality ratio was nearly 900 per 100,000 live births in the United States and subsequently decreased to per 100,000 by 1991, attributed to factors such as movement of childbirth to hospitals, common use of prenatal care, aseptic technique, screening for preeclampsia,

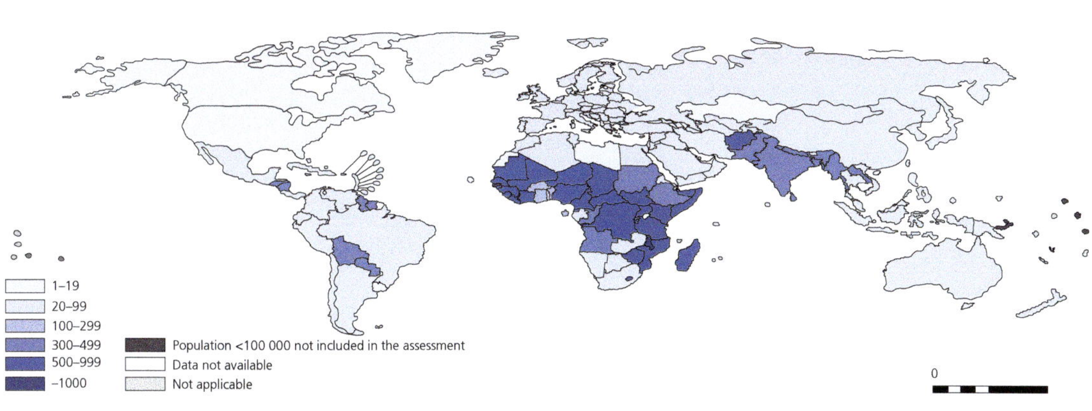

Figure 5.2 Worldwide distribution of the maternal mortality ratio (maternal deaths per 100,000 live births) in 2015
Reprinted with permission from the World Health Organization.

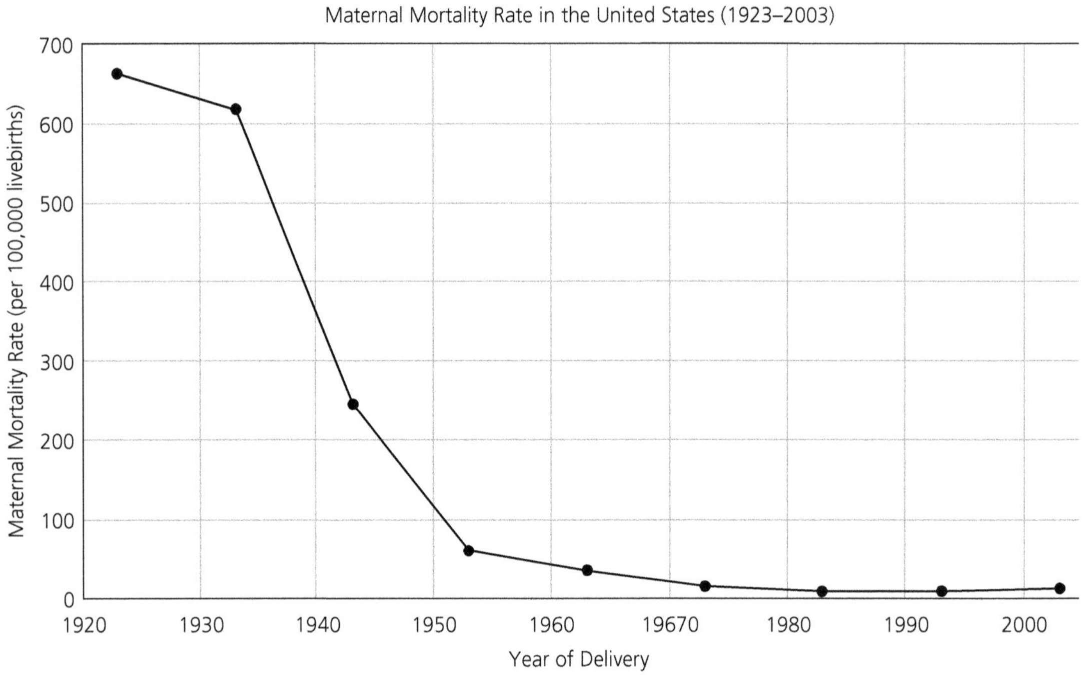

Figure 5.3 Maternal mortality rate in United States from 1923 through 2003.
Maternal mortality rate in United States from 1923

blood transfusions, antibiotics, and obstetrical anesthesia (Creasy & Resnick, 1994).

The well-known drivers of maternal mortality and morbidity include hemorrhage, preeclampsia, sepsis, thromboembolism, cardiovascular disease, cerebrovascular accidents, and suicide/homicide in both developed and developing countries. As for preventable causes of maternal mortality, hemorrhage (44–93%) and preeclampsia (50–64%) are the most preventable causes of death—at least in resource-rich countries—whereas cases of amniotic fluid embolism and cardiomyopathy are, as expected, less preventable (Berg et al., 2005; Cantwell et al., 2011).

5.14 Obstetric hemorrhage

Hemorrhage is the leading cause of death worldwide, and one of the top five causes in resource-rich countries, representing 10 percent of maternal deaths in the United States (MMR 1.6 per 100,000) (Knight et al., 2010). Causes of obstetric hemorrhage are multifactorial—related to both iatrogenic factors (rising Cesarean delivery rates, abnormal placentation or implantation, prolonged inductions, lacerations from delivery, instrumental delivery, and more) and obstetric factors (uterine atony, abruption, hypertensive disorders, retained placenta, macrosomic fetal birthweight, grand multiparity, and fetal demise) (Sheiner et al., 2005).

Hemostasis following delivery of the placenta requires adequate contraction of the myometrium to compress blood vessels coursing the denuded placental bed and innate decidual hemostatic factors to form a stable clot. Disturbances in either of these factors can result in massive blood loss as the term uterus receives 600 mL/min of blood. However, given advancements in obstetric anesthesia, standardization of management for postpartum hemorrhage, hospital simulations, medical therapies to prevent atony including oxytocin, methylergonovine, prostaglandin F2a, misoprostol, and mechanical/surgical techniques to control blood loss (balloon tamponade, uterine artery embolization/ligation, and hysterectomy), postpartum hemorrhage is only the fifth leading cause of maternal mortality in the United States despite occurring relatively frequently (approximately 3 percent of all deliveries) (Bateman et al., 2010). Nonetheless maternal death from postpartum hemorrhage is still a common, preventable cause of maternal mortality.

5.15 Cesarean for cephalopelvic disproportion

Cephalopelvic disproportion (CPD) or fetopelvic disproportion is the result of either small maternal pelvic capacity, large fetal size, or a combination of both leading to protracted labor or arrest disorders. From an obstetric standpoint, the pelvis is defined by three anatomic regions, the pelvic inlet, mid-pelvis, and pelvic outlet for which anteroposterior and transverse diameters have been described (Cunningham & Williams., 2014). Small or "contracted" pelvic measurements have variable predictive accuracy in determining likelihood of successful vaginal birth.

The true incidence of CPD is unknown and difficult to ascertain as it is a subjective clinical assessment which is often categorized into broader definitions including "failure to progress" or "arrest of labor." Protraction of labor disorders are rarely the result of true CPD but rather a multitude of other factors such as ineffective labor, fetal head position, and labor management patterns. Nonetheless protracted labor is associated with significant maternal and neonatal morbidities including shoulder dystocia, postpartum hemorrhage, higher degree perineal laceration, operative delivery, Cesarean delivery, and subsequent operative complications, as well as neonatal birth injury (brachial plexus injuries), asphyxiation, and rarely neonatal death.

Extensive obstetric literature has focused on identifying women at risk for CPD with the use of radiographic cephalopelvimetry as well as manual measurements of the pelvic inlet, midpelvis, and pelvic outlet (Thorp & Laughon, 2019; Cunningham & Williams. 2014). Other investigators have attempted to incorporate fetal biometry, including head circumference and abdominal circumference by ultrasound, in assessing risk for failed first and second stages of labor with mixed results. Many of these methods are no longer widely used given their poor ability to predict outcome, their inherent cost, and technical difficulty. Rather, the American Congress of Obstetrics and Gynecology recommends a trial of labor for all nulliparous women unless estimated fetal weight is predicted to be greater than 5000 grams or if pathological fetal disproportion is suspected as in the setting of uncontrolled maternal diabetes. Despite these recommendations the majority of CPD cases occur among pregnancies with normal birthweights (Practice Bulletin, 173).

5.16 Fetal growth restriction

In 1963, Lula Lubchenco and colleagues published comparisons of birthweights by gestational age to determine "expected" fetal size throughout pregnancy (Lubchenco, 1963). Small for gestational age (SGA) was subsequently classified as birthweight below the tenth percentile (Battaglia & Lubchenco, 1967). Regardless of etiology, SGA infants, as a group, were found to be at increased risk for neonatal morbidity and mortality, even at term. The severity of growth restriction also correlated with increasingly poor perinatal outcomes. SGA neonates had perinatal mortality rates (PMR) up to 100 times greater than their normal-weight counterparts. For those neonates at the extremes of low birthweight, this equated to PMR of greater than 10 percent. This finding led to decades of obstetric and pediatric research into the various underlying etiologies behind low birthweights. With the development of prenatal ultrasound in the 1980s, allowing the obstetrician to determine estimated fetal weight (EFW) through fetal biometry and, thus determine fetal growth restriction (FGR) or EFW < 10th% in utero, serving as a surrogate for SGA.

A neonate with a birthweight at the thirtieth percentile who did not reach his/her growth potential is still at increased risk for neonatal mortality but would remain undetected with the use of current population-based growth curves. A superior approach in identifying those fetuses at risk for poor perinatal outcomes secondary to fetal growth restriction would be to observe individual fetal growth velocities and placental function (Doppler assessment, amniotic fluid assessment), but this approach would certainly add to the cost of healthcare (Resnick & Creasy, 2009).

Genetic determinants of fetal growth likely represent 40% of the observed variation in birthweight while environmental factors determine the remainder. Observed constraints on fetal growth indicate strong maternal influences, such as height and other genetic factors, since FGR is observed more commonly between sisters, mothers, and daughters, and has high incidence of recurrence in the same mother (Cunningham & Williams, 2014). The Y chromosome also influences fetal weight as male infants weigh up to 200 grams more than female infants of similar gestational age (Thomson et al., 1968). Additionally, paternal birthweight appears to have a similarly moderate influence on progeny birthweight (Klebanoff et al., 1998).

Fetal growth peaks between 32–34-weeks' gestation (230–285g) and then decreases until 41–42 weeks' gestation when fetal weight gain may become neutral and even negative. Fetal growth restriction can occur at any point throughout fetal development; during the first phase which consists of cellular hyperplasia (<20 weeks' gestation), the second phase which comprises both cellular hyperplasia and hypertrophy (>20 weeks), and lastly within the final 8 weeks of pregnancy which consists of mostly hypertrophic growth (Creasy and Resnick, 2009; Williams et al., 1982). This is useful clinically when ultrasound findings indicate symmetrically growth restricted fetal biometry as opposed to asymmetrically small biometry. Symmetrical FGR represents approximately 20% of growth restriction cases and is usually secondary to disorders of cell hyperplasia such as aneuploidy, teratogens, environmental exposures (radiation), and fetal infection. The more commonly observed asymmetrical FGR often indicates placental disease affecting nutrient and oxygen delivery to the fetus (Cunningham & Williams, 2014).

The important and more commonly encountered causes of pathologic FGR—which are beyond the scope of this chapter—include aneuploidies, congenital anomalies (gastroschisis, CHD), infections (TORCH, HIV), multiple gestations, poor maternal nutrition, environmental toxins or teratogens (cigarette smoking, alcohol ingestion, opiates, etc.,) abnormal placentation (rudimentary trophoblast invasion), and maternal vascular disease (diabetes, hypertension) (Creasy & Resnick, 2009).

5.17 Implications for medical education

Few medical school curriculums include instruction on evolutionary biology as a basic science, let alone evolutionary medicine as a discipline that can enhance understanding of disease and its treatment in areas such as antibiotic resistance, infectious disease and transmission, autoimmune diseases, and, of course, reproduction and birth (Stearns et al., 2010). As Rebecca Sear says, "Evolutionary explanations do not assume that all human behavior is hardwired and functions to maximize genetic fitness; they are able to explain variation in human behavior; and they are not necessarily alternatives to social science explanations" (Sear, 2015). Therefore, adding an evolutionary biology or medicine focus to the curriculum does not negate any perspectives currently taught (Grunspan et al., 2017)

One model would be to interweave evolutionary thought throughout the curriculum. In many ways, this is not different than the rapidly advancing knowledge of genomics, which touches many medical specialties and requires medical school curriculums to be flexible and adaptable, as new insight comes to light. In arguing for its addition to medical school curriculums, Randolph Nesse states, "Evolutionary biology is a unifying principle that provides a framework for organizing medical knowledge from other basic sciences" (Nesse et al., 2010).

This is particularly critical in the area of reproductive health. The study of pregnancy and reproductive outcomes is vitally important to all societies. Incorporating an evolutionary framework into medical education in the future promises to reveal new insights into both proximate and ultimate causes of human health and pathogenesis. An evolutionary approach will augment traditional specialty and subspecialty-specific didactics by including physical and social anthropology—which is especially important for human childbirth as it is an attendant assisted event, comparative physiology, and genomics such that differences in species strategies can inform mechanisms rather than providing barriers, informatics to assemble high dimensional omic and epidemiological information, philosophy of science/biology to provide a historical and conceptual framework, and sociology/social evolution to better understand how changes in human societies contribute to racial disparities in pregnancy outcomes and how environmental changes—nutrient access, water and air pollution, and emerging infectious agents lead to maladaptations in pregnancy physiology that result in pregnancy disorders.

Although evolution influences human reproduction and birth, modern social influences need to be addressed simultaneously in medical education. Modern twenty-first-century medicine and the extraordinary advances in molecular biology and genetics have increased our capacity to define and answer with precision core epidemiologic tasks postured by society but have also broadened our understanding for how environmental exposures through evolutionary pressure affect not only an individual's health but the health of a population over time.

Premature birth can leave a baby vulnerable to disease throughout the lifespan. Even if born at term, the prenatal environment can exacerbate disease risk. From an evolutionary perspective, "human nutritional needs evolved under circumstances dramatically different from contemporary diets and lifestyles" (Trevathan, 2007). For example, populations who went from starvation to abundance, may have developed a gene with the ability to store fat and mobilize insulin. In today's society, that may be a reason for the increase in gestational diabetes and obesity (Trevathan, 2007). Understanding this from an evolutionary perspective changes the medical perspective on how to intervene in issues, such as obesity, that have been traditionally thought of as "lifestyle" or "behavioral." Peter Gluckman and Mark Hanson state, "Improving maternal and fetal health will allow humans to cope better with current postnatal nutritional conditions—conditions the we did not evolve to inhabit" (Gluckman & Hanson, 2004). This could someday change clinical treatment of metabolic and cardiovascular disease—a concept critical for emerging clinicians to understand.

One element that is essential in incorporating evolutionary thought into medical education going forward is the concept of pleiotropy: that a single genetic mutation can impact multiple traits. Although the concept of pleiotropy is more than a century old—first used in the medical literature by German developmental geneticist Ludwig Plate in 1910—it takes on new importance as we advance our genomic knowledge. Understanding evolutionary dynamics through the lens of pleiotropy could lead to discoveries on genetic disorders, even ultimately helping us to find treatments (Stearns, 2010). Likewise, it could help answer questions about fetal development and birth outcomes—and discover ways to prevent preterm delivery and other diseases of pregnancy in the future.

Applying an evolutionary view allows us to view pregnancy both in physiological and social contexts. Pregnancy is both a unique and personal experience for a woman to which society invests appreciably in the outcome. For instance, maternal and infant mortality are the most widely used health indicators of a nation and the incidence of low birthweight is a common indicator of a nation's nutritional health (Robbins et al., 2014). Not only are we as a society innately fascinated with pregnancy and childbirth, but reproduction exemplifies the core biological theme of evolution to optimize the fitness of a species and to ensure future procreation. Many of the adverse outcomes of pregnancy—preeclampsia, gestational diabetes, and preterm birth—may represent maladaptive complications in attempting to achieve this balance. What evolution does not do is provide selection for optimizing long-term health or longevity.

Human reproduction is a delicate biological balance between the interests of the mother and fetus—whether it is controlled by a molecular "tug-of-war" or regulatory signals; the mother allocating enough resource to maintain her own health and to propagate her own genome to both the index and subsequent offspring, and the fetus maximizing and exploiting available resource to ensure survival, which, at times, can be at the expense of maternal health. Index pregnancy—over-allocation by the mother—presumably leads to either future pregnancy failure (miscarriage or stillbirth) or, at worst, maternal loss of life and the ability to propagate current or future reproductive cycles. This balance has been described as both harmonious and conflicting in nature, but it is likely a combination of both equilibrium and competition. As Nesse says, "The power of evolutionary thinking in medicine comes from its ability to foster integrative thinking about our bodies as products of evolutionary processes" (Nesse et al., 2010).

References

Adams, M. M., Elam-Evans, L. D., Wilson, H. G., & Gilbertz, D. A. (2000). Rates of and factors associated with recurrence of preterm delivery. *Jama*, 283(12): 1591–6. ncbi.nlm.nih.gov/pubmed/2292982.

Aluvihare, V. R., Kallikourdis, M., & Betz, A. G. (2004). Regulatory T cells mediate maternal tolerance to the fetus. *Nature Immunology*, 5(3): 266–71.

Amini, P., Michniuk, D., Kuo, K., Yi, L., Skomorovska-Prokvolit, Y., Peters, G. A., Tan, H., Wang, J. Malemud, C. J., & Mesiano, S. (2016). Human parturition involves phosphorylation of progesterone Receptor-A at Serine-345 in myometrial cells. *Endocrinology*, 157(11): 4434–45.

Ananth, C. V., Keyes, K. M., Hamilton, A., Gissler, M., Wu, C., Liu, S., Luque-Fernandez, M. A., SkjÃ¦rven, R., Williams, M. A., Tikkanen, M., & Cnattingius, S. (2015). An international contrast of rates of placental abruption: An age-period-cohort analysis. *Plos One*, 10(5): 1–15.

Andersson, S., Minjarez, D., Yost, N. P., & Word, R. A. (2008). Estrogen and progesterone metabolism in the cervix during pregnancy and parturition. *The Journal of Clinical Endocrinology & Metabolism*, 93(6): 2366–74.

Apps, R., Murphy, S. P., Fernando, R., Gardner, L., Ahad, T., & Moffett, A. (2008). Human leucocyte antigen (HLA) expression of primary trophoblast cells and placental cell lines, determined using single antigen beads to characterize allotype specificities of anti-HLA antibodies. *Immunology*, 127(1): 26–39.

Barker, E. D., Mcauliffe, F. M., Alderdice, F., Unterscheider, J., Daly, S., Geary, M. P., Kennelly, M. M., O'Donoghue, K., Hunter, A., Morrison, J. J., Burke, G., Dicker, P., Tully, E. C., & Malone, F. D. (2013). The role of growth trajectories in classifying fetal growth restriction. *Obstetrics & Gynecology*, 122(2, part 1): 248–54.

Bateman, B. T., Berman, M. F., Riley, L. E., & Leffert, L. R. (2010). The epidemiology of postpartum hemorrhage in a large, nationwide sample of deliveries. *Anesthesia & Analgesia*, 110(5): 1368–73.

Battaglia, F. C. & Lubchenco, L. O. (1967). A practical classification of newborn infants by weight and gestational age. *The Journal of Pediatrics*, 71(2): 159–63.

Beagley, K. W. & Gockel, C. M. (2006). Regulation of innate and adaptive immunity by the female sex hormones oestradiol and progesterone. *Pathogens and Disease*, 38(1): 13–22.

Beekhuizen, H. V., Joosten, I., Lotgering, F., Bulten, J., & Kempen, L. V. (2010). Natural killer cells and HLA-G expression in the basal decidua of human placenta adhesiva. *Placenta*, 31(12): 1078–84.

Berg, C. J., Harper, M. A., Atkinson, S. M., Bell, E. A., Brown, H. L., Hage, M. L., Mitra, A. G., Moise, K. J., & Callaghan, W. M. (2005). Preventability of pregnancy-related deaths: Results of a state-wide review. *Obstetrics & Gynecology*, 106(6): 1228–34.

Bezold, K. Y., Karjalainen, M. K., Hallman, M., Teramo, K., & Muglia, L .J. (2013). The genomics of preterm birth: From animal models to human studies. *Genome Medicine*, 5(4): 34.

Bouteiller, P. L. & Bensussan, A. (2017). In Bouteiller, P. L. & Bensussan, A. (eds) *Up-and-Down Immunity of Pregnancy in Humans*. 6(*F1000Research*): 1216.

Boyd, H. A., Poulsen, G., Wohlfahrt, J., Murray, J. C., Feenstra, B., & Melbye, M. (2009). Maternal contributions to preterm delivery. *American Journal of Epidemiology*, 170(11): 1358–64.

Brown, J. S. & Crombleholme, W. R. (1993). Hypertensive disorders of pregnancy. In: *Handbook of Gynecology & Obstetrics*, pp. 405–16. Norwalk, CT: Appleton & Lange.

Buchbinder, A., Sibai, B. M., Caritis, S., Macpherson, C., Hauth, J., Lindheimer, M. D., Klebanoff, M., Vandorsten, P., Landon, M., Paul, R., Miodovnik, M., Meis, P., & Thurnau, G. (2002). Adverse perinatal outcomes are significantly higher in severe gestational hypertension than in mild preeclampsia. *American Journal of Obstetrics and Gynecology*, 186(1): 66–71.

Burton, G. J. & Fowden, A. L. (2015). The placenta: A multifaceted, transient organ. *Philosophical Transactions of the Royal Society B: Biological Sciences*, 370(1663): 20140066.

Burton, G.J. & Jauniaux, E. (2017). The cytotrophoblastic shell and complications of pregnancy. *Placenta*, 60: 134–9.

Carreno, C., Blackwell, S., Sibai, B., & Kase, B. (2013). The impact of medically indicated and spontaneous preterm birth among hypertensive women. *American Journal of Perinatology*, 30(10): 843–8.

Cartwright, J. E., Fraser, R., Leslie, K., Wallace, A. E., & James, J. L. (2010). Remodeling at the maternal-fetal interface: Relevance to human pregnancy disorders. *Reproduction*, 140(6): 803–13.

Castro, M., Morales, P., Rojo-Amigo, R., Martinez-Laso, J., Allende, L., Varela, P., Garcia-Berciano, M., Guillen-Perales, J., & Arnaiz-Villena, A. (2000). Homozygous HLA-G*0105N healthy individuals indicate that membrane-anchored HLA-G1 molecule is not necessary for survival. *Tissue Antigens*, 56(3): 232–9.

Cha, J., Sun, X., & Dey, S. K. (2012). Mechanisms of implantation: Strategies for successful pregnancy. *Nature Medicine*, 18(12): 1754–67.

Chang, C., Cai, J., Cheng, P., Chueh, H., & Hsu, S. (2011). Identification of metabolic modifiers that underlie phenotypic variations in energy-balance regulation. *Diabetes*, 62(12): 4284–4.

Chavan, A. R., Griffith, O. W., &Wagner, G. Ã. P. (2017). The inflammation paradox in the evolution of mammalian pregnancy: Turning a foe into a friend. *Current Opinion in Genetics & Development*, 47: 24–32.

Chen, C., Opazo, J. C., Erez, O., Uddin, M., Santolaya-Forgas, J., Goodman, M., Grossman, L. I., Romero, R., & Wildman, D. E. (2008). The human progesterone receptor shows evidence of adaptive evolution associated with its ability to act as a transcription factor. *Molecular Phylogenetics and Evolution*, 47(2): 637–49.

Chen, D.-B. & Zheng, J. (201). Regulation of placental angiogenesis. *Microcirculation*, 21(1): 15–25.

Chen, J. Z.-J., Sheehan, P. M., Brennecke, S. P., & Keogh, R. J. (2012). Vessel remodeling, pregnancy hormones and extravillous trophoblast function. *Molecular and Cellular Endocrinology*, 349(2): 138–44.

Clark, A., Glanowski, S., Nielsen, R., Thomas, P., Kejariwal, A., Todd, M., Tanenbaum, D., Civello, D., Lu, F., Murphy, B., Ferriera, S., Wang, G., Zheng, X., White, T., Sninsky, J., Adams, M., and Cargill, M. (2003). Positive selection in the human genome inferred from human-chimp-mouse orthologous gene alignments. *Cold Spring Harbor Symposia on Quantitative Biology*, 68: 479–86.

Condon, J. C., Jeyasuria, P., Faust, J. M., Wilson, J. W., & Mendelson, C. R. (2003). A decline in the levels of progesterone receptor coactivators in the pregnant uterus at term may antagonize progesterone receptor function and contribute to the initiation of parturition. *Proceedings of the National Academy of Sciences*, 100(16): 9518–23.

Coumans, B., Thellin, O., Zorzi, W., Melot, F., Bougoussa, M., Melen, L., Zorzi, D., Hennen, G., Igout, A., & Heinen, E. (1999). Lymphoid cell apoptosis induced by trophoblastic cells: A model of active foeto-placental tolerance. *Journal of Immunological Methods*, 224(1–2): 185–96.

Creasy, R. K. & Resnik, R. (1994). Maternal mortality. In: *Maternal-Fetal Medicine: Principles and Practice*, (pp. 795–801). Philadelphia: Saunders.

Creasy, R. K., Lockwood, C. J., Resnik, R., Moore, T. R., Iams, J. D., & Greene, M. F. (2014). Clinical aspects of normal and abnormal labor. In: R. K. Creasy & C. J. Resnik (eds), *Maternal-Fetal Medicine: Principles and Practice*. Philadelphia: Elsevier Saunders.

Csapo, A. (1956). Progesterone block. *American Journal of Anatomy*, 98(2): 273–91.

Cunnane, S. C., Harbige, L. S., and Crawford, M. A. (1993). The importance of energy and nutrient supply in human brain evolution. *Nutrition and Health*, 9(3): 219–35.

Cunningham, F. G., & Williams, J. W. (2014). Fetal growth disorders. In: *Williams Obstetrics*. New York: McGraw-Hill.

Dall, T. M., Yang, W., Halder, P., Pang, B., Massoudi, M., Wintfeld, N., Semilla, A. P., Franz, J., & Hogan, P. F. (2014). The economic burden of elevated blood glucose levels in 2012: Diagnosed and undiagnosed diabetes, gestational diabetes mellitus, and prediabetes. *Diabetes Care*, 37(12): 3172–9.

Diaz, P., Powell, T. L., & Jansson, T. (2014). The role of placental nutrient sensing in maternal-fetal resource allocation. *Biology of Reproduction*, 91(4).

Dimasuay, K. G., Boeuf, P., Powell, T. L., & Jansson, T. (2016). Placental responses to changes in the maternal environment determine fetal growth. *Frontiers in Physiology*, 7.

Dobzhansky, T. (1973). Nothing in biology makes sense except in the light of evolution. *The American Biology Teacher*, 35: 125–9.

Dunsworth, H. M., Warrener, A. G., Deacon, T., Ellison, P. T., & Pontzer, H. (2012). Metabolic hypothesis for human altriciality. *Proceedings of the National Academy of Sciences*, 109(38): 15212–16.

Ellish, N., Saboda, K., O'connor, J., Nasca, P., Stanek, E., & Boyle, C. (1996). A prospective study of early pregnancy loss. *Human Reproduction*, 11(2): 406–12.

Erlebacher, A. (2013). Immunology of the maternal-fetal interface. *Annual Review of Immunology*, 31: 387–411.

Evans, R. M. (1988). *The steroid and thyroid hormone receptor superfamily*. [online] Science. 240(4854): 889–95. Available at: <http://science.sciencemag.org/content/240/4854/889>

Evans, R. M. (2005). The nuclear receptor superfamily: A Rosetta Stone for physiology. *Molecular Endocrinology*, 19(6): 1429–38.

Fisher, S. & Damsky, C. (1993). Human cytotrophoblast invasion. *Seminars in Cell Biology*, 4(3): 183–8.

Flint, A. P. F., Anderson, A. B. M., Steele, P. A., & Turnbull, A.C. (1975). The mechanism by which foetal cortisol controls the onset of parturition in the sheep. *Biochemical Society Transactions*, 3(6): 1189–94.

Fowden, A. & Moore, T. (2012). Maternal-fetal resource allocation: Co-operation and conflict. *Placenta*, 33(Supplement 2): e11–e15.

Fragouli, E. & D. Wells (2012). Aneuploidy screening for embryo selection. *Semin Reprod Med*, 30(4): 289–301.

Gabbe, S. G., Niebyl, J. R., Simpson, J. L., & Anderson, G. D. (1991). *Obstetrics: Normal and Problem Pregnancies*. New York: Churchill Livingstone.

Garcia-Enguidanos, A., Calle, M. E., Valero, J., Luna, S., & Dominguez-Rojas, V. (2002). Risk factors in miscarriage: A review. *European Journal of Obstetrics and Gynecology and Reproductive Biology*, 102(2): 111–19.

Gellersen, B. & Brosens, J. J. (2014). Cyclic decidualization of the human endometrium in reproductive health and failure. *Endocrine Reviews*, 35(6): 851–905.

Gluckman, P. D. & Hanson, M. A. (2004). Developmental origins of disease paradigm: A mechanistic and evolutionary perspective. *Pediatric Research*, 56(3): 311–17.

Gluckman, P. D. & Hanson, M. A. (2008). Developmental and epigenetic pathways to obesity: An evolutionary-developmental perspective. *International Journal of Obesity*, 32(S7).

Gluckman, P. D. & Hanson, M. A. (2010). The conceptual basis for the developmental origins of health and disease. *Developmental Origins of Health and Disease*, 1(1):33–50.

Godfrey, K. M., Lillycrop, K. A., Burdge, G. C., Gluckman, P. D., & Hanson, M. A. (2007). Epigenetic mechanisms and the mismatch concept of the developmental origins of health and disease. *Pediatric Research*, 61(5): 5R—10R.

Godfrey, K. M., Gluckman, P. D., & Hanson, M. A. (2010). Developmental origins of metabolic disease: Life course and intergenerational perspectives. *Trends in Endocrinology & Metabolism*, 21(4): 199–205.

Haig, D. (1993). Genetic conflicts in human pregnancy. *The Quarterly Review of Biology*, 68(4): 495–532.

Haig, D. (1996). Altercation of generations: Genetic conflicts of pregnancy. *American Journal of Reproductive Immunology*, 35(3): 226–32.

Haig, D. (1996). Gestational drive and the green-bearded placenta. *Proceedings of the National Academy of Sciences*, 93(13): 6547–51.

Haig, D. (1997). Maternal-fetal interactions and MHC polymorphism. *Journal of Reproductive Immunology*, 35(2): 101–9.

Haig, D. (2007). Putting up resistance: Maternal fetal conflict over the control of uteroplacental blood flow. *Endothelial Biomedicine*, (4 Suppl): 135–41.

Halasz, M. & Szekeres-Bartho, J. (2013). The role of progesterone in implantation and trophoblast invasion. *Journal of Reproductive Immunology*, 97(1): 43–50.

Hammer, A. (2011). Immunological regulation of trophoblast invasion. *Journal of Reproductive Immunology*, 90(1): 21–8.

Hardy, D. B., Janowski, B. A., Corey, D. R., & Mendelson, C.R. (2006). Progesterone receptor plays a major anti-inflammatory role in human myometrial cells by antagonism of nuclear factor-KappaB activation of cyclooxygenase 2 expression. *Molecular Endocrinology*, 20(11): 2724–33.

Hartling, L., Dryden, D. M., & Guthrie, A. (2012). Screening and diagnosing gestational diabetes mellitus. *Evidence Report/Technology Assessment*. 1–327.

Hauth, J., Ewell, M. G., Levine, R. J., Esterlitz, M.S., Sibai, B., Curet, L. B., Catalano, P. M., & Morris, C. D. (2000). Pregnancy outcomes in healthy nulliparas who developed hypertension. *Obstetrics & Gynecology*, 95(1): 24–8.

Hirsch, E., Saotome, I., & Hirsch, D. (1995). A model of intrauterine infection and preterm delivery in mice. *American Journal of Obstetrics and Gynecology*, 172(5): 1598–603.

Howard, J., (2017). *Born before 22 weeks, "most premature" baby is now thriving*. [online] CNN. Available at: <https://www.cnn.com/2017/11/08/health/premature-baby-21-weeks-survivor-profile/index.html>

Horikoshi, M., Beaumont, R. N., & Freathy, R. M., (2016). Genome-wide associations for birth weight and correlations with adult disease. *Nature News*: www.nature.com/articles/nature19806.

James, J., Carter, A., & Chamley, L. (2012). Human placentation from nidation to 5 weeks of gestation. Part II: Tools to model the crucial first days. *Placenta*, 33(5): 335–42.

Jiang, S. P. & Vacchio, M. S. (1998). Cutting edge: Multiple mechanisms of peripheral T cell tolerance to the fetal "allograft." *Journal of Immunology*, 160(7): 3086–3090.

Kalantaridou, S., Zoumakis, E., Makrigiannakis, A., Godoy, H., & Chrousos, G. (2007). The role of corticotropin-releasing hormone in blastocyst implantation and early fetal immunotolerance. *Hormone and Metabolic Research*, 39(6): 474–7.

Kalkhoven, E., Wissink, S., Saag, P. T. V. D., & Burg, B. V. D. (1996). Negative interaction between the RelA(p65) subunit of NF-B and the progesterone receptor. *Journal of Biological Chemistry*, 271(11): 6217–24.

Kastner, P., Krust, K., B Turcotte, Stropp, U., Tora, L., Gronemeyer, H., & Chambon, P. (1990). Two distinct estrogen-regulated promoters generate transcripts encoding the two functionally different human progesterone receptor forms A and B. *EMBO Journal*, 9: 1603–14.

Ke, W., Chen, C., Luo, H., Tang, J., Zhang, Y., Gao, W., Yang, X., Tian, Z., Chang, Q., & Liang, Z. (2016). Histone deacetylase 1 regulates the expression of progesterone receptor a during human parturition by occupying the progesterone receptor a promoter. *Reproductive Sciences*, 23(7): 955–64.

Kim, C., Newton, K. M., & Knopp, R. H. (2002). Gestational diabetes and the incidence of Type 2 Diabetes: A systematic review. *Diabetes Care*, 25(10): 1862–8.

King, A., Boocock, C., Sharkey, A., Gardner, L., Beretta, A., Siccardi, A., & Wai, L. Y. (1996). Evidence for the expression of HLA-C class I mRNA and protein by human first trimester trophoblast. *Journal of Reproductive Immunology*, 31(3): 232–3.

King, A., Burrows, T., Hiby, S., Bowen, J., Joseph, S., Verma, S., Lim, P., Gardner, L., Bouteiller, P. L., Ziegler, A., Uchanska-Ziegler, B., & Loke, Y. (2000). Surface expression of HLA-C antigen by human extravillous trophoblast. *Placenta*, 21(4): 376–87.

Klebenanoff, M. A., Mednick, B. R., Schulsinger, C., Secher, N. J., & Shiono, P. H. (1998). Father's effect on infant birth weight. *American Journal of Obstetrics and Gynecology*, 178(5): 1022–6.

Knight, M., Callaghan, W. M., Berg, C., Alexander, S., Bouvier-Colle, M. -H., Ford, J. B., Joseph, K. S., Lewis, G., Liston, R. M., Roberts, C. L., Oats, J., & Walker, J. (2010). Trends in postpartum hemorrhage in high resource countries: A review and recommendations from the international postpartum hemorrhage collaborative group. *Obstetrical & Gynecological Survey*, 65(4): 211–12.

Kosova, G. Ã. Ã., Stephenson, M. D., Lynch, V.J., & Ober, C. (2015). Evolutionary forward genomics reveals novel insights into the genes and pathways dysregulated in recurrent early pregnancy loss. *Human Reproduction*, 30(3): 519–29.

Lamarck, J. B. D. & Elliott, H. S. R. (2011). General principles concerning animals. *Zoological Philosophy*, 47–55.

Leavey, K., Benton, S. J., Grynspan, D., Bainbridge, S. A., Morgen, E. K., & Cox, B. J. (2017). Gene markers of normal villous maturation and their expression in placentas with maturational pathology. *Placenta*, 58: 52–9.

Lewis, R., Cleal, J., & Hanson, M. (2012). Review: Placenta, evolution and lifelong health. *Placenta*, 33.

Lim, Y. C., Li, J., Ni, Y., Liang, Q., Zhang, J., Yeo, G. S. H., Lyu, J., Jin, S., & Ding, C. (2017). A complex association between DNA methylation and gene expression in human placenta at first and third trimesters. *Plos One*, 12(7).

Liu, K. -J., Wang, C. -J., Chang, C. -J., Hu, H. -I., Hsu, P. -J., Wu, Y .-C., Bai, C. -H., Sytwu, H. -K., & Yen, B. L. (2011). Surface expression of HLA-G is involved in mediating immunomodulatory effects of placenta-derived multipotent cells (PDMCs) towards natural killer lymphocytes. *Cell Transplantation*, 20(11–12): 1721–30.

Lubchenco, L. O., Hansman, C., Dressler, M., & Boyd, E. (1963). Intrauterine growth as estimated from liveborn birth-weight data at 24 to 42 weeks of gestation. *Pediatrics*, 32(5): 793–800.

Lynch, V. J., Leclerc, R. D., May, G., & Wagner, G. Ã. P. (2011). Transposon-mediated rewiring of gene regulatory networks contributed to the evolution of pregnancy in mammals. *Nature Genetics*, 43(11): 1154–9.

Mahendroo, M. S., Porter, A., Russell, D. W., & Word, R. A. (1999). The parturition defect in steroid 5Î±-Reductase Type 1 knockout mice is due to impaired cervical ripening. *Molecular Endocrinology*, 13(6): 981–92.

Maiti, K., Sultana, Z., Aitken, R. J., Morris, J., Park, F., Andrew, B., Riley, S. C., & Smith, R. (2017). Evidence that fetal death is associated with placental aging. *American Journal of Obstetrics and Gynecology*, 217(4).

Makrigiannakis, A., Zoumakis, E., Kalantaridou, S., Coutifaris, C., Margioris, A. N., Coukos, G., Rice, K. C., Gravanis, A., & Chrousos, G. P. (2001). Corticotropin-releasing hormone promotes blastocyst implantation and early maternal tolerance. *Nature Immunology*, 2(11): 1018–24.

Maltepe, E. & Fisher, S. J. (2015). Placenta: The forgotten organ. *Annual Review of Cell and Developmental Biology*, 31(1): 523–52.

Mao, G., Wang, J., Kang, Y., Tai, P., Wen, J., Zou, Q., Li, G., Ouyang, H., Xia, G., & Wang, B. (2010). Progesterone increases systemic and local uterine proportions of CD4 CD25 Treg Cells during midterm pregnancy in mice. *Endocrinology*, 151(11): 5477–88.

Martin, B., & Sacks, D. A. (2018). The global burden of hyperglycemia in pregnancy – Trends from studies in the last decade. *Diabetes Research and Clinical Practice*, 145: 17–19.

Mayne, B. T., Bianco-Miotto, T., Buckberry, S., Breen, J., Clifton, V., Shoubridge, C., & Roberts, C. T. (2016). Large scale gene expression meta-analysis reveals tissue-specific, sex-biased gene expression in humans. *Frontiers in Genetics*, 7(183): 1–10.

Mayne, B. T., Leemaqz, S. Y., Smith, A. K., Breen, J., Roberts, C. T., & Bianco-Miotto, T. (2017). Accelerated placental aging in early onset preeclampsia pregnancies identified by DNA methylation. *Epigenomics*, 9(3): 279–89.

Medawar, P. B. (1953). Some immunological and endocrinological problems raised by the evolution of viviparity in vertebrates. *Symposia of the Society for Experimental Biology*, 7: 320–37.

Mellor, A. L., Sivakumar, J., Chandler, P., Smith, K., Molina, H., Mao, D., & Munn, D. H. (2001). Prevention of T cell-driven complement activation and inflammation by tryptophan catabolism during pregnancy. *Nature Immunology*, 2(1): 64–8.

Merlino, A. A., Welsh, T. N., Tan, H., Yi, L. J., Cannon, V., Mercer, B. M., & Mesiano, S. (2007). Nuclear progesterone receptors in the human pregnancy myometrium: Evidence that parturition involves functional progesterone withdrawal mediated by increased expression of progesterone receptor-A. *The Journal of Clinical Endocrinology & Metabolism*, 92(5): 1927–33.

Mesiano, S. (2002). Progesterone withdrawal and estrogen activation in human parturition are coordinated by progesterone receptor A expression in the myometrium. *Journal of Clinical Endocrinology & Metabolism*, 87(6): 2924–30.

Moorjani, P., Amorim, C. E. G., Arndt, P. F., & Przeworski, M. (2016). Variation in the molecular clock of primates. *Proceedings of the National Academy of Sciences*, 113(38): 10607–12.

Morgan, T. (2016). Role of the placenta in preterm birth: A review. *American Journal of Perinatology*, 33(03): 258–66.

Mori, M., Bogdan, A., Balassa, T., Csabai, T., & Szekeres-Bartho, J. Ã. (2016). The decidua-the maternal bed embracing the embryo-maintains the pregnancy. *Seminars in Immunopathology*, 38(6): 635–49.

Moyer, V. A. (2014). *Screening for Gestational Diabetes Mellitus:* U.S. Preventive Services Task Force Recommendation Statement.

Munn, D. H. Zhou, M., Attwood, J. T., Bondarev, I., Conway, S.J., Marshall, B., Brown, C., Mellor, A.L. (1998). Prevention of allogeneic fetal rejection by tryptophan catabolism. *Science*, 281(5380): 1191–3.

Nadeem, L., Shynlova, O., Matysiak-Zablocki, E., Mesiano, S., Dong, X., & Lye, S. (2016). Molecular evidence of functional progesterone withdrawal in human myometrium. *Nature Communications*, 7: 11565.

National Research Council (US) Committee to Update Science, Medicine, and Animals. (2004), A Theory of Germs. *Science, Medicine, and Animals*. Washington (DC): National Academies Press (US); Available from: https://www.ncbi.nlm.nih.gov/books/NBK24649/

Nesse, R. M., Bergstrom, C. T., Ellison, P. T., Flier, J. S., Gluckman, P., Govindaraju, D. R., Niethammer, D., Omenn, G. S., Perlman, R. L., Scwartz, M. D., Thomas, M. G., Stearns, S. C., & Valle, D. (2010). Making evolutionary biology a basic science for medicine. *PNAS*, 107(1): 1800–7.

O'Malley, B. W. and Tsai, M. J. (1992). Molecular pathways of steroid receptor action. *Biology of Reproduction*, 46(2): 163–7.

Oreshkova, T., Dimitrov, R., & Mourdjeva, M. (2012). A Cross-Talk of Decidual Stromal Cells, Trophoblast, and Immune Cells: A Prerequisite for the Success of Pregnancy. *American Journal of Reproductive Immunology*, 68(5): 366–73.

Osborn, J., Cattaruzza, M. S., & Spinelli, A. (2000). Risk of spontaneous abortion in Italy, 1978–1995, and the effect of maternal age, gravidity, marital status, and education. *American Journal of Epidemiology*, 151(1): 98–105.

O'Tierney-Ginn, P. F. & Lash, G. E. (2014). Beyond pregnancy: Modulation of trophoblast invasion and its consequences for fetal growth and long-term children's health. *Journal of Reproductive Immunology*, 104–5.

Parham, P. & Moffett, A. (2013). Variable NK cell receptors and their MHC class I ligands in immunity, reproduction and human evolution. *Nature Reviews Immunology*, 13(2): 133–44.

Peters, G. A., Yi, L., Skomorovska-Prokvolit, Y., Patel, B., Amini, P., Tan, H., & Mesiano, S. (2016). Inflammatory stimuli increase progesterone receptor-A stability and transrepressive activity in myometrial cells. *Endocrinology*, 158(1): 158–69.

Phillippe, M. & Phillippe, S. M. (2017). Birth and death: Evidence for the same biologic clock. *American Journal of Reproductive Immunology*, 77(5).

Phillips, J.B., Abbot, P., & Rokas, A. (2015). Is preterm birth a human-specific syndrome? *Evolution, Medicine, and Public Health*, 2015(1): 136–48.

Platek, S. M., & Shackelford, T. K. (2006). *Female Infidelity and Paternal Uncertainty: Evolutionary Perspectives on Male Anti-Cuckoldry Tactics*. Cambridge: Cambridge University Press.

Porter, T., Fraser, A., Hunter, C., Ward, R., & Varner, M. (1997). The risk of preterm birth across generations. *Obstetrics & Gynecology*, 90(1): 63–7.

Resnik, R. & Creasy, R. K. (2009). Intrauterine growth restriction. *Creasy and Resnik's Maternal- Fetal Medicine: Principles and Practice*, pp. 635–50.

Robbins, C. L., Zapata, L., Farr, S., Kroelinger, C., Morrow, B., Ahluwalia, I., . . . Barfield, W. (2014). *Core State Preconception Health Indicators—Pregnancy Risk Assessment Monitoring System and Behavioral Risk Factor Surveillance System*. 3rd ed. Surveillance Summaries. Atlanta: National Center for Chronic Disease Prevention and Health Promotion, CDC, 1–62.

Roberts, R. M., Green, J. A., & Schulz, L. C. (2016). The evolution of the placenta. *Reproduction*, 152(5): R179.

Robillard, P. -Y., Hulsey, T. C., Dekker, G. A., & Chaouat, G. (2003). Preeclampsia and human reproduction. An essay of a long term reflection. *Journal of Reproductive Immunology*, 59(2): 93–100.

Rosenberg, K. & Trevathan, W. (2002). Birth, obstetrics and human evolution. *BJOG: An International Journal of Obstetrics and Gynaecology*, 109(11): 1199–206.

Rouas-Freiss, N., Goncalves, R. M. -B., Menier, C., Dausset, J., & Carosella, E. D. (1997). Direct evidence to support the role of HLA-G in protecting the fetus from maternal uterine natural killer cytolysis. *Proceedings of the National Academy of Sciences*, 94(21): 11520–5.

Rowe, J. Â. H., Ertelt, J. Â. M., Aguilera, M. Â. N., Farrar, M. Â. A., & Way, S. Â. S. (2011). Foxp3 Regulatory T Cell expansion required for sustaining pregnancy compromises host defense against prenatal bacterial pathogens. *Cell Host & Microbe*, 10(1): 54–64.

Ruiter, L., Ravelli, A. C.J., de Graaf, I. M., Mol, B. W. J., Pajkrt, E. (2015). Incidence and recurrence rate of placental abruption: A longitudinal linked national cohort study in the Netherlands. *American Journal of Obstetrics and Gynecology*, 213(4): 573.e1–573.e8.

Sakaguchi, S., Sakaguchi, N., Asano, M., Itoh, M., & Todah, M. (1995). Immunologic self-tolerance maintained by activated T cells expressing IL-2 receptor alpha-chains (CD25). Breakdown of a single mechanism of self-tolerance causes various autoimmune diseases. *Journal of Immunology*, 155(3): 1151–64.

Sear, R. (2015). Evolutionary contributions to the study of human fertility. *Population Studies*, 69(sup1).

Semmelweis, I. Ì. F. Ì. Ì. & Carter, K. C. (1983). *The Etiology, Concept, and Prophylaxis of Childbed Fever*. Madison, WI: University of Wisconsin Press.

Sferruzzi-Perri, A. N., & Camm, E. J. (2016). The programming power of the placenta. *Frontiers in Physiology*, 7.

Sheiner, E., Sarid, L., Levy, A., Seidman, D. S., & Hallak, M. (2005). Obstetric risk factors and outcome of pregnancies complicated with early postpartum hemorrhage: A population-based study. *The Journal of Maternal-Fetal & Neonatal Medicine*, 18(3): 149–54.

Sibai, B., Dekker, G., & Kupferminc, M. (2005). Pre-eclampsia. *The Lancet*, 365(9461): 785–99.

Smith, R. (2007). Partuition. *New England Journal of Medicine*, 356(3): 271–83.

Stearns, F. W. (2010). One hundred years of pleiotropy: A retrospective. *Genetics*, 186(3): 767–73.

Stearns, S. C. & Medzhitov, R. (2016). *Evolutionary Medicine*. Sunderland, Massachusetts: Sinauer Associates, Inc., Publishers.

Stearns, S. C., Nesse, R. N., Govindaraju, D. R., & Ellison, P. T. (2010). Evolutionary perspectives on health and medicine. *PNAS*, 107: 1691–5.

Sunderland, C. A., Redman, C. W., & Stirrat, G. M. (1981). HLA A, B, C antigens are expressed on nonvillous trophoblast of the early human placenta. *Journal of Immunology*, 127(6): 2614–15.

Tafuri, A., Alferink, J., MÂ, Ller, P., HÂ Mmerling, G. Â. N. J., & Arnold, B. (1995). T Cell awareness of paternal alloantigens during pregnancy. *Science*, 270(5236): 630–3.

Tan, H., Yi, L., Rote, N. S., Hurd, W. W., & Mesiano, S. (2012). Progesterone Receptor-A and -B have opposite effects on proinflammatory gene expression in human myometrial cells: Implications for progesterone actions in human pregnancy and parturition. *The Journal of Clinical Endocrinology & Metabolism*, 97(5).

Thellin, O., Coumans, B., Zorzi, W., Igout, A., & Heinen, E. (2000). Tolerance to the foeto-placental "graft": Ten ways to support a child for nine months. *Current Opinion in Immunology*, 12(6): 731–7.

Thomson, A. M., Billewics, W. Z., & Hytten, F. E. (1968). The assessment of fetal growth. *The Journal of Obstetrics and Gynaecology of the British Commonwealth*, 75(9): 903–16.

Thornton, J. W. (2001). Evolution of vertebrate steroid receptors from an ancestral estrogen receptor by ligand exploitation and serial genome expansions. *Proceedings of the National Academy of Sciences*, 98(10): 5671–6.

Thorp, J. M., & Laughon, S. K. (2019). Clinical Aspects of Normal and Abnormal Labor. *Creasy and Resnik's Maternal-Fetal Medicine: Principles and Practice*, by Robert Resnik et al., Philadelphia: Elsevier.

Trevathan, W. (2007). Evolutionary medicine. *Annual Review of Anthropology*, 36(1): 1–49.

Tsai, M., & O'Malley, B. W., (1994). Molecular mechanisms of action of steroid/thyroid receptor superfamily members. *Annual Review of Biochemistry*, 63(1): 451–86.

Turner, J., (2010). Diagnosis and management of pre-eclampsia: An update. *International Journal of Women's Health*, 327.

Ven, H. H., Al-Hasani, S., Diedrich, K., Hamerich, U., Lehmann, F., and Krebs, D. (1985). Polyspermy in in vitro fertilization of human oocytes: Frequency and possible causes. *Annals of the New York Academy of Sciences*, 442(1 In Vitro Fert): 88–95.

Wagner, G. Ã. P., Kin, K., Muglia, L., & Pavli, M. (2014). Evolution of mammalian pregnancy and the origin of the decidual stromal cell. *The International Journal of Developmental Biology*, 58(2–3–4): 117–26.

Walsh, S. W., Kittinger, G. W., & Novy, M. J. (1979). Maternal peripheral concentrations of estradiol, estrone, cortisol, and progesterone during late pregnancy in rhesus monkeys (Macaca mulatta) and after experimental fetal anencephaly and fetal death. *American Journal of Obstetrics and Gynecology*, 135(1): 37–42.

Wang, X., Chen, C., Wang, L., Chen, D., Guang, W., & French, J. (2003). Conception, early pregnancy loss, and time to clinical pregnancy: A population-based prospective study. *Fertility and Sterility*, 79(3): 577–84.

Wells, J. C. K., Nesse, R. M., Sear, R., Johnstone, R. A., & Stearns, S. C. (2017). Evolutionary public health: Introducing the concept. *The Lancet*, 390(10093): 500–9.

Wilcox, A. J. and Weinberg, C. R. (1988). Incidence of early loss of pregnancy. *New England Journal of Medicine*, 319(22): 1483–4.

Williams, K. C., Renthal, N. E., Condon, J. C., Gerard, R. D., & Mendelson, C. R. (2012). MicroRNA- 200a serves a key role in the decline of progesterone receptor function leading to term and preterm labor. *Proceedings of the National Academy of Sciences*, 109(19): 7529–34.

Williams, K. C., Renthal, N. E., Gerard, R. D., and Mendelson, C. R. (2012). The microRNA (miR)- 199a/214 Cluster mediates opposing effects of progesterone and estrogen on uterine contractility during pregnancy and labor. *Molecular Endocrinology*, 26(11): 1857–67.

Williams, R. L., Creasy, R. K., Cunningham, G.C., Hawes, W. E., Norris, F. D., & Tashiro, M. (1982). Fetal growth and perinatal viability in California *Obst Gynecol*, 59(5): 624–32.

Wilmoth, J., (2012. *Statistics, Politics, and Policy: A New Method for Deriving Global Estimates of Maternal Mortality*. [online] World Health Organization. Available at: <http://www.who.int/reproductivehealth/publications/monitoring/newMethod_deriving_globalEstimates.pdf>

Winkvist, A. (1998). Familial patterns in birth characteristics: Impact on individual and population risks. *International Journal of Epidemiology*, 27(2): 248–54.

World Health Organization (2013). *Born Too Soon: The Global Action Report on Preterm Birth*. [online] World Health Organization. Available at: <http://www.who.int/pmnch/media/news/2012/preterm_birth_report/en/>

World Health Organization (2014). *Trends in maternal mortality: 1990 to 2010*. [online] World Health Organization. Available at: <http://www.who.int/reproductivehealth/publications/monitoring/9789241503631/en/>

World Health Organization (2016). Fetal macrosomia. *Trends in maternal mortality: 1990 to 2015*. [online] World Health Organization. Available at: <http://www.who.int/reproductivehealth/publications/monitoring/maternal-mortality-2015/en/>

Yoder, S. R., Thornburg, L. L., & Bisognano, J. D. (2009). Hypertension in pregnancy and women of childbearing age. *The American Journal of Medicine*, 122(10): 890–5.

Young, I. R., Renfree, M. B., Mesiano, S., Shaw, G., Jenkin, G., & Smith, R. (2011). The comparative physiology of parturition in mammals: Hormones and parturition in mammals. *Hormones and Reproduction of Vertebrates*, 17(1): 95–116.

Younge, N., Goldstein, R. F., Bann, C. M., Hintz, S.R., Patel, R. M., Smith, P. B., Bell, E. F., Rysavy, M. A., Duncan, A. F., Vohr, B. R., Das, A., Goldberg, R. N., Higgins, R. D., & Cotten, C. M. (2017). Survival and neurodevelopmental outcomes among periviable infants. *Obstetrical & Gynecological Survey*, 72(7): 401–3.

Zhang G., Feenstra B., Bacelis J., Liu X., Muglia LM., Juodakis J., Miller DE., Litterman N., Jiang P., Russell L., Hinds DA., Hu Y., Weirauch MT., Chen X., Chavan A., Wagner GP., Pavlicev M., Nnamani M., Maziarz J., Karjalainen MK., Ramet M., Sengpiel V., Geller F., Boyd HA., Palotie A., Momany A., Bedell B., Ryckman KK., Huusko JM., Forney CR., Kottyan LC., Hallman M., Teramo K., Nohr EA., Davey-Smith G., Melbye M., Jacobsson B., Muglia LJ., Genetic associations with gestational duration and spontaneous preterm birth. N Engl J Med 2017; 377: 1156–1167.

Zinaman, M. J., Clegg, E. D., Brown, C. C., O'Connor, J., & Selevan, S. G. (1996). Estimates of human fertility and pregnancy loss. *Fertility and Sterility*, 65(3): 503–9.

CHAPTER 6

An Evolutionary View of Homeostasis: Bioenergetics, Life History Theory, and Responses to Pregnancy

Robert Perlman

Claude Bernard and Charles Darwin, two of the founding figures of modern biology, epitomize two distinct approaches to understanding biological phenomena. In Ernst Mayr's terminology, Bernard was interested in the proximate causes of these phenomena, the physiological mechanisms that operate during the lifetime of an organism, while Darwin sought to understand ultimate causes, the processes that acted during the evolutionary history of a population (Mayr, 1988). Although the rigid distinction between proximate and ultimate causes has recently been criticized (Laland, Sterelny, et al., 2011), it does capture an important difference between the perspectives of these two preeminent scientists. This chapter will integrate the insights of evolutionary biology into our understanding of physiological regulatory mechanisms, an area that has traditionally been the purview of the intellectual descendants of Bernard.

Bernard (1813–1878) and Darwin (1809–1882) were contemporaries; although they never met, they were well aware of each other's work. Darwin spoke of Bernard in laudatory terms. In *The Expression of the Emotions in Man and Animals*, Darwin refers to Bernard as "The great physiologist" (Darwin, 1872/1965). Bernard, in contrast, was dismissive of Darwin. In *An Introduction to the Study of Experimental Medicine*, Bernard belittles Darwin's ideas as "hypothetical contemplation" and goes on to say that if we pursued these ideas "we should soon turn our backs on reality; and, in my opinion, we should misunderstand true scientific philosophy"(Bernard, 1865/1957).

Bernard's disparaging view of Darwin's work was regrettable; it promoted the separation of physiology from the study of evolution and it probably contributed to the protracted exclusion of evolutionary biology from medical research and education. At the time of the Flexner Report, *Medical Education in the United States and Canada*, physiology was a thriving discipline with obvious importance for medicine, while the theory of evolution by natural selection was not yet widely accepted in the scientific community and only a handful of physicians thought about the application of evolutionary concepts to disease (Zampieri, 2009). Not surprisingly, Flexner made physiology the centerpiece of the preclinical medical curriculum and ignored evolutionary biology (Flexner, 1910). Flexner's decision was ironic, since Darwin himself used examples from human biology and medicine to illustrate and buttress his theory (Bynum, 1983).

Perlman, R., *An Evolutionary View of Homeostasis: Bioenergetics, Life History Theory, and Responses to Pregnancy*
In: *Integrating Evolutionary Biology into Medical Education—for maternal and child healthcare students, clinicians, and scientists*. Edited by: Jay Schulkin and Michael L. Power, Oxford University Press (2020). © Oxford University Press.
DOI: 10.1093/oso/9780198814153.003.0007

Nonetheless, after the Flexner Report, medical education became more heavily grounded in physiology and other basic medical sciences, and there were few if any evolutionists on medical school faculties. There was little opportunity for communication or collaborative research between evolutionary scientists and biomedical scientists, and these fields developed in isolation (Perlman, 2013).

Fortunately, a few physicians and biologists were interested in integrating the perspectives of evolutionary biology and physiology. One of the most prominent of these scientists was the physiologist Walter Cannon (1871–1945). Cannon admired Darwin as well as Bernard, and hung portraits of both of these scientists in his office (Hagen, 1996). Cannon is best known for his studies of the sympathetic nervous system and for developing Bernard's recognition of the importance of the *milieu intérieur*, or internal environment, into the concept of physiological homeostasis. Bernard had famously written, "All the vital mechanisms, however varied they might be, always have one purpose, that of maintaining the integrity of the conditions for life within the internal environment" (Bernard, 1878/1974). The function of Bernard's "vital mechanisms," which we now call homeostatic mechanisms, is to maintain more or less constant conditions in the blood and extracellular fluids, conditions that enable the ongoing functioning of living organisms (Cannon, 1929; Cannon, 1939). In Cannon's words, homeostasis "does not imply something set and immobile, a stagnation. It means a condition—a condition which may vary, but which is relatively constant" (Cannon, 1939). To emphasize this idea, Cannon quoted the French physiologist Charles Richet, who wrote

> The living being is stable. It must be so in order not to be destroyed, dissolved, or disintegrated by the colossal forces, often adverse, which surround it. By apparent contradiction it maintains its stability only if it is excitable and capable of modifying itself according to external stimuli and adjusting its response to the stimulation. In a sense it is stable because it is modifiable—the slight instability is the necessary condition for the true stability of the organism." **(Cannon, 1929), p. 399)**

Cannon recognized that homeostatic control mechanisms evolved because they enhanced success in the struggle for existence, or as we would now say, they enhanced fitness. He had a view of evolution as progressive—a common view at the time—and believed that evolutionary progress consisted of an increasing control of the internal environment and a concomitant independence from the external one. Like Bernard, Cannon emphasized the role of homeostatic mechanisms in promoting survival and did not consider the relationships between homeostasis and reproduction. The modern synthesis of evolutionary biology with population genetics, which was being developed at the time Cannon wrote, led to a clearer appreciation that fitness entails "success in producing offspring" (Paul, 1992). We now understand the importance of reproductive success rather than survival alone in shaping our biology, including our homeostatic mechanisms.

Although Cannon's concept of homeostasis has provided a framework for much research in physiology, his interest in connecting evolution and physiology lay dormant for a long time. Physiologists made great progress in clarifying the workings of homeostatic control mechanisms. Some were interested in the phylogenetic history of these mechanisms, such as the evolution of homeothermy in birds and mammals. By and large, though, physiologists took for granted that homeostatic mechanisms promoted survival and didn't think about them as evolutionary adaptations. And evolutionists were more interested in studying natural selection in the wild than in physiology. Evolutionists focused on the responses of populations to external challenges while physiologists focused on the responses of organisms to internal challenges. Over the last twenty years, however, stimulated in large part by the publication of *Why We Get Sick* (Nesse & Williams, 1994), there has been increasing recognition that physiological and evolutionary causes of biological phenomena are complementary and that bringing these two causal pathways together enhances our understanding of these processes. In particular, there has been growing interest in bringing evolutionary concepts and approaches into medical thinking in order to enrich our current physiological or biomedical approach to health and disease. Several conceptual advances, most importantly recognition of the central role of energy in biology, the development of evolutionary life history theory, and the elaboration of engineering control theory, have helped to integrate evolution and physiology. These concepts provide a framework in which to understand both physiological and evolutionary aspects of homeostatic regulatory mechanisms. Evolutionary insights can help us understand the ways that natural selection has shaped these mechanisms.

6.1 Bioenergetics

Soon after the laws of thermodynamics were formulated in the mid-nineteenth century, thermodynamics, and especially the concept of energy as the ability to do work, were applied both to physiology and to evolution. In the 1840s, Robert Mayer and Hermann von Helmholtz proposed that physiological processes obeyed the first law of thermodynamics, the conservation of energy. Their work ushered in the field of animal energetics, the study of heat production as a measure of metabolism. The physiologist Max Rubner (1854–1932), another pioneer in the study of biological energetics, perfected the method of animal calorimetry, the measurement of heat production by animals, and showed that the heat produced by an organism was equal to the energy that would have been released by the oxidation of the substrates that were being metabolized by the organism. Rubner demonstrated what Mayer and Helmholtz had hypothesized, "the validity of the law of the conservation of energy in the animal body" (Lusk, 1932). Rubner studied energetics and metabolism in humans and also carried out comparative studies of heat production in other animals; his work led to an understanding of the role of the nervous system and of hormones in regulating energy metabolism. He also quantified the caloric content of foodstuffs and so made possible indirect calorimetry, the measurement of O_2 consumption, and CO_2 production. Indirect calorimetry is now often used as a convenient proxy for heat production in studies of the energetic costs of physiological processes.

At about the same time, the great German physicist Ludwig Boltzmann (1844–1906), who was an early champion of Darwin's work, recognized that "the fundamental object of contention in the life-struggle, in the evolution of the organic world, is available energy" (Lotka, 1922). Since that time, understanding how organisms have evolved to harvest, store, and utilize energy has been a central topic in evolution. Echoing Theodosius Dobzhansky, Nick Lane has written "nothing in evolution makes sense except in the light of energy" (Lane, Martin et al., 2013). The role of energy in evolution is sometimes expressed in the saying "Organisms are machines that convert energy into offspring; natural selection increases the efficiency of these machines." Although there are limitations to the metaphor of organisms as machines and the word efficiency might be better replaced by effectiveness, this aphorism does convey an important insight into biology. Since the nineteenth century, the study of the acquisition and utilization of metabolic energy, and the regulation of these processes, has been a central concern in both physiology and evolutionary biology.

6.2 Life history theory

The modern synthesis of evolutionary biology and genetics stressed the fitness benefits of the phenotypes of adult organisms and the spread of those traits in populations but paid relatively little attention to the development of these traits. Life history theory is a branch of evolutionary biology which recognizes that natural selection shapes the entire life course of organisms in ways that enhance evolutionary fitness. Life history theory builds on Boltzmann's appreciation of the fundamental role of energy in natural selection and provides a valuable framework for integrating evolutionary and physiological perspectives on bioenergetics. Organisms require sources of energy (nutrients) and other resources—oxygen, water, heat, shelter, etc.—to survive and reproduce. They can acquire these resources directly from their environments or indirectly from others, most often their parents or other family members. They then must allocate their metabolic energy and other resources, including their time, among a handful of physiological functions, most importantly growth and development, reproduction, the prevention and repair of bodily damage, responses to both internal and extrinsic stresses, external work, and storage. Life history theory is an account of the ways in which natural selection has shaped the physiological and behavioral mechanisms that direct the acquisition and allocation of energy and other resources over the life course. Natural selection has shaped these mechanisms in ways that enhance fitness. In demographic terms, organisms use resources to increase survival and to increase fertility. Because resources are finite and energy (and time) that is used for one purpose is not then available for other uses, life history theory highlights the importance and ubiquity of tradeoffs in both evolution and physiology.

6.2.1 The human life course

The human life course is conventionally divided into a half-dozen more-or-less distinct stages—prenatal, infant (nursing), post-weaning dependent child or juvenile, puberty and adolescence, sexual and reproductive maturity, and post-reproductive life—that are distinguished by the ways in which we acquire and utilize energy. Natural selection has adjusted the timing and duration of these stages as well as our activities in them. During the period of prenatal development, we acquire resources from our mothers via our placentas and use them almost exclusively for growth and development. In infancy, we continue to acquire resources largely if not exclusively from our mothers but now, at least in traditional societies, by breastfeeding. The human life course is characterized by a long period of post-weaning childhood, during which we are still nutritionally dependent on others but have time to begin to acquire the knowledge and skills that we will need to become successful adults. Puberty marks the period when we begin to devote significant resources to reproduction, initially to the development of secondary sex characteristics and to behaviors that will enable us to find and be chosen by potential mating partners. It is also a period during which we become increasingly independent in acquiring nutrients and other resources for ourselves. As adults, we can have children; if we do, we must acquire resources to support them as well as ourselves. In other words, we can use energy directly for reproduction and indirectly to nourish our offspring. Finally, the human life course is characterized by a relatively long period of post-reproductive life, when we are no longer having our own children but can continue to acquire resources and use them to enhance our fitness by promoting the survival of our descendants.

6.2.2 The prevention and repair of bodily damage

Throughout our lives our bodies are subject to damage of various sorts and we confront many challenges to maintaining homeostasis. The prevention and repair of bodily damage comprises responses to a number of disparate problems and processes, including the intracellular biochemical damage that is a consequence of cell metabolism, the physiological changes that accompany our daily activities, and the pathophysiological disturbances that result from extrinsic sources. All of this somatic damage, whatever its source and its manifestations, are stresses that disturb homeostasis and that require energy for their prevention, amelioration, or repair.

First, consider intracellular biochemical damage. Ions leak across cell membranes down their electrochemical gradients and must be pumped out across these membranes. Proteins undergo conformational changes, become unfolded and non-functional, and may form aggregates. These unfolded proteins must either be refolded or degraded before they form stable aggregates; if they are degraded, they must be replaced by new proteins. A significant amount of our resting energy expenditure—our resting metabolic rate or resting rate of heat production—is used for these two purposes, ion transport and protein turnover. Energy expenditure requires nutrients and oxygen, and produces carbon dioxide. Another significant fraction of our metabolic energy is used for the cardiovascular work required to pump blood through the circulatory system to deliver oxygen and nutrients to cells and remove carbon dioxide and other metabolic products from them.

Some intracellular damage is the inevitable byproduct of our metabolic processes. Reactive oxygen species such as superoxide anion and hydrogen peroxide are formed in the course of our oxidative metabolism. These reactive oxygen species can oxidize and damage virtually all cellular components, including nucleic acids, proteins, and lipids. If damaged deoxyribose nucleic acid (DNA) is not repaired before it is replicated, this damage may result in somatic mutations. Oxidized proteins may become non-functional and are prone to aggregate, and so oxidative damage contributes to the burden of protein turnover mentioned previously. Oxidized lipids may disrupt the function of cell membranes. If this panoply of biochemical damage is not repaired, it will interfere with cellular functions and will lead to cell death and ultimately to organ failure.

Other stresses are the inevitable result of our activities of daily living, including sleeping and waking, eating and defecating, postural changes, changes in activity, and exposure to changes in ambient temperatures. Changes in posture or activity cause changes in blood flow, blood pressure, blood glucose, metabolic rate and heat production, and other variables. Two essential activities, eating and working, require energy and increase our metabolic rate, i.e., heat production, above its resting value. Something on the order of 5–10% of the caloric content of food is used for the processes of digestion, absorption, and the conversion of foodstuffs into products that can be utilized or stored. This thermic effect of food is a necessary part of our metabolism. Physical labor—work—is also a major component of energy usage. Our evolutionary ancestors may have engaged in play but most of their physical activity must have involved searching for and acquiring food, and avoiding or escaping from predators. We think of physical activity as exercise, as though it were disembodied and separate from our lives, but of course we are doing external work all the time as part of our daily living. The metabolic rates of sedentary people are on the order of 50% higher than their resting metabolic rates, and the metabolic rates of manual laborers and athletes can be many times higher than that.

In addition, we face all kinds of external threats to homeostasis. These stresses include situations we all agree are dangerous, such as exposure to radiation (including sunburn) or toxic chemicals, contact with or infection by pathogens, or physical trauma. The built environments we humans have constructed to protect ourselves from environmental challenges contribute to these stresses by altering our diets and increasing our exposure to toxins and pathogens. Living and working in confined spaces increases the risk of myopia. Artificial lighting, shift work, and long-distance air travel may disrupt our circadian rhythms. Finally, the psychic stresses of daily living and of interacting with other people, especially in hierarchical societies such as ours, evoke internal defensive responses that can be damaging in the long run.

All of these challenges to our bodily integrity activate compensatory mechanisms that help to maintain homeostasis. For some purposes, it may be useful to distinguish external threats to homeostasis from the internal stresses but there is no satisfactory way of doing this. Moreover, the same physiological mechanisms act to maintain homeostasis regardless of the source of the stress. As we have seen, some of the bodily damage we suffer comes from intracellular processes; this damage is prevented or repaired by intracellular mechanisms. Some of these mechanisms, such as the P53-based mechanisms that detect damaged DNA and activate pathways that repair this DNA or lead to apoptosis, are autonomous cellular mechanisms (Williams & Schumacher, 2016), while others, such as the heat shock response, are coordinated with whole-body responses to increased temperature or other stresses (Richter, Haslbeck et al., 2010). Although much of the interest in physiological homeostasis has focused on maintaining the properties of extracellular fluids, we now recognize that intracellular fluids are the environments in which cellular organelles function and cellular metabolism takes place. Following Bernard and Cannon, we can say that physiological homeostatic mechanisms maintain the conditions of fitness—survival and fertility—in both the extracellular and intracellular fluid environments.

6.2.3 Aging

Life history theory provides an evolutionary understanding of aging, which may be thought of as a progressive decline in physiological functions and in responses to stress, an increase in the risk of mortality, and, in humans as well as many other species, a decrease in fertility (Kirkwood, 1996). These manifestations of aging result from the accumulation of unrepaired somatic damage. Homeostatic regulatory mechanisms themselves decline in function, because the cellular components of these mechanisms also suffer from unrepaired damage. In principle, organisms could acquire sufficient energy and allocate this energy to repair all of the somatic damage they suffer. Indeed, some organisms, notably hydra, are said not to age in nature; if this is true, it would suggest that they do repair their bodily damage completely. Hydra and related organisms aside, both theoretical analyses and empirical studies of life histories have shown that prevention or repair of somatic damage is generally imperfect. Life history strategies evolve in ways that enhance fitness, not health, or longevity. Fitness requires that we survive long enough and healthy enough to bear and raise children. Beyond this, however, for the vast majority of species including our own, fitness is evidently enhanced by using energy for growth and reproduction rather than for complete bodily repair and the maintenance of homeostasis.

6.3 Feedback and feedforward in physiology

Shortly after Cannon formulated the concept of homeostasis, Norbert Wiener coined the term "cybernetics" to denote "the entire field of control and communication theory, whether in the machine or in the animal" (Wiener, 1961). Cannon and Wiener were colleagues and friends; Cannon's ideas of homeostasis influenced Wiener's work on cybernetics and Wiener's work stimulated the application of engineering control theory to physiology. An analysis of the various types of homeostatic control mechanisms will clarify the benefits and drawbacks of different mechanisms and will help us understand why we have evolved a plethora of seemingly redundant control mechanisms. Two types of regulatory mechanisms play prominent roles in homeostasis—negative feedback mechanisms and feedforward mechanisms (Houk, 1988). Negative feedback mechanisms are much better known in the medical community; they are widely discussed in physiology texts and are emphasized in medical education. Negative feedback mechanisms contribute to homeostasis of blood pressure, blood glucose, and many other physiological variables. Thermoregulation is commonly used to illustrate the properties of negative feedback mechanisms. Central thermoreceptors in the hypothalamus and elsewhere in the central nervous system sense and respond to our core body temperature. Under usual conditions, the "set point" of these thermoreceptors is approximately 37°C. If the temperature falls below the set point, these thermoreceptors activate a suite of compensatory responses, including peripheral vasoconstriction, an increase in metabolic rate through shivering, and the stimulation of lipolysis in brown fat, and behavioral adjustments such as moving to warmer surroundings and putting on warmer clothes. Together, these processes increase heat production and reduce heat loss, and thereby act to return the central temperature closer to its set point. Conversely, if the temperature rises above the set point, the central thermoreceptors activate a different set of responses, including peripheral vasodilation, sweating, and cooling behaviors, all of which bring the central temperature back toward its desired value.

It is not surprising that thermoregulation is often used as a model of negative feedback mechanisms. The operation of our physiological thermoregulatory system is easy to understand, at least in outline; we're aware of many of our thermoregulatory responses, such as sweating and shivering, from our own experience. We can readily appreciate the analogy between physiological thermoregulation and temperature regulation in our homes, where a thermostat responds to deviations from a set point and activates either a heater or an air conditioner to maintain the ambient temperature near the temperature we desire. Temperature regulation in our homes can also involve behavioral responses, such as opening or closing windows or shades, etc. The analogy of our thermoregulatory system to a thermostat is also instructive because it corrects the mistaken idea that the set point is some objective reference value and that physiological regulation involves comparison of a variable to this reference value. Just as the responses of a thermostat depend on the thermal properties of the materials it comprises, so do the responses of our central thermoreceptors depend on the temperature sensitivity of their firing rates, not on a comparison of the temperature of the blood to some fictive reference temperature.

Negative feedback mechanisms have many advantageous features. Even though most of our negative feedback systems, like thermoregulation, are complex and have multiple interacting components, they are conceptually simple; in principle they require only a few elements—a sensor whose activity or output depends on the value of a regulated variable (e.g., central thermoreceptors), an effector whose activity or output changes the value of that variable (sweat glands), and a communication channel by which the activity of the sensor affects the activity of the effector (the sympathetic nervous system). Because they are comparatively simple, negative feedback mechanisms are presumably easy to evolve. We certainly have evolved a lot of them! Negative feedback mechanisms respond to all types of stresses or perturbations. The same temperature regulatory mechanisms become activated in response to changes in ambient temperature, to changes in heat production as in exercise, and to changes in heat loss due to changes in clothing. Likewise, the negative feedback mechanisms that regulate blood respond to pressure changes caused by changes in posture, changes in activity, dehydration, and hemorrhage, among other factors.

But negative feedback control has limitations as well as advantages. Negative feedback mechanisms do not become activated until a disturbance has occurred. In addition, correction of the deviation from the set point is usually incomplete. In a steady state, the attenuation (A) of a deviation in a linear control system is determined by a parameter known as the loop gain (G); formally, $A = 1/(1 + G)$ (Houk, 1988). Engineers design control systems with sufficient loop gain to obtain the error attenuation they desire. The gain of physiological control mechanisms has been produced by natural selection rather than by engineers. Natural selection does not optimize control mechanisms or other traits. Rather, it results in "good enough" solutions. It "satisfices"; it results in solutions that on average work better than the available alternatives (Simon, 1956). We can infer the gain of a physiological feedback mechanism by estimating the error attenuation achieved by this mechanism. Although homeostatic control mechanisms are rarely in a steady state and their responses are not necessarily linear, we can still understand the attenuation of an error, the degree to which the deviation caused by stress is corrected, as the result of the finite gain of the feedback mechanism.

We might expect natural selection to maximize the loop gain of homeostatic control mechanisms, because this would minimize deviations from the desired values or set points of regulated variables. There is a tradeoff, however, because of the potential for instability. As the gain of a feedback system increases, there is increased likelihood of feedback oscillations. As Cannon noted, physiological regulation is typically maintained by the operation of opposing homeostatic mechanisms (Cannon, 1929). We have already discussed the opposing mechanisms that maintain core temperature in the face of stresses that raise or lower this temperature. High gain of negative feedback mechanisms may lead to overshoots that activate opposing feedback mechanisms. Feedback oscillations play a role in some pathological situations, perhaps most notably in Cheyne–Stokes respiration, a cyclical variation in the depth of respiration and often also in respiratory rate, that results from an instability of the respiratory control mechanisms (Berger, Mitchell et al., 1977). Similarly, some neuromuscular tremors are caused by instability of the negative feedback control mechanisms that regulate muscular position and movement.

The instability of negative feedback mechanisms is also influenced by the time course of their responses; time lags in response increase the likelihood of instability. (Standard block diagrams of negative feedback systems in physiology implicitly assume that these systems are in a steady state and so ignore the kinetics of these responses.) By and large it seems that natural selection has produced homeostatic control mechanisms that balance the benefits of minimizing error against the problem of instability, which has in turn entailed balancing the tradeoff between loop gain and response speed. Finally, in the context of life history theory, the gain of negative feedback mechanisms may be limited by the energetic costs and tradeoffs of increasing gain. Thus, for example, sweating and shivering are energetically expensive, and sweating results in salt and water loss; the strength of these homeostatic mechanisms may be limited by these other considerations.

The time course of homeostatic mechanisms must be matched to the time course of the stresses that activate them. Temperature regulation is unusual in that the perturbations that affect body temperature are typically sustained for long periods; in other words, the stress of temperature change is often close to a steady state stress. If we go, say, from inside our homes to outside, we experience an abrupt change in temperature but then the new temperature is relatively constant. Not surprisingly then, our homeostatic responses to temperature change, sweating and shivering, begin rapidly after the onset of change and can be maintained for long times provided that they are not curtailed by salt and water loss or lack of available energy. The relative stability of temperature change contrasts with blood pressure and blood glucose, which fluctuate rapidly during postural changes or after eating.

The speed of homeostatic mechanisms depends on the communication pathways between the sensors and effectors in these mechanisms. The nervous system and the endocrine system are the two main communication channels in homeostatic control systems but the immune system plays an important role in reactions to the stress of infections and other pro-inflammatory stimuli; as a rule, neural control is faster than endocrine or immune responses. More importantly, the speed of regulation depends on the biochemical processes involved in these control processes. The fastest responses are typically changes in the activity of ion channels, which can produce effects in the millisecond time scale. Regulation of the activity of other proteins is typically slower, because there are metabolic pathways that connect neural or endocrine signals with changes in enzyme activity and it takes time—perhaps minutes—for changes in enzyme activity to alter cellular functions. Changes in gene expression are even slower, taking minutes to hours to change protein levels sufficiently to produce physiological responses. And changes that entail the production of new cells are slower yet, and may take days or weeks to become important. But whatever the mechanisms involved, the speed of onset of a homeostatic response as well as its persistence must be scaled to the rate of change of the disturbance that elicits that response.

Our infatuation with negative feedback reflects the unfortunate persistence of a Cartesian view of animals as machines that denies or minimizes the role of agency in the animals themselves. But now that ideas of niche construction and developmental bias are emphasizing the agency of animals in their own evolution, it is fitting to pay more attention to the role of animals in their own physiological regulation. Regulation is not simply the passive response of animals to environmental stresses. Instead, animals engage in activities that lead to perturbations and they have evolved feedforward regulatory mechanisms that anticipate and minimize the changes caused by these stresses. Feedforward control mechanisms play crucial but underappreciated roles in physiology. Feedforward mechanisms may be triggered by animals' own activities or by environmental changes that have led to the evolution of circadian and annual rhythms. For example, in addition to the central thermoreceptors discussed previously, we have peripheral or cutaneous thermoreceptors in the skin. If animals experience a change in ambient temperature, skin temperature changes before the core temperature does. Peripheral thermoreceptors respond to changes in skin temperature by activating thermoregulatory mechanisms before there is a change in core temperature. The behavioral thermoregulatory mechanisms, which also illustrate the active participation of animals in their own physiological regulation, are feedforward responses that respond to changes in skin temperature. More dramatically, changes in day length herald changes in environmental temperature and are used by many animals as feedforward signals to prepare for seasonal changes in temperature as well as in the availability of food. Responses to these signals include migration, dormancy or hibernation, and changes in fur or other forms of insulation.

Many feedforward mechanisms are linked to animal behavior, especially to eating and to physical activity, which, as discussed, are important determinants of metabolic rate. Eating leads to the cephalic phase of insulin secretion, during which the pancreas secretes insulin before there is a rise in blood glucose (Power & Schulkin, 2008). In addition to being physiologically important, feedforward mechanisms are conceptually interesting because of their connection to animal behavior. Feedforward mechanisms have properties—advantages and limitations—that complement those of negative feedback processes. They can forestall deviations of regulated variables and can prevent deviations completely, and they don't run the risk of inducing oscillations. But they are computationally complicated and because these mechanisms may not be able to predict the size of disturbances in advance, they may undershoot or overshoot their targets.

6.4 Other aspects of homeostasis

6.4.1 Physiological adaptation of homeostatic mechanisms

Both negative feedback and feedforward mechanisms can acclimate, adapt, and be subject to physiological regulation; they can also be manipulated by pathogens. Neither the set points nor the magnitude of homeostatic responses are fixed. Although we use the term set point, we should remember that homeostatic responses maintain regulated variables within a range that is consistent with physiological function rather than at a fixed value. The set point for temperature regulation has a circadian rhythm and varies by about 1°C from a low in the early morning to a high in the afternoon. In the presence of an infection, cytokines released from macrophages and other inflammatory cells raise the set point of the thermoregulatory system and thereby lead to an increase in our core temperature. This physiological adaptation of the thermoregulatory system is an evolutionary adaptation that increases our resistance to pathogens and minimizes the fitness costs of the infection (Kluger, Kozak et al., 1996). But some pathogens, notably malaria parasites, hijack this regulatory mechanism to enhance their own fitness. The periodic fever spikes caused by *Plasmodium falciparum* is evidently such a manipulation. Elevation of body temperature and debilitation of victims occur at just the time of day when mosquitoes are most active and are able to transmit the parasite to new hosts (Perlman, 2013).

6.4.2 Development and aging of homeostatic mechanisms

Homeostatic regulatory mechanisms, like other traits, develop over the life course and decline in aging because they too are affected by the consequences of unrepaired bodily damage. For example, eccrine sweat glands develop during the prenatal period. People in different human populations have roughly the same number of sweat glands, on the order of several million, at birth. At birth, these sweat glands can respond to pharmacological stimuli but not to changes in body temperature; the reflexes that lead to thermoregulatory sweating develop over the first several weeks of life. Thermal stimulation in infancy leads to the activation of sweat glands, to their ability to respond to increases in body temperature at the moment and later in life. People who grow up in tropical climates develop a greater number of active sweat glands than do people who mature in colder climates and are thereby acclimated to the hot environments they are likely to experience later (Diamond, 1991). Sweat output from this apparently fixed number of active sweat glands increases until puberty, is maintained through adulthood, and then declines in the seventh or eighth decade of life. The ability to increase blood flow to the skin also decreases late in life. Because of these decreases in cutaneous blood flow and in sweating, elderly people have a decreased ability to respond to heat stress (Nagashima, Tokizawa et al., 2012). Less is known about the development and decline of shivering and of brown fat activity but they presumably follow roughly the same pattern of changes during the life course as does sweating, since the responses to cold develop during infancy and childhood, and the elderly become more sensitive to cold as well as to heat.

6.4.3 Safety factors

Jared Diamond has discussed the evolution of reserve capacities, or safety factors, in physiology (Diamond, 2002). Diamond was concerned primarily with the strength of bones and other structures, the activities of enzymes and transporters, and organ functions such as pancreatic enzyme secretion. To this list we can add other safety factors, including genetic redundancy that buffers the effects of somatic mutations, and stem cells, which provide a reservoir to replace dead or non-functional cells. These reserve capacities, which are built up early in development, enable organisms to maintain homeostasis and to survive (and reproduce) in the face of individual variation, incomplete responses to stress, unusually large demands, and deterioration in aging (Diamond, 2002). The economist Robert Fogel referred to these safety factors as physiological capital that can be expended in response to stresses or other challenges (Fogel, 2003). As he noted, the size of these reserves must reflect tradeoffs between their benefits and their costs, which are primarily energetic or opportunity costs, i.e., the costs of not using the energy or the crowded intracellular space for other purposes (Zimmerman & Minton, 1993). The metabolic scope of mammals—the ratio of their maximal metabolic rate to their resting rate—is about ten (Schmidt-Nielsen, 1984). Together, the processes involved in the transport of oxygen from the atmosphere to tissues and its use by mitochondria must have an overall safety factor of that magnitude.

Safety factors are also built into the set points of homeostatic control mechanisms. The levels at which regulated variables are maintained are presumably determined by the population-wide costs of increases and decreases in them. Temperature regulation is unusual in that the regulation of core body temperature is very tightly regulated with only a small margin of safety. Rises or falls in core temperature of more than a few degrees away from the set point of 37°C lead rapidly to physiological dysfunction and to death. Virtually all cellular processes are temperature dependent and their coordinated activities depend on precise temperature regulation. Elevation of body temperature also entails energetic costs; a one-degree rise in body temperature increases the resting metabolic rate by about 10%. The physiological stability of body temperature is related to its phylogenetic stability. The body temperature of monotremes is about 31°C and that of marsupials is about 35°C (Schmidt-Nielsen, 1984). There must be fitness advantages to offset the energy costs of the higher body temperatures of eutherian mammals.

Most other regulated variables, such as plasma glucose, are maintained at levels that confer large margins of safety. It is difficult to quantify these safety factors because they depend heavily on the kinetics of changes in regulated variables but it is clear, for example, that resting plasma glucose is maintained at a value that enables most people to accommodate significant (albeit gradual) increases or decreases in glucose to perhaps double or half its resting level without harm. Maintenance of resting plasma glucose at 80 or 100 mg/dl rather than 90 mg/dl would probably not entail meaningful energetic or opportunity costs. Plasma glucose is presumably regulated at a level at which the risks of hypoglycemia balance the risks of hyperglycemia.

6.4.4 Storage

Homeostatic mechanisms regulate supplies, such as glucose, or processes, such as blood pressure. Homeostatic mechanisms that regulate supplies include intake, storage, and excretion. Cannon distinguished two types of storage, which he referred to as storage by inundation and storage by sequestration (Cannon, 1929). Because we focus on the homeostatic regulation of the properties of extracellular fluids, it is easy to lose sight of the role of extracellular fluids themselves in maintaining homeostasis. The large and expandable volume of extracellular fluids provides a reservoir that stores water and low molecular weight solutes and buffers changes in their intake, metabolism, or excretion. This is storage by inundation. Storage by sequestration refers to the storage of glucose and fatty acids as glycogen and triglycerides, respectively, and the storage of calcium in bone. These materials may be stored and released in response to changes in their concentrations in extracellular fluids.

6.4.5 Long-term side effects of homeostatic mechanisms

Many homeostatic mechanisms have been designed to respond to acute, albeit repeated, stresses. Since responses to stress typically involve the brain and nervous system, the endocrine system, and the immune system, homeostatic responses are accompanied by changes in the secretion and the plasma concentrations of neurotransmitters, hormones, cytokines, and other biologically active molecules.

If the responses to stress are exaggerated, prolonged, or otherwise inappropriate, these bioactive substances may have unwanted pathophysiological effects such as increases in blood pressure and in inflammatory processes. They increase cellular damage and so increase the risk of a host of chronic diseases, or diseases of aging. Bruce McEwen and colleagues have coined the term allostatic load to describe the accumulated bodily damage that results from heightened or chronic homeostatic responses to stress (McEwen, 2015). The chronic stress of being socioeconomically disadvantaged and living in a hierarchical society is a major contributor to allostatic load and to socioeconomic disparities in health. The short-term fitness benefits of our homeostatic regulatory mechanisms evidently outweigh their long-term detrimental effects.

6.5 Metabolic regulation and life history tradeoffs in pregnancy

The physiological adjustments in pregnancy illustrate many of the concepts discussed in this chapter. Pregnancy requires major changes in the allocation of maternal resources. A pregnant woman must provide energy not only to support the growth of the fetus and placenta, but also for the growth of maternal tissues, for fat storage in adipose tissue, and for the increased circulatory, respiratory, and renal work needed to supply metabolites to the fetus and to maintain homeostasis in both the mother and her fetus. Not surprisingly, a woman's resting metabolic rate increases during pregnancy. But the metabolic and energetic changes during pregnancy depend heavily on the nutritional state of the mother. The resting metabolic rates of well-nourished mothers in economically developed countries rise progressively through pregnancy and increase to about 20–25% above pre-pregnancy levels in the third trimester (Jasienska, 2013). The increase in metabolic rate is closely correlated with the pre-pregnancy fat mass or energy stores. Many poorly nourished women as well as some who are better nourished exhibit a decline in metabolic rate in early pregnancy and a smaller increase by term. The decreased metabolic rate in poorly nourished women is likely to be an evolutionary adaptation that increases the availability of energy to the growing fetus and conserves energy for lactation; as yet, however, the physiological mechanisms that underlie this interesting and important phenomenon are poorly understood.

The changes in metabolic rate during pregnancy are accompanied by changes in core temperature (Hartgill, Bergersen et al., 2011). The slightly elevated core temperature characteristic of the luteal phase of the menstrual cycle (less than 1°C) is maintained during roughly the first trimester of pregnancy, presumably because of the continued effects of the high levels of ovary-derived progesterone on central thermoreceptors. After that, core temperature decreases gradually until it is roughly 36.5°C, or about 0.5°C below pre-pregnancy levels, at the end of pregnancy. This fall in core temperature is likely to be an adaptation that protects the fetus from heat stress; intrauterine and fetal temperatures are slightly higher than the maternal core temperature and so are maintained at close to 37°C throughout pregnancy. Surprisingly, maternal temperatures remain low, in the range of 36.5°C, for several months after delivery (Hartgill, Bergersen et al., 2011). Neither the evolutionary nor the physiological reason for this continued decrease in core temperature is well understood; it might protect new mothers from the effects of febrile illnesses in the neonatal period but it might simply be a holdover from the lower temperature characteristic of pregnancy rather than an evolutionary adaptation. To maintain maternal core temperature approximately constant, however, the increased heat production during pregnancy must be compensated by an increase in heat loss. Most pregnant women do this easily by increasing cutaneous blood flow and thereby increasing heat loss by conduction and radiation to the environment.

Given that the cardiovascular system must supply cells with oxygen and nutrients and that allometric changes in metabolic rate with body mass among mammals are roughly paralleled by changes in cardiac output (Schmidt-Nielsen, 1984), we can appreciate that physiological changes in metabolic rate will also be accompanied by changes in cardiac output. Norwitz and Robinson have recently reviewed the effects of pregnancy on the cardiovascular system (Norwitz & Robinson, 2018). Cardiac output increases to about 30–50% above pre-pregnancy levels by the end of the second trimester and is maintained at these levels for the duration of pregnancy. Although measurements of metabolic rate and of cardiac output were carried out on different populations of pregnant women, it is plausible that the increases in cardiac output are greater than the increases in metabolic rate. Systemic vascular resistance decreases by about 20% during the first two trimesters of pregnancy. This decrease in vascular resistance is due in part to the expansion of the uteroplacental circulation and in part to the action of endocrine or paracrine vasodilators. A specific decrease in the resistance of cutaneous vascular beds, as previously noted, is also important. This decrease in vascular resistance leads to a fall in blood pressure and to the activation of homeostatic mechanisms that help to restore blood pressure by increasing cardiac output (through a combination of increased heart rate and increased stroke volume) and by increasing renal retention of sodium and water. Sodium and water retention in turn leads to an increase in plasma and extracellular fluid volume and a dilutional anemia that also increases cardiac output. As we would expect, however, these negative feedback mechanisms do not completely correct the fall in blood pressure; blood pressure decreases by roughly 10–15 mm Hg during the first two trimesters before returning toward pre-pregnancy levels by the end of gestation. The limited gain of these feedback mechanisms and the resulting decrease in blood pressure may be an adaptation that decreases cardiovascular work during pregnancy and conserves energy for other needs.

The allocation of metabolic energy to fetal growth and related needs during pregnancy necessitates tradeoffs with other uses of the available energy and is likely to entail a decrease in somatic maintenance and/or in immune mechanisms. These tradeoffs will be more severe for poorly nourished women than for well-nourished ones. When they have the luxury, pregnant women can decrease the amount of physical labor they do. Unfortunately, in many economically underdeveloped societies, women must continue to work to acquire resources for themselves and their families. Although the immunological changes that accompany pregnancy are complex, there is a decrease in cell-mediated immunity during pregnancy; energy is apparently diverted from the immune system to provide some of the needs of reproduction. Recently, reproduction was found to be associated with shorter telomeres and with evidence of accelerated aging in DNA methylation patterns (Ryan, Hayes et al., 2018). This is direct evidence that reproduction decreases somatic maintenance.

The effects of reproduction on health and longevity are complex. Reproduction and lactation protect women from breast and uterine cancer but increase the risk of obesity and cardiovascular disease (Jasienska, 2013). Protection against reproductive cancers is more significant in well-nourished populations, in which women have higher levels of estrogens and in which there is a higher incidence of these cancers than in economically deprived populations. The relationship between reproduction and life expectancy is further complicated by the fact that some infertile women also suffer from other health problems that are associated with or cause their infertility. Despite these complications, examination of number of progeny and age of death in historical data from the British aristocracy provided direct evidence that reproductive effort decreases women's life expectancy (Westendorp& Kirkwood, 1998; Doblhammer & Oeppen, 2003). The detrimental effect of reproduction on longevity is exacerbated by poverty, a finding consistent with the idea that reduced longevity results from tradeoffs in the use of limited resources for reproduction or for somatic maintenance (Lycett, Dunbar et al., 2000). In summary, the physiology of pregnancy illustrates the changes in resource allocation in response to changes in physiological need, the costs engendered by life history tradeoffs, and the ways in which regulatory mechanisms maintain homeostasis in the face of these tradeoffs.

Acknowledgments

The ideas discussed in this chapter were presented at a Darwin Day Seminar at Kalamazoo College in February 2018 and at a meeting of the International Society for Evolution, Medicine, and Public Health in August 2018. I thank the participants at these events, especially Santiago Salinas and the students at Kalamazoo College, for their thoughtful critiques. I also thank Alan Schechter and Ted Steck for their helpful comments and suggestions.

Bibliography

Berger, A. J., R. A. Mitchell, & J. W. Severinghaus (1977). Regulation of respiration (third of three parts). *N Engl J Med*, 297(4): 194–201.

Bernard, C. (1865/1957). *An Introduction to the Study of Experimental Medicine* (pp. 69, 92). New York: Dover.

Bernard, C. (1878/1974). *Lectures on the Phenomena of Life Common to Animals and Plants* (p. 89). Springfield, IL: Charles C Thomas.

Bynum, W. F. (1983). Darwin and the doctors: Evolution, diathesis, and germs in 19th-century Britain." *Gesnerus*, 40(1–2): 43–53.

Cannon, W. B. (1929). Organization for physiological homeostasis, *Physiol Rev*, 9(3): 399–431.

Cannon, W. B. (1939). *The Wisdom of the Body* (p. 24). New York: Norton.

Darwin, C. (1872/1965). *The Expression of the Emotions in Man and Animals* (p. 68). Chicago: University of Chicago Press.

Diamond, J. (1991). Pearl Harbor and the Emperor's physiologists. *Nat Hist*, (12): 2–7.

Diamond, J. (2002). Quantitative evolutionary design. *J Physiol*, 542(Pt 2): 337–45.

Doblhammer, G. & J. Oeppen (2003). Reproduction and longevity among the British peerage: The effect of frailty and health selection. *Proceedings of the Royal Society B: Biological Sciences*, 270(1524): 1541–7.

Flexner, A. (1910). *Medical Education in the United States and Canada: A Report to the Carnegie Foundation for the Advancement of Teaching*. New York: The Carnegie Foundation.

Fogel, R. W. (2003). Secular trends in physiological capital: Implications for equity in health care. *Perspect Biol Med*, 46(3(Suppl.)): S24–38.

Hagen, J. B. (1996). Walter Cannon & self-regulation in animals. In *Doing Biology* J. B. Hagen & D. Allchin (eds.) (pp. 95–103). Genview, IL: Harper Collins.

Hartgill, T. W., T. K. Bergersen, & J. Pirhonen (2011). Core body temperature and the thermoneutral zone: A longitudinal study of normal human pregnancy. *Acta Physiol (Oxf)*, 201(4): 467–74.

Houk, J. C. (1988). Control strategies in physiological systems. *FASEB J*, 2(2): 97–107.

Jasienska, G. (2013). *The Fragile Wisdom: An Evolutionary View of Women's Biology and Health*. Cambridge, MA, Harvard University Press.

Kirkwood, T. B. L. (1996). Human senescence. *BioEssays*, 18(12): 1009–16.

Kluger, M. J., W. Kozak, C. A. Conn, L. R. Leon, & D. Soszynski (1996). The adaptive value of fever. *Infect Dis Clin North Am*, 10(1): 1–20.

Laland, K. N., K. Sterelny, J. Odling-Smee, W. Hoppitt, & T. Uller (2011). Cause and effect in biology revisited: Is Mayr's proximate-ultimate dichotomy still useful? *Science*, 334(6062): 1512–16.

Lane, N., W. F. Martin, J. A. Raven, & J. F. Allen (2013). Energy, genes and evolution: Introduction to an evolutionary synthesis. *Philos Trans R Soc Lond B Biol Sci*, 368(1622): 20120253.

Lotka, A. J. (1922). Contribution to the energetics of evolution. *Proceedings of the National Academy of Sciences*, 8(6): 147–51.

Lusk, G. (1932). A tribute to the life and work of Max Rubner. *Science*, 76(1963): 129–35.

Lycett, J. E., R. I. Dunbar, & E. Voland (2000). Longevity and the costs of reproduction in a historical human population. *Proc Biol Sci*, 267(1438): 31–5.

Mayr, E. (1988). Cause and effect in biology. In *Toward a New Philosophy of Biology*, (pp. 24–37). Cambridge, MA, Harvard University Press.

McEwen, B. S. (2015). Biomarkers for assessing population and individual health and disease related to stress and adaptation. *Metabolism*, 64(3 Suppl 1): S2–10.

Nagashima, K., K. Tokizawa, Y. Uchida, M. Nakamura-Masuda, & C.-H. Lin (2012). Exercise and thermoregulation. *J Phys Fitness Sports Med*, 1(1): 73–82.

Nesse, R. M., & G. C. Williams (1994). *Why we Get Sick: The New Science of Darwinian Medicine*. New York, Times Books.

Norwitz, E. R. & J. N. Robinson (2018). Pregnancy-induced physiological alterations. In *Critical Care Obstetrics* J. P. Phelan, L. D. Pacheco, & M. R. Foley (eds.) (pp. 41–67). Hoboken, NJ: John Wiley & Sons.

Paul, D. (1992). Fitness: Historical perspectives. In *Keywords in Evolutionary Biology*. E. F. Keller & E. A. Lloyd, (pp. 112–14). Cambridge, MA: Harvard University Press.

Perlman, R. L. (2013). *Evolution and Medicine*. Oxford: Oxford University Press.

Power, M. L. & J. Schulkin (2008). Anticipatory physiological regulation in feeding biology: Cephalic phase responses. *Appetite*, 50(2–3): 194–206.

Richter, K., M. Haslbeck, & J. Buchner (2010). The heat shock response: Life on the verge of death. *Mol Cell*, 40(2): 253–66.

Ryan, C. P., M. G. Hayes, N. R. Lee, T. W. McDade, M. J. Jones, M. S. Kobor, C. W. Kuzawa, & D. T. A. Eisenberg (2018). Reproduction predicts shorter telomeres and epigenetic age acceleration among young adult women. *Sci Rep*, 8(1): 11100.

Schmidt-Nielsen, K. (1984). *Scaling: Why is Animal Size so Important?* (p. 198). Cambridge: Cambridge University Press.

Simon, H. A. (1956). Rational choice and the structure of the environment. *Psychological Review*, 63(2): 129–38.

Westendorp, R. G., & T. B. Kirkwood (1998). Human longevity at the cost of reproductive success. *Nature*, 396(6713): 743–6.

Wiener, N. (1961). *Cybernetics*, (p. 11). Cambridge, MA: MIT Press.

Williams, A. B., & B. Schumacher (2016). p53 in the DNA-damage-repair process. *Cold Spring Harb Perspect Med*, 6(5).

Zampieri, F. (2009). Medicine, evolution, and natural selection: An historical overview. *Q Rev Biol*, 84(4): 333–55.

Zimmerman, S. B. & A. P. Minton (1993). Macromolecular crowding: Biochemical, biophysical, and physiological consequences. *Annu Rev Biophys Biomol Struct*, 22: 27–65.

CHAPTER 7

The Functions of MicroRNA in Female Reproduction

Michael L. Power, Caroline W. Quaglieri, Eda G. Reed, and Jay Schulkin

7.1 Introduction

The topic of this chapter may be a canonical example of the inherent dilemma in whether a firm grasp of evolutionary principles is necessary in medical training for women's health care providers. The use of micro RNAs (miRNAs) for screening and diagnosing disease in and out of pregnancy is showing promise in many areas. We contend that an understanding of evolutionary principles would enhance clinical interpretation of any miRNA diagnostic panels; however, it could be argued that most clinicians would only need to know the sensitivity and specificity of the tests and wouldn't even have to understand the underlying proximate biology of miRNAs, let alone the evolutionary implications and complexity. We emphatically contend that biomedical researchers would gain from viewing miRNAs from an evolutionary perspective, especially for research involving potential therapeutic interventions that either suppress or replace miRNAs. Translation of miRNA research in animal models also would benefit from an evolutionary perspective, in our opinion. Regardless, miRNAs are an emerging hot topic in women's health, both on the clinical and research sides, worthy of review in this volume.

The haploid human genome contains approximately 3 billion base pairs of DNA. Only about 2% of that DNA codes for the approximately 20,000 genes that produce human proteins. This led to a hypothesis that much of the genome was "junk" DNA (Ohno, 1972), continually inherited from generation to generation, but performing no adaptive function. However, modern research has determined that as much as 80% of the human genome has biochemical activity (Kellis et al., 2014), and thus could have biological function. Minimal estimates are on the order of 10% of the human genome has true biological function that affects fitness, while some argue that more than half may have relevant biological activity (Kellis et al., 2014). Some of the non-protein-coding DNA code for important RNA molecules such as ribosomal RNA (rRNA) and transfer RNA (tRNA), and thus can be considered genes under a broad definition, producing functional nucleic acid molecules rather than amino acid molecules. Other sections of non-protein-coding DNA serve regulatory functions for gene expression, for example, as binding sites for enhancers and suppressors of gene expression. The conception of non-coding DNA as mostly "junk" has been replaced with a conception of non-coding DNA performing important roles in cellular physiology and especially in regulation of gene expression.

In this chapter, we focus on a subset of small RNA molecules of about 22 nucleotides in length known as microRNA or miRNA. miRNA are transcribed from non-coding DNA, but obviously are not trans-

Power, M. L., Quaglieri, C. W., Reed, E. G., and Schulkin, J., *The Functions of MicroRNA in Female Reproduction* In: *Integrating Evolutionary Biology into Medical Education—for maternal and child healthcare students, clinicians, and scientists*. Edited by: Jay Schulkin and Michael L. Power, Oxford University Press (2020). © Oxford University Press.
DOI: 10.1093/oso/9780198814153.003.0008

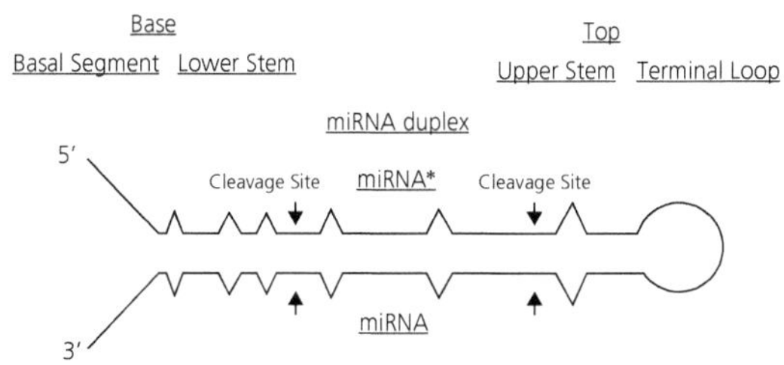

Figure 7.1 Structure of the pre-miRNA with cleavage sites

lated into a peptide. Rather, these small, non-coding strands of RNA regulate gene expression post-transcription primarily by binding to complementary strands of messenger RNA (mRNA) and act to repress the production of the peptide associated with that mRNA either by blocking translation or facilitating degradation of the mRNA. Our review will focus on the role miRNA may play in various cancers of the female reproductive system (including breast), in pathologies of pregnancy such as preeclampsia and preterm birth, and the possible functions of miRNA excreted into milk. Specifically, we examine the evidence regarding the extent to which miRNAs affect maternal, fetal, and neonatal health and whether miRNAs have utility in screening, diagnosing and treating disorders of the female reproductive system.

7.1.1 miRNAs

In the early 1990s, small non-coding RNAs (lin-4 and let-7) were discovered to play important roles in the timing of development in the roundworm *Caenorhabditis elegans* (Lee, Feinbaum & Ambros, 1993; Wightman, Ha, & Ruvkin, 1993; Reinhart et al., 2000). Subsequent research identified numerous of these miRNAs in the *C. elegans* genome and found orthologs in the *Drosophila* and human genomes, indicating that miRNAs likely had important regulatory functions in all animals (Lau et al., 2001; Lee & Ambros, 2001). They represent an important evolutionary process, providing a complex and potentially highly variable mechanism to regulate gene expression on which selection can act.

Since that time our understanding of the importance of small non-coding RNAs in the regulation of gene expression has exploded. Small interfering RNA (siRNA), piwi-interacting RNA (piRNA), and miRNA are all species of RNA transcribed from the DNA that serve to regulate the translation of messenger RNA (mRNA) into a peptide, via both pre- and post-transcriptional mechanisms. This review will focus on miRNA, which are short (usually 20–25 nucleotides) RNA fragments that are produced from a longer transcribed segment of DNA that folds back on itself to form a precursor miRNA (pre-miRNA) that has a distinctive hairpin structure (Figure 7.1). Names of miRNAs follow a basic convention of a three-letter species code (e.g., hsa for human, bta for bovine), a dash and then miR followed by a dash and a number (e.g., hsa-miR-21). Pre-miRNAs are named in the same manner except the r is lower case (e.g., the pre-miRNA hsa-mir-21 is cleaved to form hsa-miR-21). There are miRNAs for which there are multiple pre-miRNAs produced by different RNA genes that produce the identical mature miRNA. These pre-miRNAs are identified by a dash followed by 1, 2, 3, and so forth at the end of the name. For example, the pre-miRNAs hsa-mir-7-1, hsa-mir-7-2, and hsa-mir-7-3 all produce hsa-miR-7. If different pre-miRNAs produce almost identical products (e.g., differing by only one or two nucleotides) a letter is added after the number (e.g., miR-181a, miR-181b, miR-181c, and miR-181d). Both conditions can exist, for example, mir-26a-1 and mir-26a-2 both produce miR-26a, while miR-26b differs from miR-26a by two nucleotides and is produced by a single RNA gene that produces the pre-miRNA mir-26b.

There are several pathways to forming a pre-miRNA. The most common mechanism starts with the transcription of a primary miRNA (pri-miRNA) with the characteristic hairpin. The pri-miRNA is then cleaved by the microprocessor complex composed of the ribonuclease enzyme DROSHA and the RNA binding protein DCGR8 (known as PASHA in non-human animals) into the shorter (about 70 nucleotides) pre-miRNA (Figure 7.1). The pre-miRNA is exported from the nucleus into the cytoplasm where the hairpin is cleaved off by the ribonuclease DICER, leaving two complementary strands of RNA. As in most biological processes there are exceptions to the rule. For example, pre-mir-451 is cleaved by argonaute protein 2, not by DICER (Dueck et al., 2012). There are also DROSHA/PASHA-independent pre-miRNAs that are transcribed directly from an intron, which are termed mirtrons. Mirtrons were originally identified in *Drosophila* and *C. elegans*, but recently hundreds have been identified in the human and mouse genomes (Ladewig et al., 2012; Wen et al., 2015). Mirtrons and canonical miRNAs (miRNA processed by DROSHA/PASHA) appear to have different evolutionary origins despite their functional similarities. So far no mirtrons have been shown to be conserved across invertebrates and vertebrates, while many canonical miRNAs are conserved, indicating a truly ancient origin for canonical miRNAs (Westholm & Lai, 2011).

After the hairpin has been cleaved off the pre-miRNA, two complementary strands of RNA remain, usually referred to as miRNA and miRNA* (Figure 7.1). In most cases, the miRNA strand is the functional strand that will be loaded into the RNA-induced silencing complex (RISC) while the miRNA* strand is degraded. For example, miR-124 derives from the 3-prime side and miR-124* from the 5-prime side of the same pre-miRNA, but in general much more miR-124 will be found in any body fluid or tissue. In some cases both strands of the cleaved pre-miRNA appear to be functional, or at least are found in roughly equal quantities, in which case the nomenclature is to designate one as -3p and the other as -5p, referencing the 3-prime and the 5-prime sides of the pre-miRNA, respectively (e.g., miR-590-3p and miR-590-5p).

RISC is a multi-protein complex that usually includes an argonaute protein. RISC uses the loaded miRNA strand as a guide strand to identify the appropriate complementary 3 prime-untranslated regions (3′-UTR) of mRNA molecules. The bound miRNA prevents translation of the mRNA either by initiating cleavage and degradation of the mRNA or by blocking the translation machinery. Degradation of the mRNA requires a near-perfect match of the miRNA with the segment of the 3′-UTR of the mRNA and an argonaute protein to cleave the mRNA. This is what is typically found in plants; so far most plant miRNAs have been found to be a perfect match for their target mRNA sequences and to initiate mRNA degradation (Axtell, Westholm, & Lai, 2011). In animals incomplete matching appears to be more typical, and miRNA regulatory action is more through blocking translation, which does not require as close a match (Axtell et al., 2011).

Any particular miRNA may be able to bind to the 3′-UTR of many mRNAs produced by multiple genes, and multiple miRNAs may act on any particular mRNA. Thus, miRNA represent an extremely complex and powerful evolved mechanism to regulate gene expression. Advances in gene sequencing have allowed scientists to categorically analyze many of the miRNA sequences found throughout the body and various bodily fluids (e.g., Weber et al., 2010). Many of these sequences have been associated to known functions, but further analysis and designation is necessary to paint a complete picture of the possible effects of miRNAs. At least half of mammalian mRNA has been under selective pressure to conserve pairing with miRNAs (Friedman et al., 2009), indicating that the expression of most genes may be under some level of regulatory control from miRNAs and implying that the role of miRNAs in regulating gene expression has strong selective advantage. For more detailed descriptions of miRNA biogenesis and evolution the reader is directed to excellent reviews (Westholm & Lai, 2011; Ha & Kim, 2014).

The role of miRNAs in regulating gene expression is more like a rheostat than an on-off switch, modulating the amount of peptide produced. The functions of miRNA generally fall into two broad categories: fine-tuning of gene expression in response to specific cell requirements, and robustness in cellular responses, especially related to cellular differentiation state. Changes in cellular miRNA expression can be diagnostic for changes in cell function, both adaptive and pathological. For example, in some cancers the expression of miRNAs associated with terminal differentiation is reduced, promoting a proliferative state (Hayes, Peruzzi, & Lawler, 2014).

7.1.1.1 Exosomes

Despite high levels of endogenous nucleases, RNA of all types (e.g., mRNA, siRNA and miRNA) have been found to exist outside of the cell, in the intercellular fluid. Some are bound to argonaute proteins, but often they are packaged in microvesicles or exosomes. Exosomes are produced by cells via several processes (Alenquer & Amorim, 2015), but they will always include a lipid membrane composed of fragments of the cellular membrane. RNA packaged into exosomes is resistant to RNase and other enzymatic degradation, which allows these exosomes to travel through the circulatory system to be eventually absorbed into another cell, deliver its contents, and thus act as an intercellular mechanism for cell-to-cell communication, including regulating gene expression (Valadi et al., 2007; Chen et al., 2014). In addition to miRNA, mRNA and proteins have been also found to be encapsulated in exosomes in blood, and in mouse and human mast cells both mRNA and miRNA have been detected within exosomes (Valadi et al., 2007). Exosomes are able to cross biological barriers, including the blood-brain barrier (Alvarez-Erviti et al., 2011). Thus, miRNA from one set of cells (in a particular organ, for example) can enter the blood stream in a protected vesicle and be delivered to cells in other organs as an endocrine signal; miRNAs can act locally or systemically. Researchers are beginning to explore the use of exosomes as a delivery device for novel therapeutic biochemicals. Recent work has shown that bovine milk-derived exosomes can be manipulated into becoming a potential drug delivery vehicle (Munagala et al., 2016).

In summary, the miRNA biochemical system represents an ancient, evolved mechanism for complex gene expression regulation and cell-to-cell communication which was likely important for producing complex organisms with multiple cell types with different functions, all with the same DNA. The pattern of miRNA expression of a cell can tell much about its function and status. This appears especially true for pathology, such as cancer. Tumor cells often have distinctive and predictive miRNA expression signatures (Hayes et al., 2014). The use of miRNA for therapeutic interventions also has generated cautious interest, though the difficulty of delivering the miRNA intervention into the targeted cells (and only those cells) is a major stumbling block at present. In the next sections we review some of the recent advances in screening and detection of cancers of the female reproductive tract using miRNA.

7.1.2 miRNAs and cancer

The study of miRNAs related to cancer screening and treatment is perhaps the most developed of the clinical applications regarding miRNA. In 2002, a study demonstrated an association between the deletion of miR15 and miR16 and chronic lymphocytic leukemia (Calin et al., 2002). By 2011, more than 200 miRNAs had been linked to the development and progression of various cancers (Kayani et al., 2011). Recent work has focused on the miRNome, defined as the specific miRNA signature of a tumor, which is associated with many clinical states of a tumor such as its malignancy, aggressiveness, vascular invasion, and proliferation index (Di Leva, Garofalo, & Croce, 2014). In general, miRNAs are down regulated in tumors compared to normal tissue, indicating a reduction of gene expression regulation. However, there are also miRNA "oncogenes," where over expression is associated with tumors, often by suppressing expression of tumor suppressor genes (Di Leva et al., 2014). Whether a particular miRNA is considered a tumor suppressor or an "oncomiRNA" can depend on the tissue and the specific tumor, due to the large number of target genes each miRNA has the potential to affect.

The use and study of miRNAs as biomarkers of cancer shows great promise. Assessing the miRNome of a tumor is less invasive and generally more accurate than other techniques (Cheng, 2015). The miRNome of a tumor can be indicative of its malignancy and proliferation potential. The miRNA profile also can show where tumors originated (tissue of origin) (Hayes et al., 2014), which can be useful when determining the course of treatment. Cancer detection and early diagnosis by assessing circulating miRNAs or miRNAs in other body fluids may be especially valuable, as early detection of tumors is critical for survival in many cancers (Di Leva et al., 2014).

7.1.2.1 Breast cancer

The role of miRNA in breast cancer may be the best studied so far, with the most published clinical trials (Hayes et al., 2014). Dysregulation of miRNAs in breast cancer appears to be a critical aspect of the disease, and characterizing the miRNome can help stratify patient risk level and potential response to therapy (Mulrane et al., 2013). Many miRNA appear to be involved in breast cancer, including but not limited to: miR-221 and miR-222, which are upregulated in breast cancer, miR-182, miR-27a, and miR-96 which act as oncogenes, and miR-146a which downregulates BRCA1 (Kayani et al., 2011). The use of miRNAs for diagnosis shows promise. For example, miR-155 has potential to be used as a biomarker for breast cancer diagnosis, and as an indicator of response to therapy. Others are also significant, but miR-155 is the most prominent (Cheng, 2015). The future should focus on a panel of miRNAs (e.g., the miRNome), including miR-155, to identify breast cancer and type of breast cancer (Cheng, 2015).

Treatments for breast cancer using miRNAs are in the early development stage. There are two types of proposed treatments, depending on the miRNome of the cancer cells: either suppressing the expression of oncomiRNAs or providing replacement miRNAs for tumors where miRNA expression is suppressed. For either therapy to be effective, screening and diagnosis of the tumor "type" based on its miRNome has to be accurate and comprehensive. Breast cancer is heterogeneous, which raises the complexity that miRNA-based therapy may have to be personalized for almost every patient. The other major challenge is directing the therapy to the cancer cells without affecting other cells. This is especially important as each miRNA has the potential to affect multiple genes, not just the targets of interest. The potential for unanticipated consequences to miRNA therapy are very high.

7.1.2.2 Ovarian cancer

Epithelial ovarian cancer has a low (25–30%) 5-year survival rate. It is asymptomatic and difficult to diagnose until the tumors have spread beyond the ovaries, and once that has happened survival rates decrease quickly. Thus, methods for early detection, such as miRNA based diagnosis, would enhance outcomes.

Iorio et al. (2007) conducted the first study identifying different miRNA expression in normal ovarian tissue vs carcinomas of the ovaries. Ovarian tissue can be used for miRNA diagnosis, but the procedure is invasive and often only pursued during later stages of the cancer when it becomes symptomatic. Plasma miRNA is better suited for screening and early diagnosis due to ease of collection. Research to detect ovarian cancer via circulating miRNAs in body fluids is in the beginning phase. Most studies have small sample sizes and therefore need further validation, but the results are encouraging. In a retrospective study, 30 miRNAs were identified as differentially expressed in plasma between cases and controls. Of those 30, a panel of two (let-7f, downregulated and miR-205, upregulated) were diagnostic of ovarian cancer with high accuracy even for stage 1 disease (Zheng et al., 2013). In another study, a suite of miRNAs were found to predict ovarian cancer when used in a variety of models. The best model used 7 miRNAs and a neural network model, which achieved a positive predictive value greater than 90% (Elias et al., 2017). However, the study was restricted by the limited availability of early stage and pre-detection samples to test, the time period when plasma screening would be most important.

Although epithelial ovarian cancer is rarer than other cancers (it is the eighth most common cancer in women), its mortality rate is highest out of all gynecologic cancers (Iorio & Croce, 2016). This duality presents a unique challenge for researching appropriate diagnostic and therapeutic methods. Testing the general public for ovarian cancer via tissue miRNA detection is neither practical nor cost effective, but early detection is necessary for reducing the mortality rate. Further study of plasma based miRNA screening and diagnosis in larger cohorts will be most useful in the future.

One of the reasons why epithelial ovarian cancer mortality rates are so high is the low success rate of chemotherapies. Research on ovarian tissue expression of miRNA might provide insight into the gene expression regulation which could characterize these tumors and perhaps indicate whether miRNA-based therapies could be effective (Iorio & Croce, 2016). Of course miRNA therapy is limited by all the drawbacks of miRNA treatment as described in the breast cancer section, especially the concern for unintended effects due to changes in gene expression outside of the targeted cells.

7.2 Pregnancy-related miRNAs

During pregnancy, regulation by miRNA is unusual in that there are two separate individuals, the fetus and the mother, who can and do contribute to the miRNome. Not surprisingly, you find more placenta-related circulating miRNAs in pregnant women, and the resulting miRNome differs by trimester (Cai, Kolluru, & Ahmed, 2017). The evidence that miRNAs affect pregnancy are strong, but the specifics remain poorly understood. Studies are difficult to replicate, and sample size is usually small. The use of miRNAs to diagnose and treat pregnancy complications is being pursued aggressively but is still far from a reality.

Currently, there is more information about tissue resident miRNA than circulating miRNA related to pregnancy but circulating miRNA would be more useful during diagnosis due to the less invasive procedure. Another concern/complication regarding tissue resident miRNA is that often the placentas used as "normal" are placentas sent to pathology for a reason different than the case placentas (e.g., not preeclampsia versus preeclampsia), but they might not be unaffected by some pathology (Roberts et al., 2017).

There are three main types of placental miRNA involved in pregnancy: placenta-specific miRNA, placenta-associated miRNA, and placenta-derived circulating miRNA. Uterine miRNA are also involved in pregnancy, but are thought to be less important and therefore are less studied. (Cai et al., 2017).

7.2.1 Placenta-specific miRNA

Placenta-specific miRNA are largely if not uniquely expressed in mammalian placenta. More than 100 miRNAs of this type have been identified. Placenta-specific miRNAs display a strong evolutionary signal for their origin. Members of many placental miRNAs are found within clusters on the same chromosome, some even within the same intron. For example, the *Sfmbt2* miRNA cluster which codes for 36 miRNAs expressed in placenta and embryonic stem cells is located in the 10th intron of the *Sfmbt2* gene (Zheng et al., 2013). Placental miRNA families whose members are found on different chromosomes are often associated with the same transposon.

The hsa-miR-1302 family, comprised of 11 known pre-miRNA members (7 unique pre-miRNAs and 4 copies of mir-1302-2, all on different chromosomes) appears to derive from the transposon MER53 (Yuan et al., 2010). All of the 11 pre-miRNAs produce the same mature miRNA (hsa-miR-1302). Although the function of miR-1302 is not known, predicted targets have been identified and generally are implicated in cell proliferation, cell division, metabolism, development, and immune response. At least one of these RNA genes, or at least an ortholog, was found in twenty-one placental mammal species, but none were found in the opossum and platypus (Figure 7.2). The pre-miRNA hsa-mir-1302-3 appears to be a result of a duplication of one of the hsa-mir-1302-2 genes, and may be human specific, as it is not found in the other species, including apes (though a potential ortholog was detected in the alpaca; Figure 7.2). Based on these data, miR-1302 appears likely to have arisen in the placental mammal lineage soon after the split from the marsupials (Yuan et al., 2010).

The association with a transposable element explains the multiple copies and distribution of the miR-1302 family across many chromosomes (proximate mechanism). The conservation of multiple copies implies an ultimate (evolutionary) explanation. Multiple copies across many chromosomes provides a redundancy that may create a resilience against function loss. In addition, the multiple copies increase the range of expression that can be obtained, and greater flexibility in the timing of up- and downregulation of the targets of miR-1302. The fact that all the pre-miRNAs code for the same miRNA and the ubiquity of the miR-1302 family in anthropoid primates implies critical function in monkey and ape placenta and suggests that non-human primates might be good models to explore function. In contrast, rodent models would be poor choices (Figure 7.2). An evolutionary perspective provides potential insight and hypotheses beyond the proximate mechanism to explain the distribution of the placenta-specific miR-1302 family.

7.2.2 Placenta-associated miRNA

Placenta-associated miRNA are ubiquitously expressed in placenta, but also in other tissue. The pattern of placental expression varies with gestational age and responds to challenges such as inflammation and hypoxia (Cai et el., 2017). First trimester placentas are more likely to express oncogenic, angiogenic, and antiapoptotic miRNAs whereas third trimester placentas express more miRNAs related to

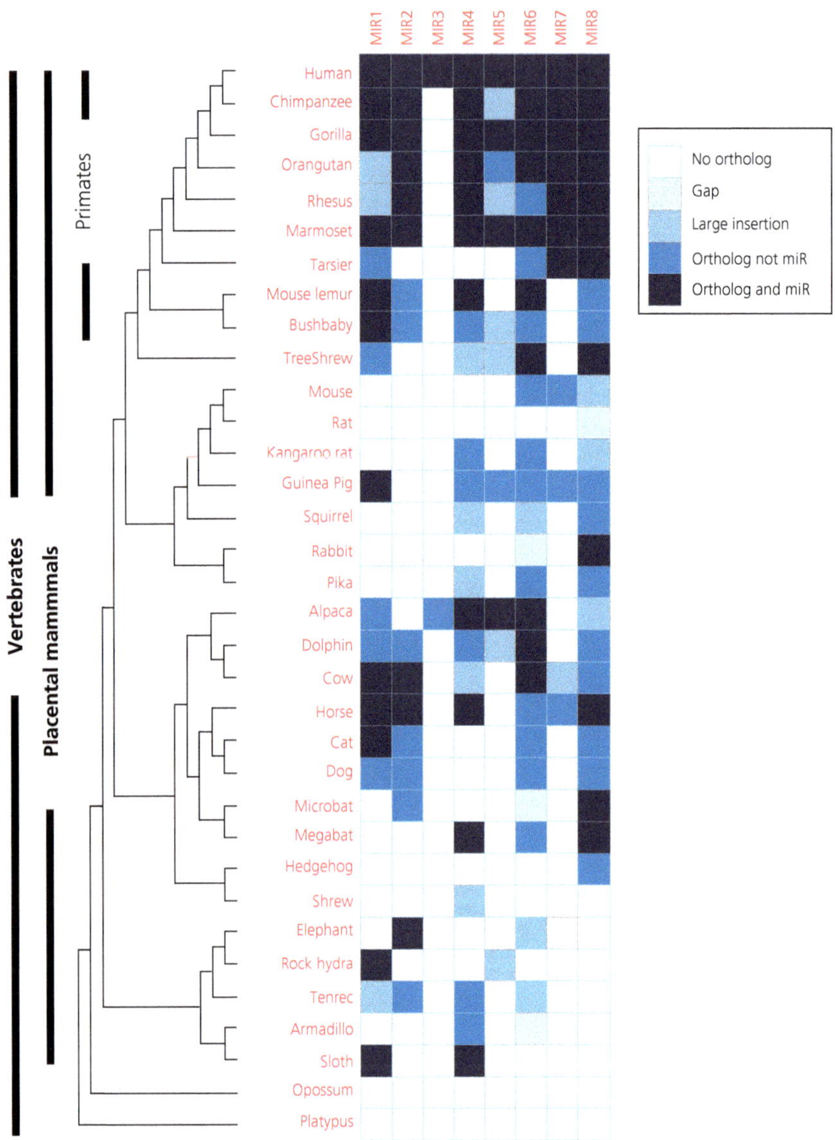

Figure 7.2 Phylogenetic distribution of orthologs of the hsa-mir-1302 family. The phylogenetic tree was modified from the UCSC Genome Browser. Although in human there are 4 identical copies of hsa-mir-1302-2, in all other the species listed there is only one. Abbreviations: miR, miRNA; Gap, a gap in the genome; Large insertion, an insertion of more than 100 bp, except for a 17 bp insertion in the ortholog of hsa-mir-1302-4 in Pika; Ortholog not miR, the ortholog is not a miRNA; Ortholog and miR, the ortholog has been validated as a miRNA; MIR1-MIR8, the orthologs of hsa-mir-1302-1, hsa-mir-1302-2, hsa-mir-1302-3, hsa-mir-1302-4, hsa-mir-1302-5, hsa-mir-1302-6, hsa-mir-1302-7, and hsa-mir-1302-8 respectively
(Yuan et al., 2010)

cell differentiation (Gu et al., 2014). Downregulation of certain miRNAs appears important for syncytiotrophoblast differentiation and upregulation of others encouraged trophoblast proliferation, migration and invasion. Trophoblast cells challenged in vitro with hypoxia exhibited altered expression of placenta-associated miRNAs (Cai et al., 2017). The placentas of women who smoked during pregnancy displayed downregulation of miR-16, miR-21, and miR-146a (Maccani et al., 2010).

7.2.3 Placenta-derived circulating miRNA

Placenta-derived miRNAs released into maternal circulation are the best candidates for biomarkers of pregnancy complications simply due to their greater ease of collection. Research on tissue expression of miRNAs, either placental tissue or in vitro studies of trophoblast cells may be the best path for understanding the function of the various placental miRNAs and the pathogenesis of pregnancy complications, but to screen and diagnose problems the advantages of using a maternal blood sample are large.

7.2.4 Preeclampsia

Preeclampsia (PE) is the pregnancy complication most studied for miRNA interactions. There is no treatment for PE (other than delivery), so an early diagnosis is imperative to try and prevent worsening of PE into severe PE or eclampsia. However, early diagnosis is challenging and recent studies have looked to miRNA to improve diagnoses. Women with preeclampsia have different placental and circulating miRNA expression than non-pregnant women (Cai et al., 2017; Barchitta et al., 2017), but there are also differences in miRNA regulation between preeclampsia and severe preeclampsia, making diagnosis an even greater challenge (Barchitta et al. 2017).

A small study (n = 4) found overexpression of miR-202-3p in the first trimester of women who developed preeclampsia compared to matched controls (Singh et al., 2017). In a slightly larger study (twenty-one cases and fifty-eight controls), concentrations of three different placenta-specific miRNAs in maternal plasma from the first trimester were higher in pregnancies that developed preeclampsia (Hromadrikova et al., 2017). Another study found an increased concentration of exosomes in the plasma of women who developed preeclampsia (Salomon et al., 2017), suggesting increased signaling from placenta faced with inadequate perfusion. This finding is consistent with earlier in vitro findings of increased release of exosomes from cytotrophoblast under conditions of hypoxia (Salomon et al., 2013). The in vivo study found eleven miRNAs that were upregulated and one that was downregulated in the placental exosomes from women with preeclampsia (Salomon et al., 2017).

7.2.5 Preterm birth

The molecular mechanisms underlying preterm labor and birth are largely unknown. It is likely that preterm labor has multiple causes, some external and some internal. Some recent studies have suggested that miRNA may be useful in screening for patients at risk (Barchitta et al., 2017). Seven miRNAs were underexpressed and one overexpressed in plasma samples of women who experienced spontaneous preterm birth (Gray et al., 2017). A retrospective study found that the relative expression of two clusters of miRNA genes were predictive of spontaneous preterm birth, suggesting that an imbalance among the miRNAs from these clusters may be associated with preterm labor (Wommack et al., 2018). Another study found that almost half of the preterm labor associated miRNAs were members of these two clusters (Fallen et al., 2018).

The process of cervical connective tissue remodeling, a necessary condition for birth, is an obvious target for investigation of association between miRNA and preterm birth. The miRNome of cervical cells from women who gave birth prematurely differed from controls, with the expression of miR-143 and miR-145 increased by more than ten-fold. Both were negatively correlated with cervical length (Barchitta et al., 2017).

Although these studies show promise, none had large sample size, and all require replication. At present there is justification for research on miRNA-based screening for preterm birth in both tissue and circulating, but clinical application is still far away.

The mammalian placenta is a profound adaptation, allowing complex signaling between mother and fetus. But in evolutionary terms, the placenta is fairly recent, probably arising only about 120 million years ago. Before the placenta, our ancestors still had a complex chemical signaling system between mother and offspring; one which is still critical to reproductive success today: milk. At birth and the breaking of the maternal-placental connection, eutherian mammal mothers continue to biochemically signal to their offspring through lactation. In the last section of this chapter we consider whether miRNA in milk may be part of that signaling.

7.3 Milk as a regulatory mechanism

Milk is the first food and for some length of time after birth the sole food for new-born mammals. As such, it must deliver all the necessary inputs for life, growth, and appropriate development to the mammalian neonate. These inputs consist of more than nutrients; milk is more than food. Milk is a biochemically complex, evolved delivery system for maternal signaling to guide and regulate mammalian offspring development (Power & Schulkin, 2013). We argue that the function of milk needs to be examined from the perspective of regulatory and developmental biology; to look beyond its nutritional importance and to begin to investigate milk for its regulatory mechanisms.

Multiple research findings have documented the importance of mother's milk in neonatal development. In several rodent cross fostering studies the phenotype of the pups appears to be more influenced by the phenotype/genotype of the nursing dam than by the genotype of the birth dam (Reifsnyder, Churchill, & Leiter, 2000; Gorski et al., 2006; Oben et al., 2010; Schroeder et al., 2010). For example, cross fostering mouse pups from lean dams to obese dams during lactation results in those pups becoming vulnerable to obesity and metabolic disease (Gorski et al., 2006; Oben et al., 2010). Cross fostering mouse pups that are genetically predisposed to obesity onto lean dams lowers the extent of obesity in the pups (Reifsnyder et al., 2000) and improves insulin sensitivity (Gorski et al., 2006). Obesity-prone and obesity-resistant rat pups displayed a phenotype consistent with the foster (nursing) dam's genotype prior to weaning rather than to their own genotype (Figure 7.3), although after weaning their food intake and weight gain more closely reflected their own genotype (Schroeder et al., 2010).

Marsupial lactation is arguably more complex than lactation in placental mammals. In marsupials, mother's milk supports growth and development in fetal-like neonates. In effect, marsupial milk accomplishes the same signaling and nurturing function as the placenta. Milk composition changes

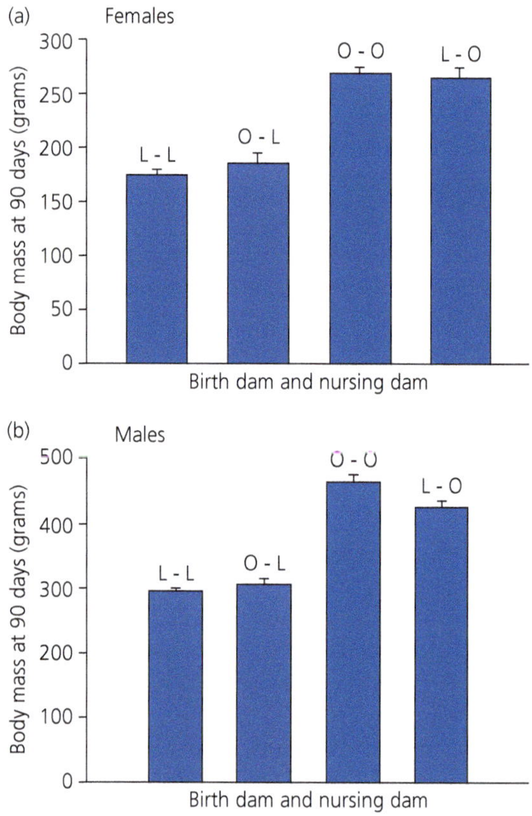

Figure 7.3 Body mass at 90 days for rat pups born to either a lean genotype dam (L) or an obese phenotype dam (O) and cross fostered to either a lean or obese genotype dam: L - L = lean-lean; O - L = obese-lean; O - O = obese-obese; L - O = lean-obese. Pup phenotype matched the nursing dam genotype, although O - O males were significantly larger than O - L male
Data from Schroeder et al., (2010).

greatly over lactation, to support the different metabolic and developmental needs of the young. A study in which tammar wallaby joeys were removed from nipples and replaced with younger joeys found that the mammary gland did not adjust to the age of the joey. In other words, the younger joeys received milk appropriate for an older joey, and not what they would have received if they had stayed on their own mother's nipple. Interestingly, the young joeys fed older joey milk grew and developed faster than controls left on their mothers (Trott et al., 2003). The fostered joeys that were 56 days younger than the joeys they replaced achieved a series of developmental markers at earlier ages than did controls. The appearance of whiskers, eyes opening, ability to stand, and having fur all over the body all occurred 35 days earlier in the fostered joeys. Fostered joeys could hop at an age 42 days younger compared to controls. By age 214 days fostered joeys had the appearance of a 242-day-old control joey. The fostered joeys also appeared fatter than controls, even at comparable developmental stages. Thus, joey age did not affect milk composition, but milk composition affected joey growth and development.

7.3.1 Are milk miRNAs a regulatory signal?

Although it is possible that these lactational effects are being mediated through nutrients and various signaling molecules in milk (e.g., hormones, growth factors and cytokines), these results strongly suggest an epigenetic potential of mother's milk; an actual change in neonatal gene expression due to ingested substances in milk. Might miRNAs in milk be part of that signaling mechanism?

Two main hypotheses regarding miRNA's role in milk exist: the functional hypothesis, maternal miRNA play a regulatory role in offspring as discussed above, and the nutritional hypothesis, miRNA provide only nutritional and not regulatory value (Melnik et al., 2016a). There is little direct evidence at present for either hypothesis, so our discussion will revolve around presenting necessary conditions for milk miRNA to be able to affect neonatal gene expression and to explore the evidence for and against these conditions holding. We ask more questions than provide answers, as what is known about miRNAs in milk is greatly outweighed by what has not yet been demonstrated. Our hope is to stimulate thought and research on the many facets of milk miRNA biology that are largely unexplored and not yet understood.

7.3.1.1 Milk miRNAs

Breast milk has been shown to contain significant concentrations of a wide variety of miRNAs. Recent studies have detected miRNA in human breast milk (Kosaka et al., 2010; Zhou et al., 2012), cow milk (Chen et al., 2010), pig milk (Gu et al., 2012), yak milk (Bai et al., 2013) and rat milk (Izumi et al., 2014), among others. The expectation is that all mammalian milks contain significant concentrations of miRNA. Human breast milk contained the highest concentration of total RNA of 12 body fluids tested, almost three times higher than seminal fluid which had the second highest concentration, and more than 80 times the concentration found in amniotic fluid (Weber et al., 2010). The number of miRNAs with known function detected in breast milk ranged from 429 (Weber et al., 2010) to more than 600 (Zhou et al., 2012).

Many of the miRNAs found in milk have immune-related function, regulating T-cells and inducing B-cell differentiation, and thus have the potential to activate and potentiate immune function. Immune cell-related miRNAs were by far the most common in cow milk (Chen et al., 2010). Some miRNA appear to affect viral replication and have other anti- and pro-viral effects. Roughly one-third of the known human pre-miRNAs were detected in breast milk, with two-thirds of the well-characterized immune-related pre-miRNAs detected (Zhou et al., 2012). Other miRNAs detected in human breast milk regulate physiological and metabolic processes and development. The potential functions of many of the common miRNAs found in milk have been determined by sequence analysis, and for some by direct experimentation/observation. Many of the milk miRNAs of highest concentration are common across multiple species. For example, miRNA-21 is one of the most common miRNA found in human, cow, and rat milk (Munch et al., 2013; Chen et al., 2010; Izumi et al., 2014). miRNA-21 is also linked to regulation of growth and adipogenesis through both upstream and downstream regulation of mTORC-1 signaling (Melnik, John, & Schmitz, 2013).

A recent study looking at select miRNAs showed that maternal pre-pregnancy BMI was negatively correlated with levels of several miRNAs (i.e., miRNA-30B, let-7a, miRNA-378) in collected colostrum. In addition, the miRNA let-7a in mature milk was negatively correlated with maternal weight in late pregnancy (Xi et al., 2015). This data suggests that maternal characteristics like weight and adiposity may play a role in miRNA secretion by the mammary gland, but future studies are needed to further elucidate the effects on a broader selection of miRNAs.

Additionally, there is evidence that maternal health status will affect the pool of miRNAs in milk.

For instance, mothers suffering from certain infections while lactating may produce milk with an influenced selection of miRNAs. In a study of bovine milk, cows suffering from *Staphylococcus aureus* infection of the mammary gland produced milk with high levels of miRNAs involved in an immune response (bta-miR-142-5p and -223; Sun et al., 2015). In another study of ruminant animals, lactating goats were studied under conditions of food deprivation or ad libitum. Certain miRNAs were up or downregulated, suggesting that the maternal nutritional status will ultimately affect the miRNA makeup of milk (Mobuchon et al., 2015).

Milk miRNAs appear resistant to degradation. Subjecting milk to repeated freeze/thaw cycles, heat, acid, and RNase does not change the assayed concentrations of miRNAs (Izumi et al., 2012). This resistance to degradation is likely due to the miRNAs being contained within exosomes (Hata et al., 2010; Munagala et al., 2016), the tiny endosome-derived membrane vesicles that are released by cells into the extracellular environment (Zhou et al., 2012). Exosomes would serve to protect the miRNAs from the acidic stomach and digestive enzymes of the stomach and small intestine (Kosaka et al., 2010). Milk exosomes appear to derive predominantly from mammary cells, as their miRNA content differs from exosomes from maternal blood (Modepalli et al., 2014; Alsaweed et al., 2015).

Breast milk miRNA encapsulated in exosomes likely survive intact into the neonatal small intestine, where they could be taken up by lumen epithelium. Research has shown that exosomes can successfully be absorbed by nearby cells (through membrane fusion) (Lotvall & Valadi, 2007). Recent in vitro research on human colon carcinoma cells and rat intestinal cells has indicated the possibility of a glycoprotein-mediated endocytic mechanism in the intestinal transport of bovine milk exosomes (Wolf, Baier, & Zempleni, 2015). Based on their endogenous actions and their probable ability to survive the neonatal digestive processes, milk miRNAs may represent a regulatory mechanism allowing the maternal genome to influence development of her offspring by affecting post transcription gene expression. However, the evidence this actually occurs is scant at present, and a number of conditions need to be met before milk miRNAs could actually affect neonatal gene expression. We consider the evidence in the next section.

7.3.1.2 Mammary or neonate?

Do miRNAs in milk have function within the mammary gland, within the neonate, or in both? Our hypothesis is to presume both; but it may depend on which fraction of milk contains the miRNAs. Milk is not a homogeneous fluid, especially after it enters the neonate's digestive tract. The action of the enzyme chymosin in the stomach breaks down the casein molecules, creating curds, formed by clumping together of casein micelles. The whey, (i.e., more liquid milk fraction), will rapidly leave the stomach while the curds will remain behind, to experience fuller digestion in the stomach. Whether milk exosomes and miRNA associate more with the curds or with the whey affects the rapidity and concentration with which the exosomes will enter the small intestine. Exosomes in the whey will flow into the small intestine relatively rapidly and in a pulse dose. Exosomes trapped in the curds will be released into the small intestine gradually, implying both a longer duration event and lower concentrations at any given time.

The miRNA of mammary tissue has been studied in dairy cattle. A diverse panel of miRNAs may play a role in the development of the mammary gland, and consequently in milk production. Of 54 miRNAs differentially expressed in non-lactating mammary tissue from a dairy breed compared to a meat breed, 52 were downregulated in the dairy breed (Wicik et al., 2016). Some of those down-regulated miRNAs were also found to be downregulated in lactating bovine mammary glands compared to non-lactating glands (Li et al., 2012). For example, miRNA-23b-3p is downregulated in both the lactating bovine mammary and also in a dairy breed relative to a meat breed outside of lactation. This miRNA has been shown to reduce expression of TGF-β, a growth factor associated with reduced milk duct growth and development (Rogler et al., 2009; Li et al., 2012).

The evidence for miRNA function in regulating gene expression in the mammary gland to transform it from a quiescent organ into a milk-secreting organ is strong. However, that does not indicate that the miRNAs in milk necessarily are acting on the mammary gland. It is reasonable to hypothesize that exosomes in milk may be both secreted and absorbed by mammary epithelial cells. The potential for milk miRNAs to be acting on mammary tissue certainly exists. But it is also possible that milk miRNAs have little effect on the mammary gland. Women with milk stasis (i.e., still had milk within the mammary years after lactation ended) differed in their milk miRNA profile by breast cancer status. Women with breast cancer had high levels of oncogenic miRNAs and low levels of tumor suppressing miRNAs (Gu et al., 2014). The presence of these milk miRNAs appears to be a better indicator of cancer than miRNAs in blood, suggesting that milk miRNAs may play a role in carcinogenesis and in the mammary tumor microenvironment (Gu et al., 2014). However, those data cannot distinguish between different miRNA profiles in milk producing differing effects on breast tissue versus the miRNA profile of milk trapped in a duct simply reflects the miRNA expression patterns of the surrounding tissue. The miRNA profile of milk trapped in the mammary gland may be a good biomarker for breast cancer, but it is not clear whether those miRNAs are cause or effect.

Milk contains cells, both sloughed epithelial cells and various immune cells. The functions of the cells found in milk are not completely understood, however, the possibility that exosomes function in cell-to-cell communication between mammary cells and the immune cells in milk, as well as between the milk immune cells themselves, should be considered. The high concentration of miRNA in milk may reflect an active (and interactive) immune system within the mammary gland and milk, which may affect both maternal and neonatal health.

7.3.2 Are milk miRNAs absorbed into offspring circulation?

The evidence that milk exosomes can be absorbed and enter neonatal circulation is not strong and mostly indirect. A study in adult humans showed that ingestion of cow milk resulted in a 2.5-fold increase in circulating miRNA-29b, one of the most common miRNA in cow milk (Baier et al., 2014). Expression of runt-related transcription factor 2 (*RUNX2*) in blood mononuclear cells was increased after milk consumption, consistent with studies that demonstrate that miR-29b is a positive regulator of *RUNX2* (Li et al., 2009). Similarly, consumption of unboiled farm milk is associated with increased T cells and demethylation of FoxP3, which have the potential to be preventative of atopic diseases, such as asthma (Melnik et al., 2016b). These findings suggest that milk miRNAs are able to be absorbed into circulation when ingested and have effects on gene expression in humans.

A study of tammar wallabies found that milk miRNAs are absorbed by the neonate and enter blood circulation, but did not provide evidence for effects on neonate regulation (Modepalli et al., 2014). The miRNAs studied were overexpressed in the mother's milk compared to the mother's serum, showing strong support that the neonates were receiving the miRNAs from the milk and not producing them endogenously (Modepalli et al., 2014). The circulating concentration of 11 of 13 miRNAs examined in piglets was directly and significantly associated with the concentrations found in the sows' milk. Half of the piglets were kept with their birth sow and half were fostered onto another sow which had given birth 8 days prior. The piglets consuming colostrum had significantly higher circulating values for 10 of the miRNAs found in higher concentration in colostrum, and a significantly lower value for the one miRNA that was lower in colostrum compared to their siblings nursing on a sow with mature milk (Gu et al., 2012). In the other two cases the circulating values for the colostrum-feeding piglets were numerically greater, but not statistically different. These data are consistent with the absorption by the neonate of milk miRNAs in biologically relevant amounts.

In contrast, a recent study found evidence of extensive degradation of milk miRNA when it is incubated in the intestinal fluid of new-born mice (Title, Denzler, & Stoffel, 2015). This study also found little evidence of any absorption of two particular miRNAs by neonatal mice null for these RNA genes. However, upon further consideration, this may be due to the knockout model mice studied and the knockout's effect on the neonate's ability to conduct endocytosis successfully (Alsaweed et al., 2015; Melnik et al., 2016a). Still, the negative results from this study provide an important cautionary note.

7.4 Conclusions

MicroRNA likely play important roles in women's and infant health, including during pregnancy, cancer, and lactation. The role of miRNA in cancer is the most well understood, and currently is the closest to clinical application. Breast cancer is the most thoroughly studied for miRNA involvement. Challenges, including pinpointing the exact genetic drivers of each cancer, require further studies and trials before miRNA can be used in diagnosis and treatment. The use of miRNAs in therapeutic treatment of cancer shows some potential promise, but a serious concern is the potential for unintended consequences if the miRNA-therapy is not well-directed only to target cells.

Likewise, more research is needed to effectively use miRNAs in the diagnosis and treatment of pregnancy complications such as preeclampsia and preterm birth. Because of the difficulty to predict or diagnose these conditions early in pregnancy, the research on miRNA-based techniques presents a novel opportunity to develop early diagnosis and treatment tools. Fulfilling these research gaps would prove incredibly valuable in improving maternal and fetal health by preventing disability and loss of life from pregnancy complications.

The concentration of miRNAs in milk exceeds that in blood in humans, yet we do not understand their function. It is not clear whether the biological actions of milk miRNAs are primarily affecting the mammary gland, the neonate, or both. However, the potential importance of milk miRNAs as a maternally-derived regulatory mechanism guiding neonatal development supports the study of miRNAs in milk.

Cancer, pregnancy, and lactation represent important times of intense physiological change in women, which can be easily disrupted by abnormal miRNA regulation. The biggest challenge in the study and use of miRNA is the lack of standardization across studies (including in sample collection procedures, methods of analysis, and sample size), which is seen in all three focus areas of this chapter. A growing body of data exists, but it is far from conclusive. Larger and longer studies with standardized methodology would benefit current knowledge of miRNAs in pregnancy, cancer, and lactation.

Acknowledgments

This work was partially supported by the Health Resources and Services Administration (HRSA) of the US Department of Health and Human Services (HHS) under cooperative agreements UA6MC19010 and UA6MC31609: Pregnancy-Related Care Research Network. This information or content and conclusions are those of the author and should not be construed as the official position or policy of, nor should any endorsements be inferred by HRSA, HHS, or the US Government.

References

Alenquer, M. & Amorim, M. J. (2015). Exosome biogenesis, regulation, and function in viral infection. Viruses 7: 5066–83.

Alsaweed, M., Hartmann, P. E., Geddes, D. T., & Kakulas, F. (2015). MicroRNAs in breastmilk and the lactating breast: Potential immunoprotectors and developmental regulators for the infant and the mother. *International Journal of Environmental Research and Public Health*, 12: 13981–4020.

Alvarez-Erviti, L., Seow, Y., Yin, H., Betts, C., Lakhal, S., & Wood, M. J. (2011). Delivery of siRNA to the mouse brain by systemic injection of targeted exosomes. *Nature Biotechnology*, 29: 341–5.

Axtell, M. J., Westholm, J. O., & Lai, E. C. (2011). Vive la difference: Biogenesis and evolution of microRNAs in plants and animals. *Genome Biology*, 12: 1–13.

Bai, W. L., Yin, R. H., Yang, R. J., Khan, W. A., Ma, Z. J., Zhao, S. J., Jiang, W. Q., Wang, Z. Y., Zhu, Y. B., Luo, G. B., & Zhao, Z. H. (2013). Technical note: Identification of suitable normalizers for microRNA expression analysis in milk somatic cells of the yak (Bos grunniens). *Journal of Dairy Science*, 96: 4529–34.

Baier, S. R., Nguyen, C., Xie, F., Wood, J. R., & Zempleni, J. (2014). MicroRNAs are absorbed in biologically meaningful amounts from nutritionally relevant doses of cow milk and affect gene expression in peripheral blood mononuclear cells, HEK-293 kidney cell cultures, and mouse livers. *The Journal of Nutrition*, 144: 1495–500.

Barchitta, M., Maugeri, A., Quattrocchi, A., Agrifoglio, O., & Agodi, A. (2017). The role of miRNAs as biomarkers for pregnancy outcomes: A comprehensive review. *Hindawi International Journal of Genomics*, 2017: e8067972.

Cai, M., Kolluru, G. K., & Ahmed, A. (2017). Small molecule, big prospects: MicroRNA in pregnancy and its complications. *Hindawi Journal of Pregnancy*, 2017: e6972732.

Calin, G. A., Dumitru, C. D., Shimizu, M., Bichi, R., Zupo, S., Noch, E., Aldler, H., Rattan, S., Keating, M., Rai, K., Rassenti, L., Kipps, T., Negrini, M., Bullrich, F., & Croce, C. M. (2002). Frequent deletions and down-regulation of micro-RNA genes miR15 and miR16 at 13q14 in chronic lymphocytic leukemia. *PNAS*, 99(24): 15524–9.

Chen, T., Xi, Q. Y., Ye, R. S., Cheng, X., Qi, Q. E., Wang, S. B., Shu, G., Wang, L. N., Zhu, X. T., Jiang, Q. Y., & Zhang, Y. L. (2014). Exploration of microRNAs in porcine milk exosomes. *BMC Genomics*, 15: 100–19.

Chen, X., Gao, C., Li, H., Huang, L., Sun, Q., Dong, Y., Tian, C., Gao, S., Dong, H., Guan, D., Hu, X., Zhao, S., Li, L., Zhu, L., Yan, Q., Zhang, J., Zen, K., & Zhang, C. (2010). Identification and characterization of microRNAs in raw milk during different periods of lactation, commercial fluid, and powdered milk products. *Cell Research*, 20: 1128–37.

Cheng, G. (2015). Circulating miRNAs: Roles in cancer diagnosis, prognosis and therapy. *Advanced Drug Delivery Reviews*, 81: 75–93.

Di Leva, G., Garofalo, M., & Croce, C. M. (2014). MicroRNAs in Cancer. *Annual Review of Pathology: Mechanisms of Disease*, 9: 287–314.

Dueck, A., Ziegler, C., Eichner, A., Berezikov, E., & Meister, G. (2012). microRNAs associated with the different human argonaute proteins. *Nucleic Acids Research*, 40: 9850–62.

Elias, K. M., Fendler, W., Stawiski, K., Fiascone, S. J., Vitonis, A. F., Berkowitz, R. S., Frendl, G., Konstantinopoulos, P., Crum, C. P., Kedzierska, M., Cramer, D. W., & Chowdhury, D. (2017). Diagnostic potential for a serum miRNA neural network for detection of ovarian cancer. *ELIFE*, 6: e28932.

Fallen, S., Baxter, D., Wu, X., Kim, T., Shynlova, O., Lee, M. Y., Scherler, K., Lye, S., Hood, L., & Wang, K. (2018). Extracellular vesicle RNAs reflect placenta dysfunction and are a biomarker source for preterm labour. *Journal of Cellular and Molecular Medicine*, 22: 2760–73.

Friedman, R. C., Farh, K. K. H., Burge, C. B., & Bartel, D. P. (2009). Most mammalian mRNAs are conserved targets of microRNAs. *Genome Research*, 19: 92–105.

Gorski, J. N., Dunn-Meynell, A. A., Hartman, T. G., & Levin, B. E. (2006). Postnatal environment overrides genetic and prenatal factors influencing offspring obesity and insulin resistance. *American Journal of Physiology-Regulatory, Integrative and Comparative Physiology*, 291: R768–R778.

Gray, C., McCowan, L. M., Patel, R., Taylor, R. S., & Vickers, M. H. (2017). Maternal plasma miRNAs as biomarkers during mid-pregnancy to predict later spontaneous preterm birth: A pilot study. *Scientific Reports*, 7: 815.

Gu, Y. Q., Gong, G., Xu, Z. L., Wang, L. Y., Fang, M. L., Zhou, H., Xing, H., Wang, K. R., & Sun, L. (2014). miRNA profiling reveals a potential role of milk stasis in breast carcinogenesis. *International Journal of Molecular Medicine*, 33: 1243–9.

Gu, Y. R., Li, M., Wang, T., Liang, Y., Zhong, Z., Wang, X., Zhou, Q., Chen, L., Lang, Q., He, Z., Chen, X., Gong, J., Gao, X., Li, X., & Lv, X. (2012). Lactation-related microRNA expression profiles of porcine breast milk exosomes. *PLoS One*, 7: e43691.

Ha, M. & Kim, V. N. (2014). Regulation of microRNA biogenesis. *Nature Reviews Molecular Cell Biology*, 15: 509–24.

Hata, T., Murakami, K., Nakatani, H., Yamamoto, Y., Matsuda, T., & Aoki, N. (2010). Isolation of bovine milk-derived microvesicles carrying mRNAs and microRNAs. *Biochemical and Biophysical Research Communications*, 396: 528–33.

Hayes, J., Peruzzi, P. P., & Lawler, S. (2014). MicroRNAs in cancer: Biomarkers, functions and therapy. *Trends in Molecular Medicine*, 20: 460–9.

Hromadnikova, I., Kotlabova, K., Ivankova, K., & Krofta, L. (2017). First trimester screening of circulating C19MC microRNAs and the evaluation of their potential to predict the onset of preeclampsia and IUGR. *PLoS ONE* 12(2): e0171756.

Iorio, M. V. & Croce, C. M. (2016). Commentary on microRNA fingerprint in human epithelial ovarian cancer. *Cancer Research*, 76(21): 6143–5.

Iorio, M. V., Visone, R., Di Leva, G., Donati, V., Petrocca, F., Casalini, P., Taccioli, C., Volinia, S., Liu, C., Alder, H., Calin, G. A., Ménard, S., & Croce, C. M. (2007). MicroRNA signatures in human ovarian cancer. *Cancer Research*, 67(18): 8699–707.

Izumi, H., Kosaka, N., Shimizu, T., Sekine, K., Ochiya, T., & Takase, M. (2012). Bovine milk contains microRNA and messenger RNA that are stable under degradative conditions. *Journal of Dairy Science*, 95: 4831–41.

Izumi, H., Kosaka, N., Shimizu, T., Sekine, K., Ochiya, T., & Takase, M. (2014). Time-dependent expression profiles of microRNAs and mRNAs in rat milk whey. *PloS One*, 9: e88843.

Kayani, M.u. R., Kayani, M. A., Malik, F. A., & Faryal, R. (2011). Role of miRNAs in breast cancer. *Asian Pacific Journal of Cancer Prevention*, 12: 3175–80.

Kellis, M., Wold, B., Snyder, M. P., Bernstein, B. E., Kundaje, A., Marinov, G. K., Ward, L. D., Birney, E., Crawford, G. E., Dekker, J., Dunham, I., Elnitski, L. L., Farnham, P. J., Feingold, E. A., Gerstein, M., Giddings, M. C., Gilbert, D. M., Gingeras, T. R., Green, E. D.,

Guigo, R., Hubbard, T., Kent, J., Lieb, J. D., Myers, R. M., Pazin, M. J., Ren, B., Stamatoyannopoulos, J. A., Weng, Z., White, K. P., & Hardison, R. C. (2014). Defining functional DNA elements in the human genome. *PNAS*, 111(17): 6131–8.

Kosaka, N., Izumi, H., Sekine, K., & Ochiya, T. (2010). microRNA as a new immune-regulatory agent in breast milk. *Silence*, 1: 7.

Ladewig, E., Okamura, K., Flynt, A. S., Westholm, J. O., & Lai, E. C. (2012). Discovery of hundreds of mirtrons in mouse and human small RNA data. *Genome Research*, 22: 1634–45.

Lau, N. C., Lim, L. P., Weinstein, E. G., & Bartel, D. P. (2001). An abundant class of tiny RNAs with probable regulatory roles in Caenorhabditis elegans. *Science*, 294: 858–62.

Lee, R. C. & Ambros, V. (2001). An extensive class of small RNAs in Caenorhabditis elegans. *Science*, 294: 862–4.

Lee, R. C., Feinbaum, R. L., & Ambros, V. (1993). The C. elegans heterochronic gene lin-4 encodes small RNAs with antisense complementarity to lin-14. *Cell*, 75: 843–54.

Li, Z., Hassan, M. Q., Jafferji, M., Aqeilan, R. I., Garzon, R., Croce, C. M., van Wijen, A. J., Stein, J. L., Stein, G. S., & Lian, J. B. (2009). Biological functions of miR-29b contribute to positive regulation of osteoblast differentiation. *Journal of Biological Chemistry*, 284: 15676–84.

Li, Z., Liu, H., Jin, X., Lo, L., & Liu, J. (2012). Expression profiles of microRNAs from lactating and non-lactating bovine mammary glands and identification of miRNA related to lactation. *BMC Genomics*, 13: 1–15.

Lotvall, J. & Valadi, H. (2007). Cell to cell signalling via exosomes through esRNA. *Cell Adhesion & Migration*, 1: 156–8.

Maccani, M. A., Avissar-Whiting, M., Banister, C. E., McGonnigal, B., Padbury, J. F., & Marsit, C. J. (2010). Maternal cigarette smoking during pregnancy is associated with downregulation of miR-16, miR-21 and miR-146a in the placenta. *Epigenetics*, 5(7): 583–9.

Melnik, B. C., John, S. M., & Schmitz, G. (2013). Milk is not just food but most likely a genetic transfection system activating mTORC1 signaling for postnatal growth. *Nutrition Journal*, 12: 103–13.

Melnik, B.C., Foteini, K., Geddes, D. T., Hartmann, P. E., John, S. M., Carrera-Bastos, P., Cordain, L., & Schmitz, G. (2016a). Milk miRNAs: Simple nutrients or systemic functional regulators? *Nutrition & Metabolism*, 13: 42.

Melnik, B. C., John, S. M., Carrera-Bastos, P., & Schmitz, G. (2016b). Milk: A postnatal imprinting system stabilizing FoxP3 expression and regulatory T cell differentiation. *Clinical and Translational Allergy*, 6: 18.

Mobuchon, L., Marthey, S., Le Guillou, S., Laloë, D., Le Provost, F., & Leroux, C. (2015). Food deprivation affects the miRNome in the lactating goat mammary gland. *PloS One*, 10: e0140111.

Modepalli, V., Kumar, A., Hinds, L. A., Sharp, J. A., Nicholas, K. R., & Lefevre, C. (2014). Differential temporal expression of milk miRNA during the lactation cycle of the marsupial tammar wallaby (Macropus eugenii). *BMC Genomics*, 15: 1–17.

Mulrane, L., McGee, S. F., Gallagher, W. M., & O'Connor, D. P. (2013). miRNA dysregulation in breast cancer. *Cancer Research*, 73(22): 6554–62.

Munagala, R., Aqil, F., Jeyabalan, J., & Gupta, R. C. (2016). Bovine milk-derived exosomes for drug delivery. *Cancer Letters*, 371: 48–61.

Munch, E. M., Harris, R. A., Mohammad, M., Benham, A. L., Pejerrey, S. M., Showalter, L., Hu, M., Shope, C. D., Maningat, P. D., Gunaratne, P. H., Haymond, M., & Aagaard, K. (2013). Transcriptome profiling of microRNA by Next-Gen deep sequencing reveals known and novel miRNA species in the lipid fraction of human breast milk. *PLoS One*, 8: e50564.

Oben, J. A., Mouralidarane, A., Samuelsson, A. M., Matthews, P. J., Morgan, M. L., Mckee, C., Soeda, J., Fernandez-Twinn, D. S., Martin-Gronert, M. S., Ozanne, S. E., Sigala, B., Novelli, M., Poston, L., & Taylor, P. D. (2010). Maternal obesity during pregnancy and lactation programs the development of offspring nonalcoholic fatty liver disease in mice. *Journal of Hepatology*, 52: 913–20.

Ohno, S. (1972). So much "junk" DNA in our genome. *Evolution of genetic systems, Brookhaven Sympoisa in Biology*, 23: 366–70.

Power, M. L. & Schulkin, J. (2013). Maternal regulation of offspring development in mammals is an ancient adaptation tied to lactation. *Applied & Translational Genomics*, 2: 55–63.

Reifsnyder, P. C., Churchill, G., & Leiter, E. H. (2000). Maternal environment and genotype interact to establish diabesity in mice. *Genome Research*, 10: 1568–78.

Reinhart, B. J., Slack, F. J., Basson, M., Pasquinelli, A. E., Bettinger, J. C., Rougvie, A. E., Horvitz, H. R., & Ruvkun, G. (2000). The 21-nucleotide let-7 RNA regulates developmental timing in Caenorhabditis elegans. *Nature*, 403: 901–6.

Roberts, J. M., Hansson, S. R., Vaiman, D., Redman, C. W. G., & Global Pregnancy Collaboration. (2017). Global Pregnancy Collaboration symposium on placental health: Summary and recommendations. *Placenta*, 52: 116–21.

Rogler, C. E., LeVoci, L., Ader, T., Massimi, A., Tchaikovskaya, T., Norel, R., & Rogler, L. E. (2009). MicroRNA-23b cluster microRNAs regulate transforming growth factor-beta/bone morphogenetic protein signaling and liver stem cell differentiation by targeting Smads. *Hepatology*, 50: 575–84.

Salomon C., Kobayashi M., Ashman K., Sobrevia L., Mitchell M. D., Rice G. E. (2013). Hypoxi-induced changes in bioactivity of cytotrophoblast-derived exosomes. *PLoS One*, 8: e79636.

Salomon, C., Guanzon, D., Scholz-Romero, K., Longo, S., Correa, P., Illanes, S. E., & Rice, G. E. (2017). Placental exosomes as early biomarker of preeclampsia: Potential role of exosomal MicroRNAs Across Gestation. *Journal of Clinical Endocrinology & Metabolism*, 102: 3182–94.

Schroeder, M., Shbiro, L., Moran, T. H., & Weller, A. (2010). Maternal environmental contribution to adult sensitivity and resistance to obesity in Long Evans rats. *PLoS One*, 5: e13825.

Singh, K., Williams III, J., Brown, J., Wang, E. T., Lee, B., Gonzalez, T. L., Cui, J., Goodarzi, M. O., & Pisarska, M. D. (2017). Up-regulation of microRNA-202-3p in first trimester placenta of pregnancies destined to develop severe preeclampsia, a pilot study. *Pregnancy Hypertension: An International Journal of Women's Cardiovascular Health*, 10: 7–9.

Sun, J., Aswath, K., Schroeder, S. G., Lippolis, J. D., Reinhardt, T. A., & Sonstegard, T. S. (2015). MicroRNA expression profiles of bovine milk exosomes in response to Staphylococcus aureus infection. *BMC Genomics*, 16: 806–16.

Title, A., Denzler, R., & Stoffel, M. (2015). Uptake and function studies of maternal milk-derived microRNAs. *Journal of Biological Chemistry*, 290: 23680–91.

Trott, J. F., Simpson, K. J., Moyle, R. L., Hearn, C. M., Shaw, G., Nicholas, K. R., & Renfree, M. B. (2003). Maternal regulation of milk composition, milk production, and pouch young development during lactation in the tammar wallaby (Macropus eugenii). *Biology of Reproduction*, 68: 929–36.

Valadi, H., Ekström, K., Bossios, A., Sjöstrand, M., Lee, J. J. & Lötvall, J. O. (2007). Exosome-mediated transfer of mRNAs and microRNAs is a novel mechanism of genetic exchange between cells. *Nature Cell Biology*, 9: 654–9.

Weber, J. A., Baxter, D. H., Zhang, S., Huang, D. Y., Huang, K. H., Lee, M. J., Galas, D. J., & Wang, K. (2010). The microRNA spectrum in 12 body fluids. *Clinical Chemistry*, 56: 1733–41.

Wen, J., Ladewig, E., Shenker, S., Mohammed, J., & Lai, E. C. (2015). Analysis of nearly one thousand mammalian mirtrons reveals novel features of dicer substrates. *PLoS Computational Biology*, 11: e1004441.

Westholm, J. O. & Lai, E. C. (2011). Mirtrons: microRNA biogenesis via splicing. *Biochimie*, 93: 1897–904.

Wicik, Z., Gajewska, M., Majewska, A., Walkiewicz, D., Osińska, E., & Motyl, T. (2016). Characterization of microRNA profile in mammary tissue of dairy and beef breed heifers. *Journal of Animal Breeding and Genetics*, 133: 31–42.

Wightman, B., Ha, I., & Ruvkin, G. (1993). Posttranscriptional regulation of the heterochronic gene lin-14 by lin-4 mediates temporal pattern formation in C. elegans. *Cell*, 75: 855–62.

Wolf, T., Baier, S. R., & Zempleni, J. (2015). The intestinal transport of bovine milk exosomes is mediated by endocytosis in human colon carcinoma Caco-2 cells and rat small intestinal IEC-6 cells. *The Journal of Nutrition*, 145: 2201–6.

Wommack, J. C., Trzeciakowski, J. P., Miranda, R. C., Stowe, R. P., & Ruiz, R. J. (2018). Micro RNA clusters in maternal plasma are associated with preterm birth and infant outcomes. *PLoS ONE*, 13(6): e0199029.

Xi, Y., Jiang, X., Li, R., Chen, M., Song, W., & Li, X. (2015). The levels of human milk microRNAs and their association with maternal weight characteristics. *European Journal of Clinical Nutrition*, 70: 445–9.

Yuan, Z., Sun, X., Jiang, D., Ding, Y., Lu, Z., Gong, L., Liu, H., & Xie, J. (2010). Origin and evolution of a placental-specific microRNA family in the human genome. *BMC Evolutionary Biology*, 10: 346–58.

Zheng, H., Zhang, L., Zhao, Y., Yang, D., Song, F., Wen, Y., Hao, Q., Hu, Z., Zhang, W., & Chen, K. (2013). Plasma miRNAs as diagnostic and prognostic biomarkers for ovarian cancer. *PLoS ONE*, 8(11): e77853.

Zhou, Q., Li, M., Wang, X., Li, Q., Wang, T., Zhu, Q., Zhou, X., Wang, X., Gao, X., & Li, X. (2012). Immune-related microRNAs are abundant in breast milk exosomes. *International Journal of Biological Science*, 8: 118–23.

CHAPTER 8

Evolutionary Medicine Viewed Through the Lens of Pregnancy and the Obesity Epidemic

Chloe Zera and Louise Wilkins-Haug

8.1 Introduction

In this chapter we focus on evolution in medical education and the role of evolutionary medicine in understanding the current obesity epidemic. Evolutionary considerations are the oxygen for understanding pathophysiology, whether assessing changes over generations or as short as a decade. With limited consideration of evolution in medical education, understanding the mechanisms behind human variation and change are ill-informed. We chose obesity as an example because it so clearly ties together clinical medicine and evolution, with population-level change occurring over a relatively short timeframe.

8.2 Evolutionary medicine informs understanding of the obesity epidemic

8.2.1 Evolution

Evolution broadly denotes the adaptation of a species to its environment. The goal is not increased survival but rather successful reproduction and the ultimate increase of an advantageous genetic combination. Evolution brings attention to genetic makeup, the associated mechanisms of gene expression, and interaction with the environment. Variation either within the genetic makeup, the associated gene expression regulating pathways or the environment (either internal or external) contributes to evolutionary change. From a population perspective, gene sequence variation affords an opportunity for differential responses to environmental challenges. These varied responses can be in a negative or positive direction—those variations increasing reproductive fitness to alter the genetic makeup of the population. Such variations on a population level are more important than those which promote longevity, which manifest benefits

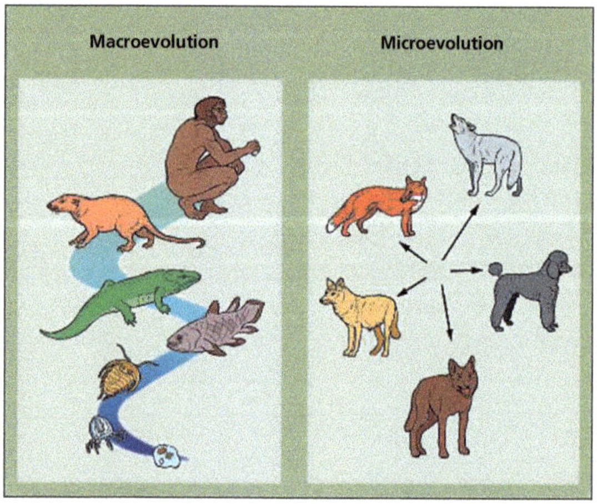

Figure 8.1 Pictorial representation of macro versus microevolution (open access)

Zera, C. and Wilkins-Haug, L., *Evolutionary Medicine Viewed Through the Lens of Pregnancy and the Obesity Epidemic* In: *Integrating Evolutionary Biology into Medical Education—for maternal and child healthcare students, clinicians, and scientists*. Edited by: Jay Schulkin and Michael L. Power, Oxford University Press (2020). © Oxford University Press.
DOI: 10.1093/oso/9780198814153.003.0009

past the optimal time for genomic transmission to the next generation.

With the completion of the human genome project over 10 years ago, the extent of variation detected within the human genome was unexpected. Those gene sequence variations that are clearly disease causing or benign are well recognized. However, many changes are rarely if ever noted in the general population. Such variation at the gene sequence level spurs a species adaptation. To change the frequency of a genetic variation typically requires multiple generations to occur, a macroevolution.

Microevolution, as compared to macroevolution, can occur within a few generations, (see Figure 8.1). Environmental interaction is key and change occurs primarily due to alterations in gene expression. Epigenetics refers to the ability to change gene expression without a change in the DNA sequence. The changes in gene expression are in part orchestrated by environmentally altered regulators of DNA translation and expression. Epigenetic mechanisms which can silence or promote a gene sequence include methylation, histone modification of the DNA coiling, and DNA packaging. Agents promoting epigenetic change are often linked to the environment external to the individual (toxins, nutrition, stress) with these external cues translated internally to the active agents altering the gene expression. Characteristic of epigenetic change is its ability to be heritable—transferred to the next generation. And importantly, many epigenetic changes are reversible (Mehler, 2008; Tang, McGowen & Reynolds, 2014; McGowan & Szyf 2010; McGowan, 2015; McGowan & Roth, 2015; McGowan & Matthews, 2018). Both epigenetics and classical genetics (DNA sequence change) are part of an individual's genome—the totality of their unique combination of DNA sequences and how such sequences, and their variations, interact with the environment.

Rapid changes over a few generations may also be an illustration of "co-evolution," defined as a population—genetic process, see Janzen (1980). Co-evolution is the term applied to the process in which our microbiome responds to the environment. The collective gene expression of microorganisms is an important component of the microbiome with connected but unique communities existing within specific organs, (see Figure 8.2). Our knowledge regarding the microbiome and its role in co-evolution has increased steadily over the past 10 years. The human body represents a well-developed, co-evolved microbial community with millions of cells comprising nearly 3% of total body weight. Advanced genetic sequencing technologies have delineated a wide range of organisms throughout the body (Davenport et al., 2017; Goodrich, Davenport, Clark, & Ley, 2017). A rapidly developing literature has demonstrated the importance of the microbiome in health and disease, as the bioactive molecular products of microbes appear to have physiologic impact on metabolism and immune function. Through co-evolution, changes in the environment are transmitted to offspring through passage of the community of microorganisms that coexist within an individual. We increasingly recognize the microbiome as an instrumental modulator of health and disease.

From an evolutionary perspective, while each individual's microbiome is unique, clustering of microbiota within populations and within species occurs. Western individuals' microbiome differs from that of non-Westerners' which has fewer overall species, and fewer species targeted to raw meat (Spor, Koren, & Ley, 2017; Goodrich et al., 2016). When compared to primates, the human gut has lower levels of microbiome diversity. Notably, the changes noted between populations within a species are similar to those seen when between species analyses are performed. This would suggest a highly versatile state able to accommodate changes in diet or changes in native resources (Davenport et al., 2017).

8.2.2 Evolutionary medicine

Evolutionary medicine theory emerged in the late twentieth century with concepts distinct from the earlier and often targeted Darwinian Theory which emphasized inherited variation and disease as a means to support social and political agenda such as eugenics (Gluckman & Bergstrom, 2011; Lamm & Jablonka, 2008). With a more balanced and analytic approach, evolutionary medicine was first introduced in writings by Williams and Nesse in 1991.

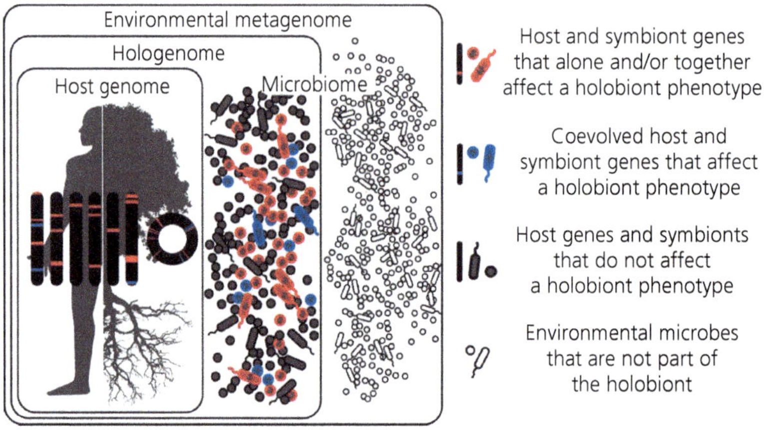

Figure 8.2 Holobionts are entities comprised of the host and its symbiotic microbes. Microbiomes include those co-evolved with the host (blue), those not coevolved (red), and those which do not affect the Holobiont's phenotype (gray). Microbes in the environment are not part of the Holobiont (white). Hologenomes then encompass the genomes of the host and all of its microbes at any given time point. Holobionts and Hologenomes are entities, whereas co-evolution or the evolution of host-symbiont interactions are processes
(Modified from *Theis*, 2016, open access)

Key to this new area of medical theory was the examination of diseases through an evolutionary perspective. Why do humans continue to succumb to diseases if evolutionary theory supports betterment of a species through reproductive fitness? Humans possess numerous features distinctive from other species such as longer life, reduced offspring numbers and most notably, unique and profound control of their environment (Gluckman & Bergstrom, 2011; Gluckman, Low, Buklijas, Hanson, & Beedle, 2011). Approaching questions of evolutionary medicine in the unique human population incorporates large, multigenerational cohorts as well as increasingly the precisions and availability of DNA molecular techniques. Detection of various mechanisms of epigenetics and recognition of the capacity for trans-generational passage of these changes further contributes to medical evolution theory (Lamm & Jablonka, 2008). An evolutionary medicine approach using such tools can then focus on the ultimate cause of disease—the why it occurs, as opposed to the proximate cause—how it occurs. Secondly, for the most part, genetic variation contributes to disease susceptibility and not necessarily the disease itself. And lastly, environments have changed dramatically overtime. These three principles frame evolutionary medicine as a study of variation rather than a state of health and disease (Gluckman & Bergstrom, 2011; Gluckman et al., 2011).

8.3 Evolutionary theory in modern medical education

Neither classic evolution nor evolutionary medicine are consistently part of medical school curricula. Neither of these principles should be considered outdated and are very much needed if the extensive variation of the human genome is understood at a pathophysiologic level. With fast and inexpensive DNA sequencing, the extensive variation across any individual's genome is now recognized as the norm. Some variants are classically pathogenic and consistently disease producing, while others can be pathogenic in one individual of a family yet benign in another relative. For many variants, their role in human physiology is unknown.

Inclusion of evolutionary theory into medical education has faced challenges. During the early twentieth century, the art of medicine transformed to the science of medical disciplines. The Flexner report of 1910 laid the foundation for science-based medical education (Pearlman, 2013). At this time evolutionary medicine was not a recognized concept, Mendelian inheritance was new and the composition of a gene unknown. Evolutionary biology emerged two to three decades later but at a time when many biologists wanted to maintain a separation from the concepts of eugenics. Efforts to incorporate evolutionary theory into medicine have often been met with resistance—no space in the curriculum, no skilled faculty, and a lack of understanding of how human beings are a part of evolution. An educational focus on the similarities between higher order primates often overshadowed closer assessment of the species differences. New concepts of evolution as explanatory for the continued existence of disease in man were met with skepticism. With time the concepts of medicine as directed toward the individual and evolution as focused on the population began to emerge as complimentary to one another, as in Pearlman, 2013. Perhaps the widest differences in the two beliefs centered on views of inherent human variation. Evolutionary medicine brings these disparate views of variation together—medicine's focus is on normal and abnormal with less recognition of variation at all levels while an evolutionary approach views variation as essential to a population's ability to respond their environment, see again Pearlman, 2013.

The formal introduction of evolutionary medicine, and integration of evolutionary concepts with the medical approach, occurred with the "Dawn of Darwinian Medicine" (Williams & Nesse, 1991). Reviews, commentary, and textbooks rapidly followed all further integrating the concepts of evolution and population theory into the medical treatment of an individual, (see Figure 8.3). Both fields are likely strengthened by shared knowledge and increased recognition of the individual as a part of a population (Stearns, Nesse, Govindaraju, & Ellison, 2010). Inclusion of evolutionary science as a core competency of premedical coursework has been supported by the American Association of Medical College since 2009 (Nesse et al., 2010).

In the twenty-first century, medical education is changing rapidly. In response to the ever-enlarging fire-hose of knowledge from which students are to drink prior to graduation, experts advocate for a transition to clinical case-based learning as well as techniques that foster independent learning such as the "flipped classroom" approach (Prober & Heath, 2010). The American Association of Medical

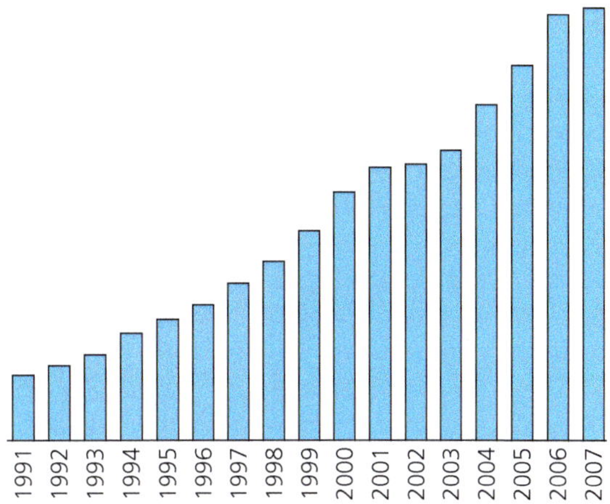

Figure 8.3 Increase in Pub Med citations for Evolutionary Medicine over 20 years (1991–2011)
(open source)

Colleges (AAMC) as well as the accrediting body of medical schools, the Liaison Committee on Medical Education (LCME) have documented major curriculum reform efforts at the majority of US medical schools since 2012. In the most recent LCME survey results, 98% of medical schools implementing change reported enhanced integration of basic science content with clinically relevant (organ-systems or case-based) learning, according to the Association of American Medical Colleges (2010).

While the AAMC has recommended evolution as a core competency for preclinical education, (AAMC, 2009), coverage of evolutionary principles in medical education remains incomplete. A recent survey of medical school deans suggested nearly half (48%) anticipated controversy if evolution were formally included in curricula (Hidaka, Asghar, & Aktipis, 2015) , emphasizing case studies as effective and well-received tools to integrate the principles of evolutionary medicine.

8.4 Obesity through the lens of evolutionary medicine

Obesity is a global epidemic, with rates increasing exponentially over a short period of time; we are in the midst of a historical shift in human health. Obesity affects nearly one third of US adults, but traditional medicine has had little to offer beyond admonitions of "eat less, move more" from physicians. However, the scientific understanding of obesity is now advancing far beyond simple caloric balance to incorporate evolutionary medicine theory. Evolutionary medicine provides a framework to understand and address the complex interplay between genetic predisposition to disease and the environment, from the cellular to the community level. In the remainder of this chapter, we look in depth at a case of a woman with obesity, focusing on her reproductive concerns. This will highlight the interconnected principles of evolutionary medicine and the contributors to rapid intergenerational change, including genetic/environmental mismatch, the influences of epigenetics and the microbiome.

Karen, a 30-year-old preschool teacher, is having her first baby. She struggled with obesity since childhood and sees a high-risk obstetrician because of the type 2 diabetes diagnosed as a teenager. This pregnancy was a surprise after she lost 20 lbs for her wedding. Although she is happy, she asks: "How will my weight affect the baby?"

Karen is not alone; this story plays out daily in the US, where an estimated 25% of women enter pregnancy with a body mass index in the "obese" category (Deputy, Dub, & Sharma, 2011). Her story also highlights the multigenerational impact of obesity, which has reached epidemic proportions globally, and the potential for pregnancy care practices to affect outcomes.

8.4.1 Weight trends over last 50 years

Obesity is best understood as an excess of adipose tissue. While controversy about the best measure of adiposity at a population level certainly exists, the broadly accepted measure of body mass index (BMI), or weight for height, is used by most authorities including the Centers for Disease Control to define obesity as a BMI greater than or equal to 30 kg/m^2. The increase in rates over time is dramatic: less than 100 years ago, fewer than 10% of adults were overweight or obese, while the most recent estimates are that approximately two out of three adults meet those criteria (Hales, Carroll, Fryar, & Ogden, 2015).

8.4.2 Obesity and evolution

A predisposition to obesity may have several benefits from an evolutionary perspective. For most of human history, food has been neither abundant nor energy dense in the way it is now. It is likely that our ancestors survived periods of relative famine interspersed with periods of abundance, and therefore the ability to store energy in the form of adiposity was beneficial. Furthermore, women with abundant energy stores would be able to support the energy requirements of pregnancy and lactation, while those without adequate energy stores would not be able to reproduce efficiently.

The link between maternal adiposity and ovulatory function is seen at the level of the hypothalamus, the brain center that regulates critical functions including energy expenditure. Gonadotropin releasing hormone (GnRH) produced by the hypothalamus controls the production of follicle stimulating hormone (FSH) and luteinizing hormone (LH), the hormones that then signal the ovaries to produce and release oocytes. Leptin produced by adipocytes is a key regulator of ovulation at the level of the hypothalamus, and in both mouse and human models, insufficient leptin stops ovulation through modulation of GnRH secretion. In the absence of normal maternal energy stores, the hypothalamus stops producing GnRH in a pulsatile fashion, leading to what is called "hypothalamic amenorrhea." As infertility caused by lack of energy would be detrimental to survival, it is likely that a predisposition to ample fat stores was a beneficial trait (Figure 8.4; Della Torre, Benedusi, Fontana, & Maggi, 2014).

Women with particularly low levels of adiposity, such as seen in anorexia nervosa or the female athlete triad, can suffer from hypothalamic amenorrhea and infertility. Conversely, women with obesity are at risk for oligoovulation due to polycystic ovarian syndrome (PCOS), a disorder characterized by high levels of insulin resistance, ovarian androgen production, and oligomenorrhea, and are also at risk for unexplained infertility (Broughton & Moley, 2017). Reproduction is thus closely related to maternal energy balance and therefore may also be particularly susceptible to modulation by gut microflora. Based on genome-wide association

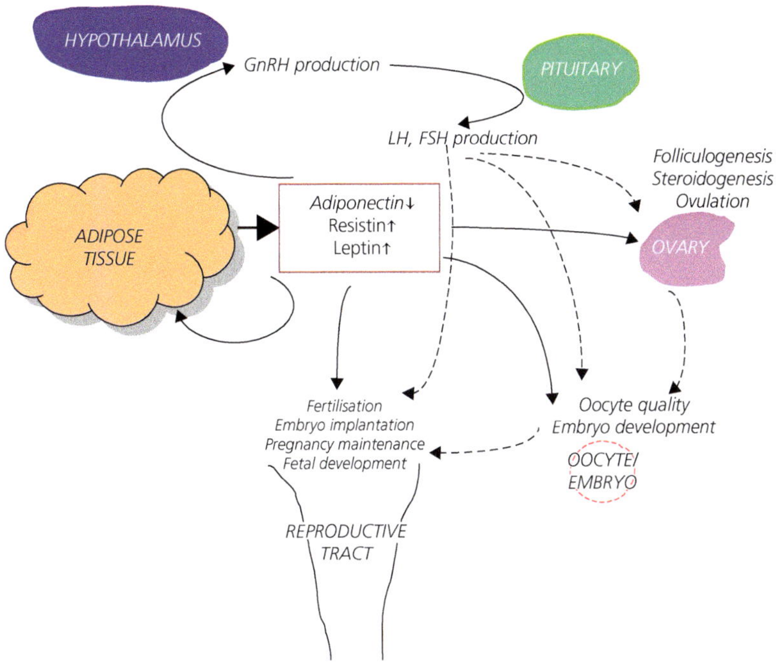

Figure 8.4 Interactions between adipose tissue, adipokines, and reproductive function (Mitchell M. (2005). *Reproduction*)

studies, while a genetic predisposition to a small amount of excess adiposity may predispose to reproductive success, the influence is considered minimal.

In the developed world, food is now both energy dense and abundant. Within one generation, immigrants from countries where traditional diets are prevalent develop obesity at rates that mirror the native-born population. The rapidity of this change is striking, suggesting that more than Mendelian genetics are at work. This rapid change also is an example of the evolutionary principle of "mismatch." Our ability to store energy in the form of adiposity is no longer an advantage, predisposing the majority of humans to the chronic disease of obesity.

At five months along, Karen is concerned her baby is too big—many people have commented that she looks to be at her due date. Her type 2 diabetes has been well controlled, but she has already gained more weight than she was intending to gain. She asks if there is anything she can do during the pregnancy or at delivery to prevent obesity in her child? She recently saw a story in the news about the importance of exposure to the maternal vaginal canal at delivery and asks: "Will a vaginal delivery prevent my baby from struggling with her weight?"

As the rates of maternal obesity have risen, obesity in children and adolescents reaches over 10% in some populations (Hales et al., 2017; Frederick, Snellman, & Putnam, 2014; Figure 8.5 a, b). Many have pointed to the role of large for gestational age (LGA) birthweight, defined as >90%ile for gestational age, as a driver of childhood obesity rates, as it can be difficult to achieve a normal weight after starting significantly above normal. However, this is likely only one driver of the epidemic. This increase in childhood obesity has significant consequences, including chronic diseases rarely before seen in children that threaten to curtail the life expectancy of this generation.

Infant weight is controlled by several factors, many of which are maternal. The risk for LGA is greatest in women who start pregnancy with a high BMI, who gain excessive weight during pregnancy and have glucose intolerance during pregnancy (Kim, Sharma, Sappenfield, Wilson, & Salihu, 2014). The placenta regulates fuel transport to the fetus

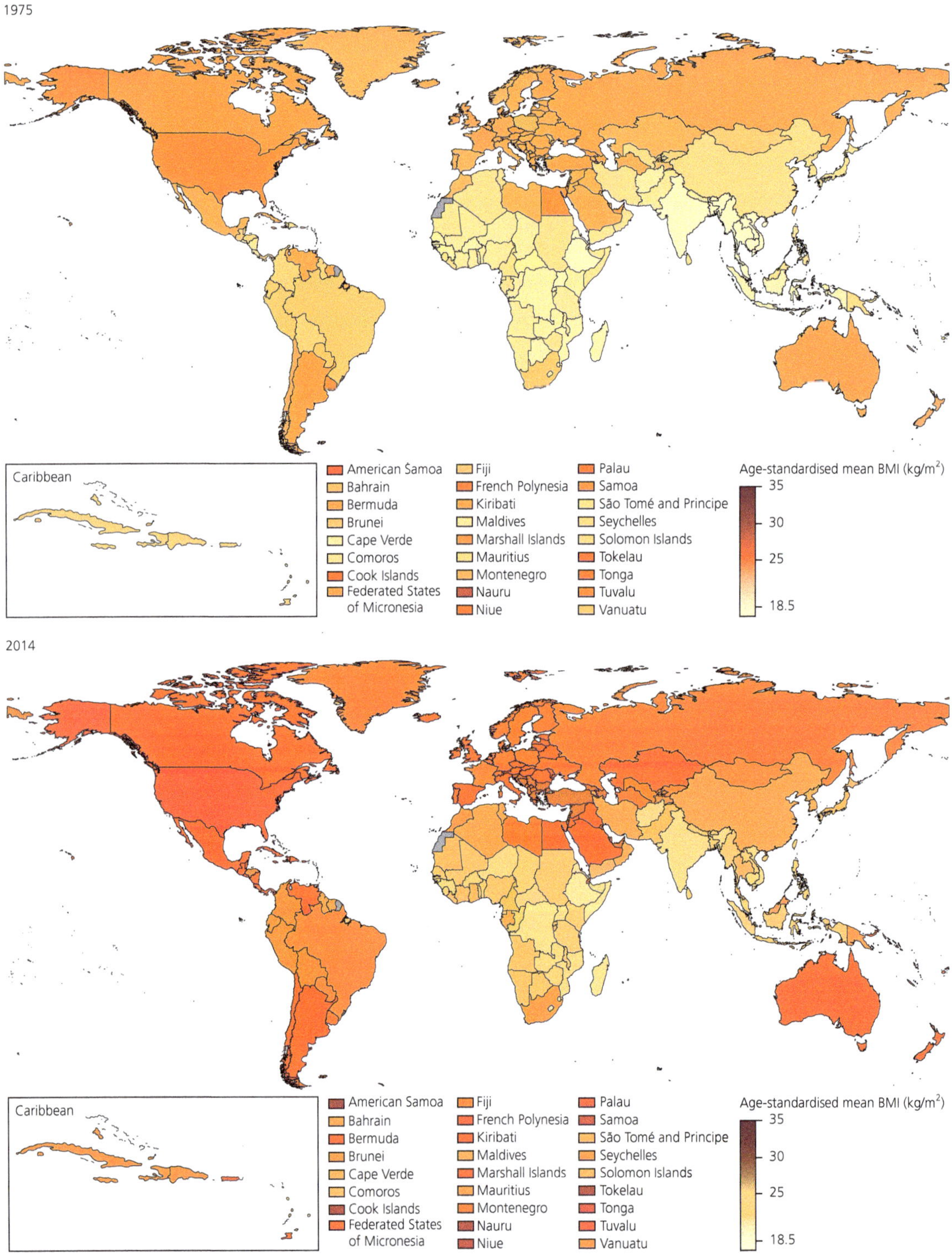

Figure 8.5a,b A) Percent of world population with obesity and the changes in 25 years (1980–2015) (open source)

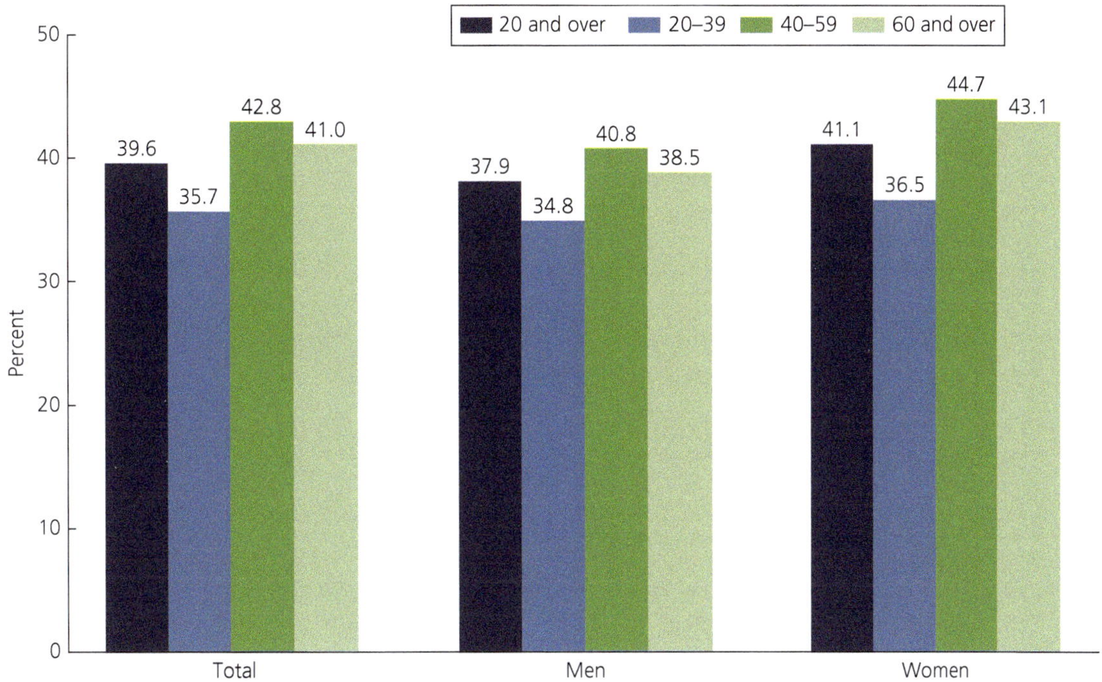

B) Prevalence of obesity among youth age 2 to 19 years, by sex, race, and Hispanic origin; United States 2015–2016 (NCHS, open source)

and therefore fetal growth. Key fetal fuel sources include glucose, which crosses the placenta through facilitated diffusion, and free fatty acids, which cross the placenta via transport proteins. Glucose and free fatty acids are primarily driven by the amount and content of dietary intake but are also high in women with peripheral insulin resistance such as seen in women with obesity.

The role of epigenetics on fetal development has substantial epidemiological support. Epigenetics is the ability of the environment to alter gene expression without altering the gene sequence itself with the potential to elicit change in one to two generations. Epigenetic control of fetal programing with a secondary effect on childhood/adult disease was initially brought forward by Barker in the 1980s (Barker, Eriksson, Forsén, & Osmond, 2002). His analysis of adults who had been born following a period of nationwide starvation found those with low birth weight in later life had higher rates of cardiovascular disease (Barker, 1997). These associations are becoming known as Developmental Origins of Health and Disease (DOaHD). Epidemiologic support further led Wells to postulate epigenetics as the pathway for the intergenerational transmission of obesity-related conditions as a reflection of social inequalities, (Wells, 2003; Wells, 2007; Wells, 2009; Wells, 2011). Through epigenetics, maternal exposures program the fetus for health inequalities. Maternal and social "capital" instruct the fetus for positive but also negative responses to their eventual environments. Most notably, these environments he characterizes as "metabolic deserts" (Wells, 2010). "Metabolic ghettos" increase marginalization of the individuals with the least positive responses to their environment as well as those with least supportive environments (Figure 8.6). These concepts continue with the work of Kuzawa with descriptions of how the pregnancy environment works with the plasticity of the fetus to program adult onset disorders (Kuzawa, 2005; Kuzawa & Sweet, 2009). In essence, the fetal environment responds by epigenetic programming using numerous alterations to gene expression to anticipate or set the groundwork for the anticipated postnatal environment. This translates to a role in fetal

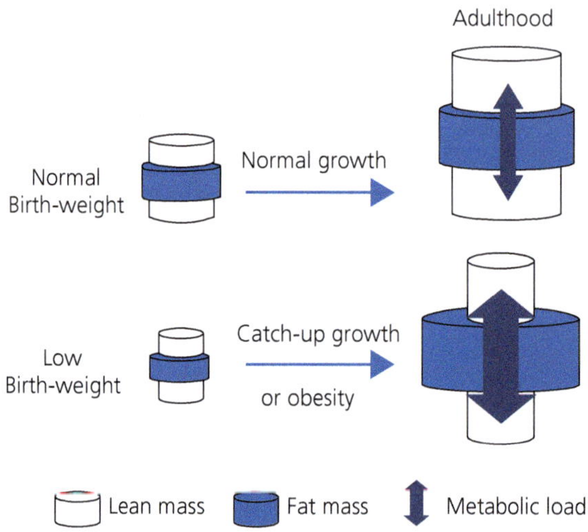

Figure 8.6 Conceptual diagram illustrating life history trajectories and their associated risk of Non-Communicable Diseases (NCDs). Each trajectory demonstrates different trade-offs, shaping both metabolic capacity and load (traits relevant to the life-course emergence of NCD risk) and the somatic traits that underlie partitioning of maternal investment between pregnancy and lactation

o: high maternal investment in pregnancy favors energy allocation to growth/maintenance, promoting longevity and health.

*: low maternal investment in pregnancy is followed by post-natal energy constraint, preventing catch-up. This leads to small adult size but low NCD risk.

●: low maternal investment in pregnancy is followed by catch-up during lactation, continuing into childhood. The extra energy accelerates maturation and increases adiposity without benefitting adult size, thus increasing NCD risk. Each of these trajectories produces in adult women the traits that favor the same trajectory in the next generation

Reproduced with permission from Wells (2018a).

determination of its eventual adult functioning. These epigenetic marks are based between generations though reflect past in utero environments. "Intergenerational inertia" may dampen a population's ability to respond to new, rapid change in their environment (Kuzawa, 2007). These thoughts were extended with a comparison with classic genetic theory of a "bottleneck"—not as a mechanism forcing selection of a subgroup of genes but a perpetual increase of the particular group in later generations (Kuzawa, 2007).

8.4.3 Obesity, energy balance, and reproduction

Karen describes her family as "big people" and feels like she has always been heavier than she wanted to be. She weighed 8 lbs when she was born a couple of weeks before her due date after her mother had her labor induced due to high blood pressure. Her mother was slightly overweight before her pregnancy, during which she gained 50 lbs. Since her diabetes diagnosis at age 16, Karen and her mom have dieted together multiple times, usually losing 10–15 lbs and then slowly regaining it; over time Karen's weight has crept up to 235 lbs, which at 5'4" makes her BMI 40 kg/m^2. She asks: "Is my child destined to struggle with her weight no matter what I do?"

We return to Karen's story to see several common themes emerge: first, that obesity is intergenerational, with roots in pregnancy that carry forward into childhood. Secondly, obesity affects all aspects of childhood including not just chronic disease risk, but relationships with family members. Finally, Karen's early experiences with dieting are not unusual. The difficulty in successfully treating obesity has physiologic roots that current science is only beginning to understand.

Body weight is set in a narrow range and regulated by a complex interplay of neuroendocrine hormones in the gut and brain. Weight loss triggers mechanisms to preserve body mass, including hunger, decreased resting energy expenditure, and increasing energy storage. Population level data demonstrates this phenomenon clearly; it is a consistent finding that even a small amount of weight loss is followed by gradual gain, often to a slightly higher weight than at baseline (Zenténius, Andersson-Assarsson, Carlsson, Svensson, & Larsson, 2018; Sjöström et al., 2004). Similarly, weight gain triggers satiety (through multiple hormones) and increased resting energy expenditure. These strategies "defend" body mass through times of relative deprivation. An increase in the body weight set point over time results in obesity. Several mechanisms are theorized to impact the set point including sleep disruption, stress, excessive weight gain at critical developmental periods (such as adolescence and pregnancy), and exposure to gut microbiota.

The gut and brain work together to regulate the body weight set point through a complicated system of hormonal regulation. Eating triggers release of hormones from the stomach, pancreas, and small intestine including cholecystokinin, amylin, glucagon-like peptide 1 (GLP-1) and peptide YY. These gut hormones act on neurons locally to connect immediate food ingestion to central signals that regulate continued energy intake both consciously and unconsciously. These gut signals can be modified through multiple mechanisms including exposure to microorganisms (Torres-Fuentesm Schellekens, Dinan, & Cryan, 2017).

An individual's microbiome, and its unique variations which respond to the environment, is best appreciated through an understanding of medical evolution theory. Ongoing major research efforts such as the NIH's Human Microbiome Project, (2018) seek to fully characterize a "healthy" microbiome and broaden our understanding of mechanisms of disease influenced by microbial communities. Technologies including metagenomic shotgun sequencing provide detailed descriptions from multiple sources including the oral, nasopharyngeal, intestinal, and urogenital tracts. However, the gut microbiome remains the most well-characterized, with a clear relationship to obesity.

The digestive tract has long been known to harbor large microbial communities. Originally, microbes were thought to be useful in digestion, in particular of fibrous foods. Recent advances have moved our understanding beyond digestive health toward an essential role in maintenance of host health including maturation of gut epithelium, immune function, and regulation of body weight. Microbes facilitate maturation of the intestinal epithelium in neonates. Human milk promotes the development of normal gut microbiota. Abnormal gut microbiota are associated with necrotizing enterocolitis in preterm infants with human milk having a lower risk than formula. In adults, the intestinal microbiome contributes to immune function by resisting colonization by pathogenic microbes; as an example, infection by *Clostridium dificile* is associated with loss of normal gut flora after recent antibiotic treatment. A normal fecal microbiome is associated with resistance to *C. dificile* colonization, and a growing treatment for *C. dificile* is fecal transplant.

Obesity is associated with alterations in the microbiome. In animal models, intestinal flora act as a vector by which obesity can be transmitted between individuals. In humans, disruptions in the proportion of firmicutes species are demonstrable in people with obesity. Furthermore, viral infections that disrupt normal intestinal flora may influence risk for obesity. Infection with adenovirus 36 (Adv36) produces obesity in animal models and is associated with risk for obesity within human populations, with children at particularly high risk (Ponterio & Gnessi, 2015). These examples represent our initial and growing understanding of the co-evolution of obesity and microbiomes as a function of individual response to their environment. The concepts of co-evolution of a specific microbiome and obesity become integral to the overarching theory of medical evolutionary medicine. One's ability to respond to their environment, and in particular a changing environment, is an orchestrated relationship at multiple levels.

Despite her best efforts to control her diabetes, Karen's baby is estimated on ultrasound to weigh nearly 9 lbs at 36 weeks (larger than 99% of fetuses at this gestational age). She is worried about her delivery, and after discussion with her obstetrician, she is planning a c-section. Although she is settled in her decision, she is worried about things she has read online suggesting that a cesarean section may increase the risk for her baby to develop obesity in later life. She has heard about something called "vaginal seeding" and wants to know if this is a common practice. She asks: "If I have a c-section, is there anything you can do to reduce my baby's risk of obesity in childhood?"

Although the gut microbiome is currently the best studied, the vaginal, placental, and milk microbiomes are also important contributors to human health including normal reproductive function and infant outcomes. Tissues previously thought to be sterile, including mammary glands and the placenta, may support microbial communities. Many of these communities are interconnected and can affect each other. For example, maternal oral and placental microbiota share many common organisms, potentially through hematogenous transmission. Furthermore, the discovery of both genital and milk microbiota supports their potential manipulation to impact the risk of fetal disease.

While the oral and lower gastrointestinal microbiota are diverse and vary between individuals, "normal" vaginal flora are relatively simple and consistent in comparison. Manipulation of these microbiomes to improve pregnancy and fetal outcomes is being investigated with antibiotics, probiotics and other means. Currently, the ability to alter the microbiome associated with obesity and reflect changes in either the mother's health or that of her infant are speculative but interesting given the concepts of co-evolution and medical evolutionary theory.

8.4.3.1 The vaginal microbiome

The vaginal microbiome was first investigated more than 100 years ago with the publication of Albert Döderlein's 1892 monograph *Das Scheidensekret*. Döderlein correctly hypothesized that the Gram-poitive, rod-shaped bacteria of the *Lactobacillus* species were the source of the lactic acid in the vagina and crucial to normal vaginal epithelial function through maintenance of healthy (acidic) vaginal pH and immunity to non-commensal microbes. The vaginal microbiome is generally dominated by *Lactobacillus* species with small numbers of other microbes though inconsistent across populations of healthy women. The vaginal microbiome varies by race and ethnicity, and even within a given woman over time, particularly by hormonal status (non-pregnant, pregnant, lactating, or postmenopausal).

Disruption of the normal vaginal flora is reflected by the common condition bacterial vaginosis (BV), an overgrowth of abnormal bacteria increasing the vaginal pH. The diagnosis is confirmed by characteristic "clue cells" (epithelial cells coated in bacteria) on microscopy, vaginal pH >4.5, and/or release of a fishy odor after 10% potassium hydroxide is added to a sample of vaginal secretions. Prevalence estimates range from 6.1% to 51.4% of women (Kenyon, Colebunders, & Crucitti, 2013) and seem to vary by race and ethnicity. BV is associated with an increased susceptibility to sexually transmitted infections, pelvic inflammatory disease (Sharma, Tal, Calrk, & Segars, 2014), premature birth, and miscarriage.

One challenge in directing therapy by altering the vaginal microbiome to improve pregnancy outcome is identifying differences between normal variations from BV. In asymptomatic women, a significant minority have vaginal microbiomes with few or no *Lactobacillus* species, characteristic of BV (Ravel et al., 2011; Anahtar et al., 2015). However, in these women, less common lactic acid-producing species (e.g., *Atopobium* and *Megasphaera*), result in a similar pH and microbiome to women with a *Lactobacillus*-predominant microbiome (Ma et al., 2012; Lamont et al., 2011). Individual-level variation in the vaginal microbiota can be dramatic and yet not result in symptoms, as long as production of lactic acid is preserved (Gajer et al., 2012).

Hormonal changes significantly impact the normal microbial communities of the vagina. High circulating estrogen, as occurs during pregnancy, is associated with high vaginal epithelial glycogen content, and promotes *Lactobacillus*-dominant microbial communities. In contrast, low circulating estrogen, as seen during the postpartum period in lactating women or after menopause, lowers the glycogen content of vaginal epithelial cells and thus is less conducive to *Lactobacillus*. Parturition also seems to be associated with an increase in the overall diversity of the vaginal microbiota which persists for up to a year postpartum (DiGiulio et al., 2015; MacIntyre et al., 2015). These shifts are not, however, necessarily associated with disease.

8.4.3.2 The upper genital tract microbiome

While the vaginal microbiome has been studied for many years, it is less clear if the upper-genital tract, previously thought to be sterile, has an active microbiome that protects against or promotes pregnancy complications (Sharma, Tal, Cark, & Segars, 2014). Several groups have established the presence of bacteria in the uterus and fallopian tubes of asymptomatic women, although these communities are less diverse than the lower genital tract (Sharma et al., 2014; Wassenaar & Panigrahi, 2014). The colonization of the upper genital tract by these bacteria does not seem to induce an inflammatory response (Mitchell et al., 2015), suggesting benign colonization rather than a pathogenic state. Furthermore, non-pathogenic microbes can be found in amniotic fluid (DiGuilio et al., 2008), and meconium of asymptomatic newborns (Ardissone et al., 2014), suggesting that the pregnant uterus is not in fact sterile. Further research is needed to understand the potential role of upper genital tract microflora in the pathogenesis of reproductive complications, particularly in preterm birth, currently a specific aim of the Human Microbiome Project.

8.4.3.3 The placental microbiome

While early understanding of the chorioamniotic unit during pregnancy was as a sterile entity, the amniotic fluid, membranes, and placenta may harbor a small but important microbial community. Nonpathogenic bacteria can be cultured from amniocentesis specimens in asymptomatic women (Horowitz, Mazor, Romero, Horowitz, & Glezerman, 1995). Intracellular bacteria have also been identified within the placental basal plate (Stout et al., 2013). Several studies have identified low biomass microbial communities within the placentas of women with pregnancy complications. While these may reflect normal findings, it is not yet well understood whether there is a naturally occurring stable placental microbial community.

The placental microbiota may originate from hematogenous spread from the maternal oral cavity (Prince, Antony, Chu, & Aagaard, 2014), as microbes in placenta appear more similar to the maternal oral flora than to the vaginal or stool flora (Aagaard et al., 2014). There is a well-established association between maternal periodontal disease and risk for spontaneous preterm birth, and common organisms have been cultured from both the oral cavity and amniotic fluid samples in women who gave birth prematurely (León et al., 2007).

Microbial organisms are identified within the placentas of women with pregnancy complications, a finding which may or may not reflect pathogenesis. Term and preterm placentas differ in their microbial communities (Doyle et al., 2014), and both preterm birth and chorioamnionitis are associated with lack of placental microbiota diversity (Prince et al., 2016). Placental bacterial colonization has been associated with extremely preterm births (23 to 27 weeks' gestation); organisms were more likely to be cultured from placentas exhibiting inflammation than those without (31% versus 68%) (Hecht et al., 2008). Bacterial species associated with BV are likely found in placentas from women with self-reported cervical/vaginal infection (Fichorova et al., 2015) and associated with an inflammatory response by the neonate. The presence of *Lactobacillus* was associated with reduced risk of an inflammatory response (Fichorova et al., 2011). Preeclampsia is less clearly associated with the placental microbiome (Amarasekara, Jyasekaram Senanayake, & Dissanayake, 2015).

The placental microbiome may also play a role in normal pregnancy. The maternal and/or fetal inflammatory response to bacteria may be an important factor in the timing of parturition (Ardissone et al., 2014). If placental bacteria are able to enter the amniotic fluid, the placental microbiota may also play an important part in early colonization of the infant gut. Preterm infants exhibit differences in gut microbiome from those born at term, suggesting the possibility of differential colonization in utero (Collado, Rautava, Aakko, Isolauri, & Salminen, 2016). One study has cultured meconium and found it more closely resembles placental than vaginal microbiota (Dong et al., 2015).

8.4.4 The maternal microbiome and the fetus

Infants are born lacking a diverse community of microorganisms, however, quickly establish a functional microbiome. Inoculation of the fetus by the mother through vaginal parturition is critical in establishing a normal population of microflora in the infant gastrointestinal tract. Both gestational age and mode of delivery are associated with establishment of the infant microbiome, suggesting the link between exposure to maternal vaginal microbiota and the infant's developing microbiome (Rutayisire, Huang, Liu, & Tao, 2016).

Current data suggest that infants born via cesarean section have different gut microbial communities than those delivered vaginally (Grönlund, Lehtonen, Eerola, & Kero, 1999). Normal vaginal flora change throughout pregnancy, with less diverse communities observed at the end of pregnancy than the beginning (DiGuilio et al., 2015). This mechanism may help limit the exposure of newborns to pathogenic bacterial strains as a co-evolved strategy to protect developing immunity. Babies delivered by scheduled cesarean section do not gain exposure to vaginal secretions during labor. A small study found more skin microbiota in the gut flora of babies delivered by cesarean than those delivered vaginally (Dominguez Bello, et al., 2010), and several groups have demonstrated babies born by cesarean section to have low or even absent levels of *Bacteriodes* species in their stool microbiota, at least over the first months of life (Penders et al., 2006; Azad et al., 2013; Madan et al., 2016; Yassour et al., 2016).

Other important factors that may affect the developing infant microbiome include gestational age at delivery, antibiotic exposure, maternal obesity, gestational weight gain, and exposure to formula (Penders et al., 2006; Yassour et al., 2016; Bäckhed et al., 2015; Bokulich et al., 2016); all of these factors are independently associated with mode of delivery. Currently available data lack specificity regarding these confounders and therefore mode of delivery should not be viewed as the most important determinant of the infant microbiome. Furthermore, a recent systematic review concluded that while differences may exist in the first three months, these differences are transient (Rutayisire et al., 2016).

Because of the data suggesting links between the infant microbiome and mode of delivery, several groups have investigated the potential to modify the infant microbiome in infants delivered by cesarean. In one recent pilot study of swabbing with maternal vaginal secretions (Dominguez Bello et al., 2016), the authors found no differences in the infant microbiotas at 30 days after birth, suggesting the potential to "normalize" the infant microbiome with exposure to maternal bacterial populations. Pilot clinical trials to investigate vaginal seeding are underway, however, the main outcomes being investigated are whether vaginal seeding can affect the neonatal microbiome and short-term clinical outcomes (ClinicalTrials.gov). There may also be risks associated with seeding (Cunnington et al., 2016); in the pilot study, women with group B strep or any evidence of other infections were excluded, however the risk of neonatal infection with invasive GBS or other pathogens exists and has been documented via other neonatal exposures including maternal placentophagy (Farr, Chervenak, McCullough, Baergen, & Grünebaum, 2018).

After a conversation with her obstetrician, Karen is no longer interested in vaginal seeding. She remains worried about how she can reduce her baby's risk for future obesity and obesity-related disease. Her OB asks her: "How are you planning to feed your baby?"

8.4.5 The milk microbiome

While modifying the infant microbiome through vaginal seeding is not a currently recommended strategy, breastfeeding may be a crucial step in establishing a normal infant microbiome. Human milk, previously thought to be relatively sterile, is likely a key driver of infant microbiome development, with its own microbiome as well as oligosaccharides which uniquely support the gut microflora. Both human milk microflora and maternal areolar skin bacteria are transmitted to infants (Pannaraj et al., 2017), with the overall composition and diversity of the infant's microbiome determined in part by breastmilk in a dose-dependent relationship. Moreover, infants who are fed with formula have distinct gut microbial populations from those who are breastfed (Bezirtzoglou, Tsiotsias, & Welling, 2011). The bacterial species found in human milk reflect the maternal gut microbiome, however, do contain unique species not seen in other sites. In addition to bacteria which colonize the infant gut, human milk contains components that can modulate the development of the microbiome. Specifically, bifidobacteria species have adapted to the specific sugars found in human milk, supporting the role of breastfeeding in the establishment of a normal infant gut microbiome (Milani et al., 2017; Turroni et al., 2018). It is not clear if probiotic supplementation helps formula-fed infants to establish a normal gut microbiome, however several trials have demonstrated safety of such an approach (Bazanella et al., 2017; Cooper et al., 2016).

The milk microbiome does vary by maternal characteristics that are tied to the maternal gut microbiome. Specifically, the microbiome of milk seems to be determined largely by maternal nutrition (Williams et al., 2017). The observed geographic diversity of the maternal microbiome likely reflects the influence of diet as well (Kumar et al., 2016). Maternal overweight and obesity have been linked to a lack of overall diversity in the milk microbiome (Cabrera Rubio et al., 2012), reflecting a potential mechanism for the transgenerational risk of obesity through the transmission of an "obesogenic" microbiome. How an obesogenic microbiome impacts the perpetuation of obesity remains investigational.

With these and other well-established benefits of breastfeeding, it is recommended as the standard for nearly all women (ACOG, 2016; Eidelman, 2012). Nonetheless, in the US only 52% of infants are breastfed at six months, and only 22% are exclusively breastfed as recommended (Breastfeeding Report Card, 2018). Women with obesity are unfortunately at particularly high risk for failure to meet their own infant feeding goals (Visram et al., 2013). While several groups have demonstrated an inverse association of maternal obesity with breastfeeding, it is unclear whether the association is causal or rather reflective of increased difficulty with breastfeeding relative to normal-weight women. Obesity is associated with delayed lactogenesis (Nommsen, Rivers, Dolan, & Huang, 2012), particularly among women with gestational diabetes (Matias, Dewey, Quesenberry, & Gunderson, 2014). Women with obesity may benefit from targeted interventions to support lactation (Carlsen et al., 2013), and providers should be aware of the potential need for additional support to meet their goals.

8.5 Disrupting the intergeneration transmission of obesity

8.5.1 Impact of obesity on human health in the modern era

The impact on health of obesity in adults and children is multifactorial and will negatively impact the health care system. Obesity is associated with several diseases affecting both quality of life and longevity, including type 2 diabetes, many cancers, and cardiovascular disease. Historically, infectious diseases and trauma were the most common causes of death. In those that survived to middle age, chronic diseases were rare. Today, cancer, cardiovascular disease, and diabetes are among the top causes of death (WHO, 2018; WHO, 2017). Rates of type 2 diabetes in particular are rising rapidly both in the US and globally; diabetes is diagnosed in an estimated 23.1 million adults (nearly 10% of the adult US population), and at least 132,000 children and adolescents have diagnosed diabetes (CDC, 2017). Adolescents with obesity and diabetes are at particularly high risk of developing fatty liver disease, which is projected to be the most common cause of liver transplant by 2030 (Shaker, Tabbaa, Albeldawi, & Alkhouri, 2014). As part of the metabolic syndrome (a constellation of complications of obesity) many adolescent girls have polycystic ovarian syndrome (PCOS), which can impact fertility. Nonetheless, the majority of women with obesity are able to successfully conceive and carry pregnancies.

Both short-term pregnancy outcomes and long-term offspring health are influenced by maternal prepregnancy BMI. Like other chronic diseases, maternal weight is best optimized prior to pregnancy to reduce the fetal and maternal impacts. As recruitment prior to pregnancy is difficult, there are few studies of weight loss prior to pregnancy to optimize outcomes (Agha, Agha, Sandall, & Sandell, 2014). Those trials that exist are primarily in a population of women with infertility (Rothberg et al., 2016), who are at higher risk for several adverse outcomes independently of their BMI. While trials of prepregnancy weight loss are therefore limited, observational data is available from women who underwent bilipancreatic diversion and lost a mean of 36% of their body weight between successive pregnancies. These studies showed a significant decrease in severe obesity in their children; only 11% had severe obesity compared to 35% of their siblings born prior to surgery (Smith et al., 2009). These results highlight the importance and potential intergenerational impact of maternal metabolic improvement through weight loss prior to pregnancy.

While many groups have studied interventions to reduce gestational weight gain, current evidence is not encouraging. Studies suggests only a modest ability to impact gestational weight gain with a mean difference of less than 1 kg between women in intervention groups compared with controls (Shepherd et al., 2017; Farpour Lambert, Ells, Martinez de Tejada, & Scott, 2018). The long-term effect of this small change in weight gain is similarly small with no overall impact on childhood adiposity. Metformin, a medication used for the treatment of type 2 diabetes is also used off-label for obesity, particularly with PCOS, and may control weight gain when administered during pregnancy. In a recent small randomized controlled trial in women with a BMI >=35 prior to pregnancy, in addition to reducing the rate of preeclampsia, metformin reduced gestational weight gain, but did not reduce the risk for LGA (Syngelaki et al., 2016). While these results are promising, metformin use for prevention of adverse outcomes in women with obesity remains investigational, due in large part to the concern for long-term effects of fetal exposure as metformin freely crosses the placenta.

Although fetal epigenetics as an alterable path to childhood obesity and chronic adult disease is well supported, opportunities to improve the prenatal environment remain challenged by societal, policy. and individual preferences. The contribution of co-evolution of the microbiome with obesity represents an exciting new frontier. Mode of delivery likely affects the infant microbiome only in the short term, with the most important determinants of the human microbiome remaining dietary factors. Critically, for infants, method of feeding is of utmost importance and therefore strategies to support women with obesity with lactation are crucial. Unraveling the potential co-evolution between gut, breast, placental, and upper genital tract microbiome and an obesogenic profile requires further work.

As obesity is increasingly recognized as a condition that should be managed comprehensively using a chronic disease model, solutions focusing on an informed activated patient within a high-functioning system of care that addresses both the social and individual biological determinants of obesity are increasingly the way forward. These solutions will require medical education to adequately address several concepts that have not traditionally been included within medical school curricula. These include a life-course approach to individuals that recognizes the crucial role of fetal development and therefore women's health, methods to study and improve the health of not just individuals but also populations, and practical skills to enable health care providers to function well within multidisciplinary teams. An understanding of evolutionary medicine with a focus moved from not how to treat but how the condition has occurred will enable new approaches to understanding innovations for disrupting the epidemic of intergenerational obesity.

8.6 Conclusions

The clinical problem of obesity highlights several core concepts of evolutionary medicine including reproductive success, epigenetics, co-evolution, environmental mismatch and cultural practices. The concepts of evolution and evolutionary medicine as described previously are essential for understanding the obesity epidemic. Evolution by gene mutation can take multiple generations to alter the human genome. However, generational change as driven by epigenetics and co-evolution, the cornerstones of micro-evolution, underpins the rapid increase in adiposity seen at the population level in the last 20 years. Evolutionary medicine provides a framework for understanding this interplay between genetics and the environment.

However, most physicians receive scant attention to this approach in medical school. Indeed, thoughtful case-based teaching that integrates evolutionary principles is crucial to adequately prepare the next generation of clinicians, investigators and policy-makers to take on the global obesity epidemic. The intersection of obesity and pregnancy provides an ideal case study on which to base new approaches to educating health professionals regarding evolutionary principles, complex microbiology and pathophysiology, concepts of population health and the potential of the health care system to influence generations while treating individuals.

References

Aagaard, K., Ma, J., Antony, K. M., Ganu, R., Petrosino, J., & Versalovic, J. (2014). The placenta harbors a unique microbiome. *Sci Transl Med*, doi:10.1126/scitranslmed.3008599

Aagaard, K., Petrosino, J., Keitel, W., et al. (2013). The Human Microbiome Project strategy for comprehensive sampling of the human microbiome and why it matters. *FASEB J*, doi:10.1096/fj.12–220,806

AAMC. (2008).Scientific foundations for future physicians. *Assoc Am Med Coll*, doi:10.1016/j.stemcr.2014.03.001

ACOG. (2016). Optimizing support for breastfeeding as part of obstetric practice. *ACOG Comm Opin*, doi:10.1016/S0140-6736(16)31898–0

Agha, M., Agha, R. A., Sandall, J., & Sandell, J. (2014). Interventions to reduce and prevent obesity in pre-conceptual and pregnant women: A systematic review and meta-analysis. *PLoS One*, 9(5): e95132. doi:10.1371/journal.pone.0095132

Amarasekara, R., Jayasekara, R. W., Senanayake, H., & Dissanayake, V. H. W. (2015). Microbiome of the placenta in pre-eclampsia supports the role of bacteria in the multifactorial cause of pre-eclampsia. *J Obstet Gynaecol Re*, doi:10.1111/jog.12619

Anahtar, M. N., Byrne, E. H., Doherty K. E., et al. (2015). Cervicovaginal bacteria are a major modulator of host inflammatory responses in the female genital tract. *Immunity*. doi:10.1016/j.immuni.2015.04.019

Ardissone,, A. N., De La Cruz, D. M., Davis-Richardson, A. G., et al. (2014). Meconium microbiome analysis identifies bacteria correlated with premature birth. *PLoS One*, doi:10.1371/journal.pone.0090784

Association of American Medical Colleges. (2018). Curriculum change in US medical schools. https://www.aamc.org/initiatives/cir/427196/27.html. Accessed November 30, 2018.

Azad, M. B., Konya, T., Maughan, H., et al. (2013). Gut microbiota of healthy Canadian infants: Profiles by mode of delivery and infant diet at 4 months. *CMAJ*, doi:10.1503/cmaj.121189

Bäckhed, F., Roswall, J., Peng, Y., et al. (2015). Erratum: Dynamics and stabilization of the human gut microbiome during the first year of life. *Cell Host Microbe*, 17(5): (690–703), doi:10.1016/j.chom.2015.05.012

Barker, D. J. P, Eriksson, J. G., Forsén, T., & Osmond, C. (2002). Fetal origins of adult disease: Strength of effects and biological basis. *Int J Epidemiol*, doi:10.1093/ije/31.6.1235

Barker, D. J. P. (1997). The fetal origins of coronary heart disease. *Eur Heart J*, doi:10.1093/oxfordjournals.eurheartj.a015368

Bazanella, M., Maier, T. V., Clavel, T., et al. (2017). Randomized controlled trial on the impact of early-life intervention with bifidobacteria on the healthy infant fecal microbiota and metabolome. *Am J Clin Nutr* doi:10.3945/ajcn.117.157529

Bezirtzoglou, E., Tsiotsias, A., & Welling, G. W. (2011). Microbiota profile in feces of breast- and formula-fed newborns by using fluorescence in situ hybridization (FISH). *Anaerobe*, doi:10.1016/j.anaerobe.2011.03.009

Bokulich, N. A., Chung, J., Battaglia, T., et al. (2016). Antibiotics, birth mode, and diet shape microbiome maturation during early life. *Sci Transl Med*, doi:10.1126/scitranslmed.aad7121

Breastfeeding Report Card Breastfeeding CDC. (2018). https://www.cdc.gov/breastfeeding/data/reportcard.htm. Accessed November 30, 2018.

Broughton, D. E., & Moley, K. H. (2017). Obesity and female infertility: Potential mediators of obesity's impact. *Fertil Steril*, doi:10.1016/j.fertnstert.2017.01.017

Cabrera-Rubio, R., Collado, M. C., Laitinen, K., Salminen, S., Isolauri, E., & Mira, A. (2012). The human milk microbiome changes over lactation and is shaped by maternal weight and mode of delivery. *Am J Clin Nutr*, doi:10.3945/ajcn.112.037382

Carlsen, E. M., Kyhnaeb, A., Renault, K. M., Cortes, D., Michaelsen, K. F., & Pryds, O. (2013). Telephone-based support prolongs breastfeeding duration in obese women: A randomized trial. *Am J Clin Nutr*. doi:10.3945/ajcn.113.059600

CDC. National Diabetes Statistics Report, (2017). Estimates of diabetes and its burden in the United States Centers for Disease Control and Prevention (CDC). *US Dep Heal Hum Serv*, doi:10.1177/1527154408322560

ClinicalTrials.gov. https://clinicaltrials.gov/ct2/results?cond=vaginal+seeding&term=&cntry=&state=&city=&dist=. Accessed December 6, 2018.

Collado, M. C., Rautava, S., Aakko, J., Isolauri, E., & Salminen, S. (2016). Human gut colonisation may be initiated in utero by distinct microbial communities in the placenta and amniotic fluid. *Sci Rep*, doi:10.1038/srep23129

Cooper, P., Bolton, K. D., Velaphi, S., et al. (2016). Early benefits of a starter formula enriched in prebiotics and probiotics on the gut microbiota of healthy infants born to HIV+ mothers: a randomized double-blind controlled trial. *Clin Med Insights Pediatr*, 2016. doi:10.4137/CMPed.S40134

Cunnington ,A. J., Sim, K., Deierl, A., Kroll, J. S., Brannigan, E., & Darby. J. (2016). "Vaginal seeding" of infants born by caesarean section. *BMJ*, doi:10.1136/bmj.i227

Davenport, E. R., Sanders, J. G., Song, S. J., Amato, K. R., Clark, A. G., & Knight, R. (2017). The human microbiome in evolution. *BMC Biol*, doi:10.1186/s12915-017-0454-7

Della Torre, S., Benedusi, V., Fontana, R., Maggi, A. (2014). Energy metabolism and fertility—A balance preserved for female health. *Nat Rev Endocrinol*, doi:10.1038/nrendo.2013.203

Deputy, N. P., Dub, B., & Sharma, A. J. (2018). Prevalence and trends in prepregnancy normal weight—48 states, New York City, and District of Columbia, 2011–2015. *MMWR Morb Mortal Wkly Rep.*, 2018. doi:10.1002/1521–3951(199706)201:2<339::AID-PSSB339>3.0.CO;2-J

DiGiulio, D. B., Callahan, B. J., McMurdie. P. J, et al. (2015). Temporal and spatial variation of the human microbiota during pregnancy. *Proc Natl Acad Sci*, doi:10.1073/pnas.1502875112

DiGiulio, D. B., Romero, R., Amogan, H. P., et al. (2008). Microbial prevalence, diversity and abundance in amniotic fluid during preterm labor: A molecular and culture-based investigation. *PLoS One*, doi:10.1371/journal.pone.0003056

Döderlein, A. (1892). *Das Scheidensekret*. Leipzig: Drack von Otto Durr.

Dominguez-Bello, M. G., Costello, E. K., Contreras, M., et al. (2010). Delivery mode shapes the acquisition and structure of the initial microbiota across multiple body habitats in newborns. *Proc Natl Acad Sci*, doi:10.1073/pnas.1002601107

Dominguez-Bello, M. G., De Jesus-Laboy, K. M., Shen, N., et al. (2016). Partial restoration of the microbiota of cesarean-born infants via vaginal microbial transfer. *Nat Med*, doi:10.1038/nm.4039

Dong, X. D., Li, X. R., Luan, J. J., et al. (2015). Bacterial communities in neonatal feces are similar to mothers' placentae. *Can J Infect Dis Med Microbiol*, doi:10.1155/2015/737294

Doyle, R. M., Alber, D. G., Jones, H. E., et al. (2014). Term and preterm labour are associated with distinct microbial community structures in placental membranes which are independent of mode of delivery. *Placenta*, doi:10.1016/j.placenta.2014.10.007

Eidelman, A. I. (2012). Breastfeeding and the use of human milk: An analysis of the American Academy of Pediatrics 2012 Breastfeeding Policy Statement. *Breastfeed Med*, doi:10.1089/bfm.2012.0067

Farpour-Lambert, N. J., Ells, L. J., Martinez de Tejada, B., & Scott, C. (2018). Obesity and weight gain in pregnancy and postpartum: An evidence review of lifestyle interventions to inform maternal and child health policies. *Front Endocrinol (Lausanne)*, 9: 546. doi:10.3389/fendo.2018.00546

Farr, A., Chervenak, F. A., McCullough, L. B., Baergen, R. N., & Grünebaum, A. (2018). Human placentophagy: A review. *American journal of obstetrics and gynecology*, *218*(4), 401-e1.

Fichorova, R. N., Beatty, N., Sassi, R. R. S., Yamamoto, H. S., Allred, & E. N., Leviton, A. (2015). Systemic inflammation in the extremely low gestational age newborn following maternal genitourinary infections. *Am J Reprod Immunol*, doi:10.1111/aji.12313

Fichorova, R. N., Onderdonk, A. B., Yamamoto, H., et al. (2011). Maternal microbe-specific modulation of inflammatory response in extremely low-gestational-age newborns. *MBio*, doi:10.1128/mBio.00280–10

Frederick, C. B., Snellman, K., & Putnam, R. D. (2014). Increasing socioeconomic disparities in adolescent obesity. *Proc Natl Acad Sci*, doi:10.1073/pnas.1321355110

Gajer, P., Brotman, R. M., Bai, G., et al. (2012). Temporal dynamics of the human vaginal microbiota. *Sci Transl Med*, doi:10.1126/scitranslmed.3003605

Geuking, M. B., Köller, Y., Rupp, S. &, McCoy, K. D. (2014). The interplay between the gut microbiota and the immune system. *Gut Microbes*, 5(3): 411–18. doi:10.4161/gmic.29330

Gluckman, P. D., Low, F. M., Buklijas, T., Hanson, M. A., & Beedle, A. S. (2011). How evolutionary principles improve the understanding of human health and disease. *Evol Appl*, doi:10.1111/j.1752–4571.2010.00164.x

Gluckman, P., & Bergstrom, C. (2011). Evolutionary biology within medicine: A perspective of growing value. *BMJ*, doi:10.1136/bmj.d7671

Goodrich, J. K., Davenport, E. R., Beaumont, M., et al. (2016). Genetic determinants of the gut microbiome in UK twins. *Cell Host Microbe*, doi:10.1016/j.chom.2016.04.017

Goodrich, J. K., Davenport, E. R., Clark, A. G. &, Ley R. E. (2017). The relationship between the human genome and microbiome comes into view. *Annu Rev Genet*, doi:10.1146/annurev-genet-110,711–155,532

Goodrich, J. K., Waters, J. L., Poole, A. C., et al. (2014). Human genetics shape the gut microbiome. *Cell*, doi:10.1016/j.cell.2014.09.053

Grönlund, M. M., Lehtonen, O. P., Eerola, E., & Kero, P. (1999). Fecal microflora in healthy infants born by different methods of delivery: Permanent changes in intestinal flora after cesarean delivery. *J Pediatr Gastroenterol Nutr*, doi:10.1097/00005176-199,901,000-00007

Hales, C. M., Carroll, M. D., Fryar, C. D., & Ogden, C. L. (2017). Prevalence of obesity among adults and youth: United States, 2015–2016. NCHS data brief, no 288. Hyattsville, MD: National Center for Health Statistics. *NCHS data brief, no 288 Hyattsville, MD Natl Cent Heal Stat*, doi:10.1017/S1368980017000088

Heard, E., Tishkoff, S., Todd, J. A., et al. (2010). Ten years of genetics and genomics: What have we achieved and where are we heading? *Nat Rev Genet*, doi:10.1038/nrg2878

Hecht, J. L., Onderdonk, A., Delaney, M., et al. (2008). Characterization of chorioamnionitis in 2nd- trimester c-section placentas and correlation with microorganism recovery from subamniotic tissues. *Pediatr Dev Pathol*, doi:10.2350/07-06-0285.1

Hidaka, B. H., Asghar, A., Aktipis, C. A., et al. (2015). The status of evolutionary medicine education in North American medical schools. *BMC Med Educ*, doi:10.1186/s12909-015-0322-5

Horowitz, S., Mazor, M., Romero, R., Horowitz, J., & Glezerman, M. (1995). Infection of the amniotic cavity with Ureaplasma urealyticum in the midtrimester of pregnancy. *The Journal of reproductive medicine*, *40*(5), 375-379.

Huttenhower, C., Gevers, D., Knight, R., et al. (2012). Structure, function and diversity of the healthy human microbiome. *Nature*, doi:10.1038/nature11234

Janzen, D. H., Janzen (1980). When is it coevolution? *Evolution*, 34(3): 611–12. doi:10.1016/j.ecolecon.2006.12.016

Kenyon, C., Colebunders, R., & Crucitti, T. (2013). The global epidemiology of bacterial vaginosis: A systematic review. *Am J Obstet Gynecol*, doi:10.1016/j.ajog.2013.05.006

Kim, S. Y., Sharma, A. J., Sappenfield, W, Wilson, H. G., Salihu, H. M. (2014). Association of maternal body mass index, excessive weight gain, and gestational diabetes mellitus with large-for-gestational-age births. *Obstet Gynecol*, doi:10.1097/AOG.0000000000000177

Kumar, H., du Toit, E., & Kulkarni, A., et al. (2016). Distinct patterns in human milk microbiota and fatty acid profiles across specific geographic locations. *Front Microbiol*, doi:10.3389/fmicb.2016.01619

Kuzawa, C. (2005). The origins of the developmental origins hypothesis and the role of postnatal environments: Response to Koletzko. *Am J Hum Biol*, doi:10.1002/ajhb.20421

Kuzawa, C. W. & Sweet, E. (2009). Epigenetics and the embodiment of race: Developmental origins of US racial disparities in cardiovascular health. *Am J Hum Biol*, doi:10.1002/ajhb.20822

Kuzawa, C. W. (2005). Fetal origins of developmental plasticity: Are fetal cues reliable predictors of future nutritional environments? *Am J Hum Biol*, doi:10.1002/ajhb.20091

Kuzawa, C. W. (2007). Developmental origins of life history: Growth, productivity, and reproduction. In: *American Journal of Human Biology*, doi:10.1002/ajhb.20659

Lamm, E. & Jablonka, E. (2008). Integrating evolution and development: From theory to practice. *Perspect Biol Med*, doi:10.1353/pbm.0.0055

Lamont, R. F., Sobel, J. D., Akins, R. A., et al. (2011). The vaginal microbiome: New information about genital tract flora using molecular based techniques. *BJOG An IntJObstetGynaecol*,doi:10.1111/j.1471–0528.2010.02840.x

León, R., Silva, N., Ovalle, A., et al. (2007). Detection of *Porphyromonas gingivalis* in the amniotic fluid in pregnant women with a diagnosis of threatened premature labor. *J Periodontol*, doi:10.1902/jop.2007.060368

Ma, B., Forney, L. J., & Ravel, J. (2012.) Vaginal Microbiome: Rethinking Health and Disease. *Annu Rev Microbiol*, doi:10.1146/annurev-micro-092611–150,157

Ma, J., Coarfa, C., Qin, X., et al. (2014). MtDNA haplogroup and single nucleotide polymorphisms structure human microbiome communities. *BMC Genomics*, doi:10.1186/1471-2164-15-257

MacIntyre D. A., Chandiramani M, Lee YS, et al. (2015). The vaginal microbiome during pregnancy and the postpartum period in a European population. *Sci Rep*, doi:10.1038/srep08988

Madan, J. C., Hoen, A. G., Lundgren, S. N., et al. (2016). Association of cesarean delivery and formula supplementation with the intestinal microbiome of 6-week-old infants. *JAMA Pediatr*, doi:10.1001/jamapediatrics.2015.3732

Matias, S. L., Dewey, K. G., Quesenberry, C. P., & Gunderson, E. P. (2014). Maternal prepregnancy obesity and insulin treatment during pregnancy are independently associated with delayed lactogenesis in women with recent gestational diabetes mellitus. *Am J Clin Nutr*, doi:10.3945/ajcn.113.073049

McGowan, P. O. & Roth, T. L. (2015). Epigenetic pathways through which experiences become linked with biology. *Dev Psychopathol*, doi:10.1017/S0954579415000206

McGowan, P. O. (2015). Epigenetic mechanisms of perinatal programming: Translational approaches from rodent to human and back. In: *Advances in Neurobiology*, doi:10.1007/978-1-4939-1372-5_17

McGowan, P. O. & Matthews, S. G. (2018). Prenatal stress, glucocorticoids, and developmental programming of the stress response. *Endocrinology*, doi:10.1210/en.2017–00896

McGowan, P. O. & Szyf, M. (2010). Environmental epigenomics: Understanding the effects of parental care on the epigenome. *Essays Biochem*, doi:10.1042/bse0480275

Mehler, M. F. (2008). Epigenetics and the nervous system. *Ann Neurol*, doi:10.1002/ana.21595

Methé, B. A., Nelson, K. E., Pop, M., et al. (2012). A framework for human microbiome research. *Nature*, doi:10.1038/nature11209

Milani, C., Duranti, S., Bottacini, F., et al. (2017). The first microbial colonizers of the human gut: Composition, activities, and health implications of the infant gut microbiota. *Microbiol Mol Biol Rev*, doi:10.1128/MMBR.00036–17

Mitchell, C. M., Haick, A., Nkwopara, E, et al. (2015). Colonization of the upper genital tract by vaginal bacterial species in nonpregnant women. *Am J Obstet Gynecol*, doi:10.1016/j.ajog.2014.11.043

Nesse, R. M. (2008). What evolutionary biology offers public health. *Bull World Health Organ*, doi:10.2471/BLT.07.049601

Nesse, R. M., Bergstrom, C. T., & Ellison, P. T., et al. (2010). Making evolutionary biology a basic science for medicine. *Proc Natl Acad Sci*, doi:10.1073/pnas.0906224106

NIH Human Microbiome Project—Home. (2018). https://commonfund.nih.gov/hmp. Accessed November 30, 2018.

Nommsen-Rivers, L. A., Dolan, L. M., & Huang, B. (2012). Timing of stage II lactogenesis is predicted by antenatal metabolic health in a cohort of primiparas. *Breastfeed Med*, doi:10.1089/bfm.2011.0007

Pannaraj, P. S., Li, F., Cerini, C., et al. (2017). Association between breast milk bacterial communities and establishment and development of the infant gut microbiome. *JAMA Pediatr*, doi:10.1001/jamapediatrics.2017.0378

Penders, J., Thijs, C., & Vink, C., et al. (2006). Factors Influencing the composition of the intestinal microbiota in early infancy. *Pediatrics*, doi:10.1542/peds.2005–824

Perlman, R. L. (2013). Evolution and medicine. *Perspect Biol Med*, 56(2): 167–83. doi:10.1353/pbm.2013.0012

Ponterio, E., & Gnessi, L. (2015). Adenovirus 36 and obesity: An overview. *Viruses*, doi:10.3390/v7072787

Prince, A. L., Antony, K. M., Chu, D. M., & Aagaard, K. M. (2014). The microbiome, parturition, and timing of birth: More questions than answers. *J Reprod Immunol*, doi:10.1016/j.jri.2014.03.006

Prince, A. L., Ma, J., Kannan, P. S., et al. (2016). The placental membrane microbiome is altered among subjects with spontaneous preterm birth with and without chorioamnionitis. *Am J Obstet Gynecol*, doi:10.1016/j.ajog.2016.01.193

Prober, C. G., & Heath, C. (2012). Lecture Halls without Lectures—A Proposal for Medical Education. *N Engl J Med*, doi:10.1056/NEJMp1202451

Ravel, J., Gajer, P., Abdo, Z., et al. (2011). Vaginal microbiome of reproductive-age women. *Proc Natl Acad Sci*, doi:10.1073/pnas.1002611107

Rothberg, A., Lanham, M., Randolph, J., Fowler, C., Miller, N., & Smith, Y. (2016). Feasibility of a brief, intensive weight loss intervention to improve reproductive outcomes in obese, subfertile women: a pilot study. *Fertil Steril*, doi:10.1016/j.fertnstert.2016.06.004

Rutayisire, E., Huang, K., Liu, Y., & Tao, F. (2016). The mode of delivery affects the diversity and colonization pattern of the gut microbiota during the first year of infants' life: A systematic review. *BMC Gastroenterol*, doi:10.1186/s12876-016-0498-0

Shaker, M., Tabbaa, A., Albeldawi, M., & Alkhouri, N. (2014). Liver transplantation for nonalcoholic fatty liver disease: New challenges and new opportunities. *World J Gastroenterol*, doi:10.3748/wjg.v20.i18.5320

Sharma, H., Tal, R., Clark, N. A., & Segars, J. H. (2014). Microbiota and pelvic inflammatory disease. *Semin Reprod Med*, doi:10.1055/s-0033–1,361,822

Shepherd, E., Gomersall, J. C., Tieu, J., Han, S., Crowther, C. A., & Middleton, P. (2017). Combined diet and exercise interventions for preventing gestational diabetes mellitus. *Cochrane Database Syst Rev*, doi:10.1002/14651858.CD010443.pub3

Sjöström, L., Lindroos, A. K. Peltonen, M., et al. (2004). lifestyle, diabetes, and cardiovascular risk factors 10 years after bariatric surgery. *N Engl J Med*, doi:10.1056/NEJMoa035622

Smith, J., Cianflone, K., Biron, S., et al. (2009). Effects of maternal surgical weight loss in mothers on intergenerational transmission of obesity. *J Clin Endocrinol Metab*, doi:10.1210/jc.2009–0709

Spor, A., Koren, O., & Ley, R. (2011). Unravelling the effects of the environment and host genotype on the gut microbiome. *Nat Rev Microbiol*, doi:10.1038/nrmicro2540

Stearns, S. C., Nesse, R. M., Govindaraju, D. R., & Ellison, P. T. (2010). Evolutionary perspectives on health and medicine. *Proc Natl Acad Sci*, doi:10.1073/pnas.0914475107

Stout, M. J., Conlon, B., Landeau, M., et al. (2013). Identification of intracellular bacteria in the basal plate of the human placenta in term and preterm gestations. *Am J Obstet Gynecol*, doi:10.1016/j.ajog.2013.01.018

Syngelaki, A., Nicolaides, K. H., Balani, J., et al. (2016). Metformin versus placebo in obese pregnant women without diabetes mellitus. *Obstet Gynecol Surv*, doi:10.1097/01.ogx.0000483242.75040.db

Tang, H., Mcgowan, O. O., & Reynolds, G. P. (2014). Polymorphisms of serotonin neurotransmission and their effects on antipsychotic drug action. *Pharmacogenomics*, doi:10.2217/pgs.14.111

Torres-Fuentes, C., Schellekens, H., Dinan, T. G., & Cryan, J. F. (2017). The microbiota–gut–brain axis in obesity. *Lancet Gastroenterol Hepatol*, doi:10.1016/S2468-1253(17)30147–4

Turroni, F., Milani, C., Duranti, S., et al. (2018). Bifidobacteria and the infant gut: An example of co-evolution and natural selection. *Cell Mol Life Sci*, doi:10.1007/s00018-017-2672-0

Visram, H., Finkelstein, S. A., Feig D., et al. (2013). Breastfeeding intention and early post-partum practices among overweight and obese women in Ontario: A selective population-based cohort study. *J Matern Neonatal Med*, doi:10.3109/14767058.2012.735995

Wassenaar, T.M., & Panigrahi, P. (2014). Is a foetus developing in a sterile environment? *Lett Appl Microbiol*, doi:10.1111/lam.12334

Wells, J. C. K. (2003). The thrifty phenotype hypothesis: Thrifty offspring or thrifty mother? *J Theor Biol*, doi:10.1006/jtbi.2003.3183

Wells, J. C. K. (2007). Environmental quality, developmental plasticity and the thrifty phenotype: A review of evolutionary models. *Evol Bioinforma*, doi:10.1177/117693430700300027

Wells, J. C. K. (2007). The thrifty phenotype as an adaptive maternal effect. *Biol Rev*, doi:10.1111/j.1469-185X.2006.00007.x

Wells, J. C. K. (2009). Thrift: A guide to thrifty genes, thrifty phenotypes and thrifty norms. *Int J Obes*, doi:10.1038/ijo.2009.175

Wells, J. C. K. (2010). Maternal capital and the metabolic ghetto: An evolutionary perspective on the transgenerational basis of health inequalities. *Am J Hum Biol*, doi:10.1002/ajhb.20994

Wells, J. C. K. (2011). The thrifty phenotype: An adaptation in growth or metabolism? *Am J Hum Biol*, doi:10.1002/ajhb.21100

Wells, J. C. K. (2012). Ethnic variability in adiposity, thrifty phenotypes and cardiometabolic risk: Addressing the full range of ethnicity, including those of mixed ethnicity. *Obes Rev*, doi:10.1111/j.1467-789X.2012.01034.x

Wells, J. C. K. (2012). Body composition in infants: Evidence for developmental programming and techniques for measurement. *Rev Endocr Metab Disord*, doi:10.1007/s11154-012-9213-9

WHO, (2017). Noncommunicable diseases, Fact Sheet 355. Update April 2017. *WHO | Noncommunicable Dis.*

WHO, (2018). Top 10 Causes of Death. http://www.who.int/gho/mortality_burden_disease/causes_death/top_10/en/. Accessed November 30, 2018.

Williams, G. C., & Nesse, R. M. (1991). The Dawn of Darwinian Medicine. *Q Rev Biol*, doi:10.1086/417048

Williams, J. E., Carrothers, J. M., Lackey, K. A., et al. (2017). Human milk microbial community structure is relatively stable and related to variations in macronutrient and micronutrient intakes in healthy lactating women. *J Nutr*, doi:10.1088/1126-6708/2009/03/056

Yassour, M., Vatanen, T., Siljander, H., et al. (2016). Natural history of the infant gut microbiome and impact of antibiotic treatment on bacterial strain diversity and stability. *Sci Transl Med*, doi:10.1126/scitranslmed.aad0917

Zenténius, E., Andersson-Assarsson, J. C., Carlsson, L. M. S., Svensson, P. A., & Larsson I. (2018). Self-reported weight-loss methods and weight change: Ten-year analysis in the Swedish obese subjects' study control group. *Obesity (Silver Spring)*, 26(7): 1137–43. doi:10.1002/oby.22200

SECTION III

PERSPECTIVES: PAST, PRESENT, AND FUTURE

CHAPTER 9

Darwin's Impact on the Medical Sciences

Fabio Zampieri

The influence of Darwinism on medical education and clinics remains an issue particularly significant today. In fact, one of the main criticisms directed against current evolutionary medicine is related to its lack of practical applications. Can evolution-based understandings of health help with medical treatment? This is the typical question that arises in current debate, and many doctors think that the answer is negative. However, the following historical sketch might prove that evolutionary concepts contributed not only to the foundation of basic sciences such as microbiology and immunology, but also gave important contributions to medical education and clinics by stimulating the debate around methods, fundamental concepts and scopes of medicine, and diagnostic reasoning.

9.1 Introduction: Pioneers of evolutionary concepts of health and disease

9.1.1 The anatomical concept of disease

For around two millennia, diseases in Western medicine were conceived as functional disorders caused by an imbalance among the four humors of the body. Thanks to the anatomical studies that flourished in the Renaissance, the concept of disease started to change, until its new foundation during the late eighteenth and the beginning of the nineteenth century. In particular, it was recognized that the majority of diseases could affect particular organs. Therefore, stable and universal characteristics of diseases were identified in organic lesions, which represented the core of the anatomo-clinical method introduced by the work of the Italian anatomist Giovanni Battista Morgagni (1682–1771). In 1761, he published *The Seats and Causes of Diseases investigated by Anatomy* (*De sedibus et causis morborum per anatomen indagatis*), by which he demonstrated a systematic correlation between clinical symptoms and lesions in the organs (Morgagni, 1761; Zampieri, 2016). The post-revolutionary medical school of Paris developed Morgagni's method, completing its translation in the clinical field. Diseases started to be diagnosed exclusively in relation with the lesions of the organs affected, to such an extent that internal diseases became equivalent to surgical ones. The stethoscope was invented by René Laennec (1781–1826) to detect internal lesions in the lungs and the heart during the life of patients, so as to formulate the exact diagnosis which could potentially indicate the right therapy. Laennec, in his *On mediated auscultation* (*De l'auscultation médiate*), stated that "The alteration of organs is, without comparison, what is more stable and positive and less variable in local diseases" (Laennec 1819).

The anatomical concept of disease represented, at the same time, a limit and a fundamental prerequisite for an evolutionary understanding of health and

Zampieri, F., *Darwin's Impact on the Medical Sciences* In: *Integrating Evolutionary Biology into Medical Education—for maternal and child healthcare students, clinicians, and scientists*. Edited by: Jay Schulkin and Michael L. Power, Oxford University Press (2020). © Oxford University Press. DOI: 10.1093/oso/9780198814153.003.0010

disease in medicine. Organic lesions correlated with clinical symptoms became the new basis for classifying diseases. The nosographic system of the French Philippe Pinel (1745–1826), published in 1797 and re-edited several times, marked a turning point in the scientific conception of disease. As written by Pinel:

> Taking, as far as possible, pathological anatomy as the basis for the methodical distribution of diseases, the lesions of specific organic functions, separately considered [...] present differences by which are reciprocally discernible. [...] After all, pathological anatomy could improve only by comparing the appearance of internal state of organs after death with the most exact and characterized history of disease's progress and termination.
>
> **(Pinel, 1797, pp. xxiij–xxiv)**

The fundamental point of Pinel's system consisted in believing that a scientific nosology would be possible only with "a permanent order in its object." According to Pinel, scientific nosology had to be based on immutable characteristics of diseases arranged into classes, genres, and species. Pinel followed the example of the first classifications of diseases proposed by the "British Hippocrates" Thomas Sydenham (1624–1689) and the French physician and botanist François Boissier de Sauvages de Lacroix (1706–1767). As plants were being divided and subdivided into various categories of the new Linnean method, so did Sydenham and Sauvages adopt a similar system for a classification of diseases based on external appearances, that is, symptoms. Pinel maintained the idea that nosology had to imitate this cataloging method of plants, which at that time was based on the fundamental theory of *fixism*, or that species alive were identical to those of the past and that evolution did not happen. Considering diseases such as immutable species, medical sciences of the early nineteenth century were closed to any idea of evolutionary changing.

At the same time, to found medicine and the concept of disease on anatomy brought some doctors to become aware that men and animals were very similar from a morphological, physiological and pathological point of view, and that this could indicate a certain degree of kinship between them. Comparative anatomy started to be studied from the sixteenth century, following the renaissance of human normal anatomy. The French traveler, naturalist and diplomat Pierre Belon (1517–1564) was the first to compare human and bird skeletons, marking the homologous bones and demonstrating the similarities between them (Belon, 1555). This is widely used as one of the earliest ideas on comparative anatomy. The Scottish surgeon John Hunter (1728–1793) was a paradigmatic naturalist and physician whose understanding of human health and disease was based on the study of organic nature as a whole. An understanding that brought him to prefigure the idea of evolution. Hunter wrote extensively on natural history and comparative anatomy. Several of his papers on these subjects were posthumously published in two volumes in 1861, under the title *Essay and Observations on Natural History, Anatomy, Physiology, Psychology, and Geology*. The first volume starts with a significant statement:

> Our ideas of the first formation of the world, of the production and distinctions of vegetables and animals, and of the alteration which has taken place in the course of time, can only arise from observation on the present state, or from history, of things.
>
> **(Hunter 1861, vol. 1, p. 1)**

Soon after, he suggested that variations of species on Earth were originated during the time in correlation with changes in the environment:

> To attempt to trace any natural production to its origin, or its first production, is ridiculous; for it goes back to that period, if ever such existed, of which we can form no idea, viz. the beginning of time. But, I think, we have reason to suppose there was a period in time in which every species of natural production was the same; there then being no variety in any species; but the variations taking place on the surface of the earth, such as the earth and water changing situations, which is obvious; as also the change in the poles or ecliptic, which I think is also obvious. The varieties [so produced] are but few and are still existing in what may be called the "Natural" Animal. Also civilization has made varieties in many species, and without number, which are the "Domesticated."
>
> **(Ibid., p. 4)**

Another example is represented by the German physician Christoph Wilhelm Hufeland (1762–1836), who can be considered as one of the fathers of comparative pathology. In his *System of Practical Medicine* (*System der practischen Heilkunde*), he proposed the use of comparative pathology to trace the history of species and human populations (Hufeland, 1800). He discussed that diseases which affect human populations in different ways represent "distinctive" physical characteristics not less important than morphological ones.

Finally, Erasmus Darwin (1731–1802), Charles' grandfather, deserves, in this perspective, a special place. Between 1794 and 1796, he published the two volumes of his *Zoonomia; or the Laws of Organic Life* in which he postulated one of the earliest ideas of organic evolution (E. Darwin 1794–1796). This work represents an attempt to explain the whole phenomenon of organic life combining comparative observations on anatomy, development, physiology, and pathology between man and animals. During the seventeenth and eighteenth centuries, in part influenced by the philosophy of Gottfried Wilhelm von Leibniz (1646–1716), physicians were looking for the fundamental morphological and functional unit of life. Before the discovery of cells and the foundation of the cell theory of living beings, the "fiber" was postulated by many anatomists such as this kind of unit. According to the Swiss anatomist and physiologist Albrecht von Haller (1708–1777), the "fiber" in medicine had the same significance and role of the "line" in geometry (Haller, 1753). Erasmus Darwin founded his theory of organic evolution on the concept of fiber and its two fundamental properties, irritability and sensitivity, advancing that all living beings originated from a unique filament of fiber possessing the faculty of continuing to improve by its own inherent activity, and of delivering down those improvements by generation to its posterity. Differences in animals' forms and qualities depended on differences in irritability and sensitivity of their fibers evolved during the time, under the pressure of three fundamental desires: sex, anger, and safety (E. Darwin, 1794–1797). Erasmus closed the first volume of his *Zoonomia* with an incontrovertible statement:

> As the habitable parts of the earth have been, and continue to be, perpetually increasing by the production of sea-shells and corallines, and by the recrements of other animals, and vegetables; so from the beginning of the existence of this terraqueous globe, the animals, which inhabit it, have constantly improved, and are still in a state of progressive improvement. **(Ibid., pp. 528–9).**

9.2 Darwin and medicine

As stated by William Bynum in a paper on *Darwin and the Doctors*: "Charles Darwin, it might be said, had medicine in his blood" (Bynum, 1983, p. 43). His grandfather and father were physicians and Charles, at his father's direction, started university at the age of sixteen in Edinburgh as a medical student. However, he showed little academic interest in medicine and was revolted by the brutality of surgery being performed without pain relief. He had to leave the operating theatre before the two operations he ventured to witness could be completed (*ibidem*). Darwin dropped out of medical school after two years of study in 1827.

However, Darwin was surely inspired by the materialistic approach of medical science, by its researches in natural history with fundamental insights on variability and heredity of living species' and humans' populations. Darwin's notebooks on *Man, Mind and Materialism*, written between July 1838 and August 1839, are filled with Darwin's father observations on heredity in human beings, on humans' adaptations to pain and diseases, and on the influence of organic diseases on the functions of mind (Barret, 1974; Bynum, 1983). For instance, Darwin wrote: "My father says, perfect deformity, as an extra number of fingers—harelip or imperfect roof to the mouth [...] are hereditary" (Barret, 1974, p. 270). Again: "My father has somewhere heard (possibly from Hunter?) that pulse of newborn babies of laboring classes are slower than those of Gentle-folks, & that peculiarities of form in trades (as sailor, tailor, blacksmiths?) are likewise hereditary, & therefore that their children have some little advantage in these trades" (ibid, p. 274).

Even in Edinburgh, Darwin enriched his knowledge on natural history thanks to a physician, Robert Edmond Grant (1793–1874), who was a noted professor of anatomy and much interested in comparative anatomy and marine biology. In his autobiography, Darwin mentioned Grant and his admiration for the evolutionary theory of Jean-

Baptiste de Lamarck (1744–1829). In the same passage, he mentioned also his grandfather's *Zoonomia*:

> He one day, when we were walking together, burst forth in high admiration of Lamarck and his views on evolution. I listened in silent astonishment, and as far as I can judge without any effect on my mind. I had previously read the "Zoonomia" of my grandfather, in which similar views are maintained, but without producing any effect on me. Nevertheless it is probable that the hearing rather early in life such views maintained and praised may have favoured my upholding them under a different form in my "Origin of Species." At this time I admired greatly the "Zoonomia"; but on reading it a second time after an interval of ten or fifteen years, I was much disappointed; the proportion of speculation being so large to the facts given. **(F. Darwin, 1887, vol. 1, p. 38)**

The heredity of diseases testified and studied by medicine gave Darwin the fundamental examples of heredity of variations in man. In his main work on heredity, *The Variation of Animals and Plants under Domestication*, the only discussion about man was focused on hereditary disease (C. Darwin, 1868).

In this book, hereditary variations and hereditary diseases were almost synonymous:

> What can be more wonderful than that some trifling peculiarity, not primordially attached to the species, should be transmitted through the male or female sexual cells, which are so minute as not to be visible to the naked eye, and afterwards through the incessant changes of a long course of development, undergone either in the womb or in the egg, and ultimately appear in the offspring when mature, or even when quite old, as in the case of certain diseases? **(Ibid., vol. II, p. 2)**

With regard to man, in fact, hereditary diseases represented the unique example of hereditary variations:

> Some writers have doubted whether those complex mental attributes, on which genius and talent depend, are inherited, even when both parents are thus endowed. But he who will read Mr. Galton's able paper on hereditary talent will have his doubts allayed. Unfortunately it matters not, as far as inheritance is concerned, how injurious a quality or structure may be if compatible with life. No one can read the many treatises on hereditary disease and doubt this. The ancients were strongly of this opinion, or, as Ranchin expresses it, *Omnes Græci, Arabes, et Latini in eo consentiunt*. A long catalogue could be given of all sorts of inherited malformations and of predisposition to various diseases. With gout, fifty per cent. of the cases observed in hospital practice are, according to Dr. Garrod, inherited, and a greater percentage in private practice. Everyone knows how often insanity runs in families, and some of the cases given by Mr. Sedgwick are awful, – as of a surgeon, whose brother, father, and four paternal uncles were all insane, the latter dying by suicide; of a Jew, whose father, mother, and six brothers and sisters were all mad; and in some other cases several members of the same family, during three or four successive generations, have committed suicide. Striking instances have been recorded of epilepsy, consumption, asthma, stone in the bladder, cancer, profuse bleeding from the slightest injuries, of the mother not giving milk, and of bad parturition being inherited.
>
> **(Ibid., pp. 7–8).**

The concept that disease represents the most striking instance of hereditary variation had a deep impact in the evolutionary conceptualization of human pathology developed by medicine between the end of the nineteenth and the first decades of the twentieth century.

9.2.1 How medicine incorporates Darwin

9.2.1.1 The concept of diathesis and "medical Darwinism"

As well known, Darwinian theory of evolution, when it appeared in 1859, immediately became one of the most popular and discussed issues not only by the scientific community, but also by public opinion. The impact of the theory of evolution in medicine was only apparently slight. From the one hand, relatively few physicians explicitly tried to combine evolution and medicine, even if an important part of the medical community demonstrated to join and support evolutionary theory. This acceptance, in fact, was principally based on a philosophical point of view, rather than on a strictly medical one. Cesare Lombroso (1835–1909) and Ernst Haeckel (1834–1919), for instance, welcomed Darwinian theory as the final victory of scientific materialism (Zampieri, 2006). As stated by John Haller:

> Doctors were disposed, by habit and temperament, to study nature: and their perception of Darwinian hypothesis – its concept of time, species, force, law, justice, and causality – indicated a deep and abiding concern for every imminent point in evolutionary speculation. [...]

Their delight for the written word brought wisdom, wit, and indignation to the fore, making them fiery partisans of the evolutionary feuds that so dominated the age.

(J. Haller, 1981, pp. 280–81).

However, it is important to note that nineteenth century physicians were supposed to be charged with the cure not only of human body, but also with human "soul." The prescriptive and proscriptive aspects of medicine at that time were most often rendered in the name of the Christian God (ibid., p. 281). Some of the doctors who entered in the Darwinian controversy accused Darwin of being in league with atheists and skeptics. Others tried to combine Darwinian Theory and theology, following the American botanist Asa Gray (1810–1888), according to whom evolution was the expression of the Divine plan (Ibid., pp. 292–3).

From the other hand, after an initial period, roughly between 1859 and the end of the 1870s, in which doctors discussed Darwinian theory along paths already traveled by theologians and philosophers, in the period between 1880 and 1940 we can find a significant number of articles (Zampieri, 2009) and books (Ross, 1872; Millican, 1883; Hutchinson, 1884; Mitchell, 1888; Bland-Sutton, 1890; Nash, 1915; Adami, 1918; Ribber, 1918; Roberts, 1926) dedicated to evolutionary issues related to medical problems. During the early years of the Darwinian controversy, few practitioners were sufficiently trained in morphology and physiology to defend or otherwise question the evidence of evolution (J. Haller, 1981). Since the 1880s, they entered in the scientific debate also because the controversy became even more focused on the origin of man (Pancaldi, 1983). It is even possible to support that evolutionary concepts contributed in a significant way to the most important medical issues, debates, and new discipline in that period, such as familial patterns of disease, diseases of civilization, germ theory, immunology, and neurology. Therefore, in the period between 1880 and 1940 we can support the existence of a movement of thought in medicine that can be labelled of "medical Darwinism," as to be clearly distinguished from current Evolutionary or Darwinian Medicine, born at the end of the 20th century on different theoretical and experimental basis (William & Nesse 1991).

The main idea based on Darwinian theory was that diseases could evolve exactly as the organisms evolved. James Paget (1814–1899), English surgeon and pathologist, in a lecture on *Some Rare and New Diseases* delivered before the *Royal College of Surgeons* in 1882, wrote:

It will be better for us if we study, in pathology as in natural history, varieties as much as species; changes as well as more stable forms. Types there are the tenacity with which they are maintained – some, even, from prehistoric times – in all the varieties of the conditions of our lives is one of the most remarkable facts in all pathology. But there are not unalterable. Types vary in diseases as in species; even in the diseases which depend least upon external conditions, and most of the qualities which are transmitted by heredity. **(Paget, 1882, p. 1190)**

Another British surgeon and pathologist, James Bland-Sutton (1855–1936), in his book on *Evolution and Disease* published few years later, reiterated the same conception: the idea of evolution of diseases was strictly related with the problem of heredity. Following Darwin's thought, hereditary diseases were viewed as hereditary variations. According to Bland-Sutton:

As with mental so with bodily conditions, it is impossible to state definitely the borderland between health and disease, either in relation with functional aberration or textural alteration. And in many instances we shall find conditions which we regard as abnormal in man, presenting themselves as normal states in other animals.

(Ibid., p. 2).

Through heredity, diseases could not only be transmitted from parents to offspring, but they could also combine and vary in many different ways. James Paget insisted particularly on that point:

We should look out for indications of the existence in the same person of two or more morbid conditions or dispositions such as many be derived from both parents or from several ancestors. For, as in plants and animals there are hybrids and mongrels, or, as in chemistry, many compounds and many mixtures, so are there in diseases.

(Paget, 1882, p. 1192).

The issue of the evolution of diseases through hereditary transmission was mainly focused on the problem of morbid disposition or liability to disease, because diseases seemed to be almost never transmitted in a linear and simple way from parents to offspring:

> An exact likeness is never transmitted by inheritance: neither an exact likeness of either parent, nor an exact composite of both. [...] The certainty and probable extent of this variation must seem the greater if we consider the mingling of diatheses and of all dispositions and liabilities to diseases in transmission from and through both parents.
>
> **(Ibid., p. 1190)**

It is very significant that Paget used in this context the concept of diathesis. Diathesis was an old medical concept which indicated a constitutional tendency toward a disease (Bynum, 1983, p. 46). It could be acquired by an individual through the influence of environment or in consequence of his habits, but its main characteristic was its hereditary nature. In fact, after being acquired, it could be transmitted through some generations undergoing variations in many ways.

It is important to underline that the concept of heredity was a product of the nineteenth century and that medical community played a major role in its articulation. If the notion of hereditary transmission had been elaborated since classical medicine, the concept of heredity itself emerged in France during the early decades of nineteenth century. In particular, there was a shift from the hereditary as an important but secondary condition, to heredity as a main cause responsible for all the natural bodily endowments of individuals (López-Beltrán, 2003, p. 10). From the one hand, the transition from old humoralizm to the solidizm endorsed by anatomo-clinical approach played an important role in promoting hereditary explanations of diseases. The concept of bodily constitution shaped by heredity replaced that of humoral complexion. On the other hand, the concept of diathesis became an alternative explanation to that of the anatomo-clinical approach, because this latter was focused on local causes of diseases, while the diathesis looked at the general state of the body which could predispose to diseases or be their direct cause. Diathesis, in fact, offered a way out of the etiological dead end into which the abandonment of the humors has led in the field of general diseases (Ackerknecht, 1982). It is worth mentioning that also the modern concept of diathesis, as well as that of heredity born in France, gained popularity in that country from the beginning of the nineteenth century. The article "Diathèse" in the popular French *Dictionnaire des science médicales* (1812–1822) stated that diathesis "is that state of the body which makes it acquire certain diseases" (Pariset & Villeneuve, 1814, p. 247). Among the fourteen diatheses mentioned, the original ones were scrophulous, herpetic, cancerous, and gouty, whereas the acquired ones were syphilitic, verminous, scorbutic, and gangrenous. There were also "combined" diatheses, such as mucous, inflammatory, bilious, and serous, and everybody had variolous and measles diatheses. The difference between diathesis and predisposition consisted in that the former always indicated a latent pathologic state, while the latter could exist within a perfect state of health (Rainard, 1840).

9.2.1.2 Diathesis, germ theory, and evolution

Even if the possibility that microorganisms existed was discussed for many centuries before their actual discovery in the seventeenth century, microbiology as a medical discipline was founded not before the work of Louis Pasteur (1822–1895) and Robert Koch. They demonstrated the validity of the *germ theory of diseases*, according to which microorganisms were the causes of many pathologies. However, the growing success of the germ theory during the late nineteenth and the beginning of the twentieth century was not a linear process without any resistance from the scientific community. On the contrary, in particular during the first decades, many physicians expressed criticisms and cautions. Most of them belonged to British medicine because, generally speaking, Victorian medicine was characterized by a conservative tendency which brought many doctors to be particularly cautious in front of the introduction of the "laboratory medicine," remaining in favor of a classic clinical approach still based on bedside observations (Bearn, 1993; Zampieri, 2006).

The concept of diathesis, founded on the Darwinian concepts of variation and evolution, was used by many physicians to counterbalance the growing importance given to germs and the external causes of diseases. James Paget and his pupil Jonathan Hutchinson were among the first to suggest that the study of individual susceptibility still remained important in medicine, despite the demonstration that many diseases were caused by a germ. Paget stated that:

Perhaps the brilliant success which has been achieved by the recent studies of disease-producing organism, or other materials acting on us from without – a success not equaled in any other field of medical enquiry – has made some think too little of those changes within ourselves which occur in such ordinary conditions of life that they may be called spontaneous. Yet these are not less important in the production of diseases, and these must be studied; just as in agriculture, souls must be studied as well as seeds. **(Paget, 1882, p. 1192)**

Hutchinson repeated the same concepts:

We define scrofula to be a state of the solid tissues, and more especially of the lymphatic system as a whole, in which there is peculiar proneness to chronic inflammations resulting in products more or less peculiar and specialized. Modern discoveries will perhaps lead us to believe that those specialized products are usually accompanied by the presence of particulate organisms which are neither part of, nor derived from, the original tissues. Be this as it may, we shall have to study the scrofulous diathesis, and the predisposing causes of scrofula, just as before. We shall still have to ask what are the conditions which favor, or otherwise, the development of the bacillus in question.

(Hutchinson, 1884, p. 127)

Other doctors shared the same idea of Hutchinson, supporting that germ theory of disease was an oversimplification, because it neglected the many other factors which have a part to play in deciding whether the host/germ/environment complex is to lead to infection (Stewart, 1968). This criticism continued to be supported also in the first years of the twentieth century. The Edinburgh doctor Sir Dyce Duckworth (1840–1928), writing on *The Diatheses: The personal factor in disease*, complained that "The doctrine of the diatheses as applied to practical medicine is little in vogue in this twentieth century" (Duckworth, 1908). This is because laboratory medicine, based on the methods of experimental physiology and bacteriology, brought physicians to neglect classic clinical approach:

The clinician is always in face of the *personal factor* in each patient. [...] a long clinical experience has only served to deepen my confidence in the certainty of the doctrine the diatheses or habit of body. The primal significance of this doctrine is that one man is not as another, that there are individual personal peculiarities relating to tendency, to vulnerability, and to immunity. [...] It becomes, therefore, a question relating to the *soil* or tissue-proclivity of the particular individual that we have to consider. The modern investigator is solely occupied with the seed which he believes to be capable of growing and developing its wide-spread powers indiscriminately on any individuality. **(Ibid., p. 700).**

A year later, the British doctor James Lindsay (1856–1931), writing on *Darwinism and Medicine*, stated that:

[...] we must recognize that in disease great secular processes, as well as more concrete causal factors, come into view. The present generation of medical workers and observers is somewhat impatient of those conceptions of "diathesis" and "temperament" which meant so much to our predecessors. [...] Yet a study of evolution in its bearing upon disease may suggest to us that these terms, vague though they are, adumbrate a great truth – namely, that in proclivity to infection and response to disease nothing is so potent as what we vaguely call "constitution", "temperament", "hereditary tendency" and the like. We have been intent upon the "search for the microbe", a search which has already yielded brilliant results, and, it is needless to say, must be continued. But the response of the organism is an element at least as important as the presence of the germ. In the future it may turn out that we shall be compelled to devote less attention to the seed, more attention to the soil.

(Lindsay, 1909, p. 1328)

Darwin's theory of evolution was also used to argue that germs evolved in the same way that all other organism did. A certain Dr James Ross from Manchester, for instance, tried to use Darwin's theory of pangenesis to explain infectious diseases, by supporting that *contagium particles* were not living such as individual organisms, but they were portions detached from a living being with similar characteristics of the *pangenes* which Darwin hypothesized as responsible of hereditary traits (Ross, 1872). The majority of physicians, in any case, used the Darwinian theory of evolution to argue that disease germs of one infection could evolve into the germs of another. The British doctor Kenneth W. Millican (1853–1915), for instance, published a book in 1883 entitled *The Evolution of Morbid Germs* (Millican, 1883). Here, he advanced the idea that the theory of evolution was fundamental to solving several difficulties of the germ theory related to the clinical field. As it could be supported that the ancestors of pathological bacteria were born as non-pathogens and only during the time evolved into infectious diseases, it was reasonable to support that infectious diseases could evolve their characteristics even within a single human host, changing into different pathologies according to different circumstances. This theory derived from the fact that bacteria seemed to be capable to modify themselves in relationship to the medium where they were cultivated and that these acquired modifications seemed to become hereditary (ibid., p. 55). Therefore, doctors advanced that bacteria could change their nature and their pathological effects according to the chemical and physical environment represented by different human hosts.

Darwin's doctrine of evolution, applied to infections, questioned in the end the doctrine of *specificity of infectious disease*, one of the theoretical basis of germ theory according to which any infectious disease was caused by just one specific germ. The Scottish pathologist Sir William Aitken (1825–1892) published a series of papers in *The Glasgow Medical Journal* in 1885 under the title *Darwin's doctrine of evolution in explanation of the coming into being of some diseases* (Aitken, 1885). Here, he stated that:

> All organic existences possess an inherent tendency to vary. [...] We must therefore recognize this fact as regards diseases, that while uniformity is general, it is not universal. Hence it comes that [...] in spite of the influence of even such precise and definite objects (seen or imagined) as "germs", "bacilli" or "bacteria", which have been identified or associated with numerous diseases, the hitherto accepted doctrine of the specificity of disease must to some extent to be modified. **(Ibid., pp. 103–4)**

Darwin's doctrine of evolution, supporting that specific infectious diseases arises from a non-specific fever as a common ancestry, seemed to be particularly valid to explain clinical phenomena which were otherwise still not possible to understand. In particular, that one form of disease could complicate or overlay another and that different infectious diseases could mingle their types and eventually produce a new type. The extreme clinical variability of infectious diseases, in sum, was explained by the idea that a single "poison" could evolve during a single patient's infection, or even during an epidemic, generating clinical phenomena so different that they could be wrongly attributed to different and unconnected poisons or germs.

In 1888, the British surgeon William Job Collins (1859–1946), writing on *Specificity and evolution in disease*, developed the same ideas of Aitken:

> [...] diseases usually accounted specific have under certain conditions arisen independently of contagion, that is to say *de novo*; [...] diverse diseases may own a common external ancestry; [...] the same poison may produce different diseases in different individuals; [...] specificity depends on the soil on which the seed is sown and on which it is subsequently cultivated.
>
> **(Collins, 1888, p. 171)**

Similarly, Chicago physician G. Frank Lydson (1858–1923), more than ten years later, supported the "[...] spontaneous generation of new and virulent properties in hitherto innocuous germs, and a natural variation of type and pathogenic effect of germs supposed to be invariably specific" (Lydson, 1902, p. 1290).

Only during the late 1930s the conflict between the concept of specificity and the theory of evolution was solved with regard to infectious diseases, when scientists such as the French Charles Nicolle (1866–1936) (Nicolle, 1937) and the Australian Frank Mcfarlane Burnet (1899–1985) (Mcfarlane Burnet, 1940) supported that germs were species which evolved exactly like other more complex organisms, that is, by natural selection of spontaneous hereditary variations.

9.2.1.3 Darwin's theories of immunology and neurology

The fundamental theory that evolution was equivalent to a progress and dissolution was the reverse process causing degeneration and disease was at the base of important achievements of late nineteenth century medical science in the fields of immunology and neurology. On the one hand, the Russian zoologist Elie Metchnikoff (1845–1916) applied Darwin's theory to explain how the immune system developed, becoming able to resists diseases. His famous *Leçons sur la pathologie comparée de l'inflammation* were opened by the following statement: "It is not necessary to give special proof for supporting that disease and pathological processes evolve such as men and superior animals" (Metchnikoff, 1892, p. 1). In this book, he advocated that immunity was acquired by a mechanism of struggle for life between leukocytes and bacteria. At the same time, he stressed the imperfection of the organisms, derived from the fact that they were product of a blind natural selection. This could explain why leucocytes, sometimes, had less defensive efficacy and they left a disease to prevail or even contributed to it.

On the other hand, the British neurologist John Hughling Jackson (1835–1911) supported the evolutionary organization of the human nervous system for which he proposed three levels: a lower, evolved for first, a middle, and a higher, most recently evolved. He explained its diseases as a process of dissolution which could affect any of the three levels, starting from the most recent in evolutionary terms and concluding with the most ancient:

> The higher nervous arrangements evolved out of the lower to keep down those lower, just as a government evolved out of a nation as well as directs that nation. If this be the process of evolution, then the reverse process of dissolution is not only 'a taking off' of the higher, but is at the very same time a 'letting go' of the lower. If the governing body of this country were destroyed suddenly, we should have two causes for lamentation: (1) the loss of services of eminent men; and (2) the anarchy of the now uncontrolled people. The loss of the governing body answers to the dissolution in our patient (the exhaustion of the highest two layers of his highest centres); the anarchy answers to the no longer controlled activity of the next lower level of evolution (third layer).
>
> **(Hughling Jackson, 1884, p. 593)**

He supported that neurological symptoms are dual in nature. Negative symptoms resulted from the loss of functions of higher levels, and positive symptoms resulted from the appearance of the function of previously inhibited lower levels. By these theories, Hughlings Jackson created the conceptual framework for clinical neurophysiology, the discipline that underlies diagnostic neurology (York & Steinberg, 2006).

9.2.1.4 Reproduction

Given the main argument of this book, few words might be appropriate about the influence of early Darwinism to midwifery, prenatal and postnatal care. In the decades between the two centuries, the birth of genetics monopolized the debate around the formation of new individuals and the determination of their morphological and mental characteristics. This new science was believed as crucial for shedding light on the mechanisms determining the individual variability considered so important in susceptibility and reaction to disease. Darwinian perspective resulted equally important for the consideration of environmental factors interplaying with hereditary mechanisms. The German doctor Julius Bauer (1887–1979), for instance, stated that "[...] hereditary units or genes themselves are merely potential energies and can act only under conditions supplied by the environment" (Bauer, 1941). He believed that "The study of identical or uniovular twins, their morphologic and functional behavior under normal and pathologic conditions, is the most reliable method of evaluating the share of both environmental and genetic influences" (Ibid., p. 5). Therefore, the Darwinian approach contributed to the study of fertilization and development by stressing the importance of taking into account environmental factors as equally important as genetic ones.

However, the Darwinian and evolutionary concepts were mostly used, in relationship with midwifery, within the eugenic debate. Sterilizations and abortions were proposed as methods to prevent the spreading of unfit individuals among the populations, as well as forbidding marriage among individuals of different "races." This "sexual hygiene" was widely discussed in Europe and the United States between the late nineteenth century and the end of the Second World War. Probably it is not by chance if the first doctor claiming that natural selections failed in the case of civilized man was Lawson Tait, considered one of the fathers of gynecology (Tait, 1966). In his famous paper of 1869, Tait advanced a correlation between the degrees of civilization, the failure of natural selection, the severity of disease and the number of births (Tait, 1869). If natural selection failed preventing deliveries in weak women or of weak individuals, the number of births might increase, producing a genetic and constitutional degeneration in the population.

Using an evolutionary approach found an important application in embryology with the famous theory of recapitulation, according to which the development of the embryo of an animal went through stages resembling or representing successive stages in the evolution of the animal's remote ancestors. Formulated for the first time well before the works of Lamarck and Darwin, this model was proposed within an evolutionary framework by the German biologist Ernst Haeckel. He formulated his theory as "Ontogeny recapitulates phylogeny" (Haeckel, 1866). The Italian clinician Achille De Giovanni (1838–1916) applied this theory, advancing that diseases could result from an interruption of development in a given anatomical part of the embryo, which maintained the feature of inferior animals (De Giovanni, 1904, pp. 59–60).

We will see (in Section 2.3.2) that constitutional medicine, focusing on human constitution, was especially interested in the issues related to the development of the individual from fecundation, to birth, to his life history. The constitution of an individual, in fact, was, at any moment, the result of his developmental history. Therefore, everything related with diseases in pregnancy, anatomical and physiological features of embryogenesis were considered crucial for the understanding of individual constitution (Castaldi, 1928). With the birth of endocrinology, the study of individual development found a new, promising perspective (Barker, 1922), which brought the Italian doctor Nicola Pende (1880–1970) to advance that any constitution was the result of a specific "hormonal constellation" (Pende, 1916; Betta, 2015). The Italian doctor Luigi Castaldi (1890–1945), follower of the constitutional school, produced significant experimental works on the influence of endocrine glands on the morphogenesis (Taccari, 1978).

9.3 Medical education and medical decision-making

The circulation of Darwin theory within the medical community did not affect medical education during the period 1880–1940. On the contrary, in the United States the teaching of evolutionary theory was even forbidden in some states, in 1926 in Mississippi and North Carolina and in 1928 in Kentucky, with attempted bans in Florida and Arkansas in 1927 (Zampieri, 2009, p. 344).

One of the most important changes in medical education was brought on by the famous "Flexner Report," a study of medical education in the United States and Canada, written by Abraham Flexner (1866–1959) and published in 1910 under the aegis of the Carnegie Foundation. By the turn of the twentieth century, a series of scientific breakthroughs had altered the values held by the public and the medical profession: clinical and laboratory research had exposed the irrationality of "heroic" treatments (such as blistering, bleeding and purging) and had proven the therapeutic efficacy and rational scientific basis of modern practices, such as antiseptic surgery, vaccination, and public sanitation. Most of the public and virtually all physicians now believed in the superiority of "scientific medicine." Educators at leading US medical schools now asserted that students should spend most of their time not with the memorization of accepted truths, but actively engaged in laboratory experimentation and hands-on care at the bedside (Beck, 2004, p. 2139).

Evolution was not mentioned in the Flexner report, but one response to the report was for medical schools to establish departments of anthropology to bring human natural history, comparative anatomy, and geographical medicine explicitly into medical training (Morton, 1926). However, evolutionary biology still was not incorporated into medical training. The reductionist scientific paradigm advocated by the Flexner report trained doctors and scientists to reject teleological notions of evolution as constantly and endlessly progressive. Teleology of this kind has been mistakenly conflated with the adaptive process of natural selection and is a common misunderstanding of evolution that still persists (Antolin, 2011). The American physician George Draper (1880–1959), in his work on *Human Constitution*, complained that:

From this point on the present and universally accepted arrangement of the medical curriculum makes it more and more difficult for the student to think of a man except in terms of disease. Whatever suggestion has come to the student of Man's relation to the past, and to more lowly forms, has been through certain smatterings of comparative anatomy. [...] In a word, he knows nothing about these specimens of humanity – whence they came, what their individual characteristics are, and, above all, what their original health potentialities have been. **(Draper, 1924, p. 21)**

Nevertheless, even if evolutionary processes could not be reproduced in the laboratory, evolutionary problems emerged, in particular related to the study of bacteria. In a paper on *The evolutionary aspect of infectious diseases*, Frank Lydson, for instance, supported that "The laboratory investigator has been witnessing the operations of the law of germ adaptation year by year" (Lydston, 1902, p. 1290). Unfortunately, the phenomenon of bacterial adaptation was interpreted, in the first period, with the Lamarckian concept of inheritance of acquired characteristics (Adami, 1918). The fact remains that Darwinian and Lamarckian concepts were important in the birth of laboratory sciences such as bacteriology and immunology.

Moreover, the concepts of "science" and "experimental method" remained at that time broader than those related exclusively to the laboratory. To them, evolutionary theory and Darwinism gave a significant contribution. In particular, diseases were considered by Darwinian doctors as "natural experiments" or "experiments of nature" though which nature manipulated organisms, eliminated the unfit, or even created new adaptations. In 1949, the famous evolutionary biologist John Burdon Sanderson Haldane (1892–1964) wrote a paper on *Disease and evolution*, supporting that infectious diseases could be an accelerator of speciation. Briefly summarized, this occurs when each parasite foments a specialized ecological niche, namely resistance to it. Haldane suggested also that thalassemia heterozygotes might be more resistant to malaria (Haldane, 1949; Lederberg, 1999).

The laboratory became the central place where not only medical education could be carried out, but also medical diagnosis could be scientifically based. From the laboratory came new theories of aetiology and new diagnostic technologies which seemed to give an unprecedented degree of certitude to medical diagnosis. However, the debate around diagnosis remained open. Some clinicians continued to see diagnostic acumen as the centerpiece of medicine, an acumen which could be reached through a continuous bed-side experience, rather than through laboratory instruments. Moreover, many physicians advocated evolutionary theory as a fundamental instrument to improve diagnostic skills and to avoid the danger that doctors would excessively focus on laboratory tests and data, rather than on patients and diseases themselves. Jonathan Hutchinson, in his *Pedigree of Disease*, wrote that:

Our forefathers, who knew far less about the details of pathology than we do, attached far more importance to such matters as temperament and diathesis. They were accustomed to prescribe for a man's temperament; we think only of his disease, and turn aside with weariness from classifications of diathesis in which the physicians of an older day delighted. Although to a large extent this change of sentiment has been the result of advance in knowledge, yet I think it might easily be shown that it has gone too far, and that we now neglect unwisely the study of those differences between man and man of which, for the most part, physiology takes no cognizance, but which may yet prove of much importance in modifying the processes of disease.

(Hutchinson, 1884, p. 1)

The concept of diathesis, in Hutchinson's mind, was crucial for clinical diagnosis, because it allowed for the understanding of individual differences that laboratory medicine failed to take into account. The work of the British neurologist John Hughling Jackson represents another important example of clinical application of the early Darwinian medicine. Starting from his evolutionary interpretation of nervous system and its diseases (as a process of dissolution), he advanced that also clinical symptoms should follow a precise order of manifestation related to the evolutionary antiquity of the nervous systems affected. As already stated, he supported that neurological symptoms could be "negative," that is, resulting from the loss of functions of higher levels, and "positive," that is, resulting from the appearance of the function of previously inhibited lower levels (York & Steinberg, 2006). For instance, Hughling

Jackson pointed out the order of the dissolution of function in the brain when under the influence of alcohol. The first to go were the highest faculties: judgement, memory, taste. In the next stage, speech became affected. Next, the emotions were involved: the individual became hilarious, maudlin, or quarrelsome. Last of all the centers of organic life—respiration, circulation, and deglutition—were affected, and death could ensue (Lindsay, 1909).

More in general, the Darwinian approach combined with the concept of human constitution was important, in that period, for maintaining the ancient idea that diagnosis and therapy had to be based not only on the anatomical lesion, but also on the consideration of the individual as a whole, alive. George Draper, father of the American constitutional medicine, stated that "[...] most doctors are good students of disease, but not of men" (Draper, 1924). In the study of evolution and human constitution, Draper indicated the possible solution for improving a better understanding on disease and therapy:

> The question arises, then, is it possible for doctors to gain a more comprehensive, more synthetic conception of the human animal? Embryology, instead of forming the first chapter in the text-book of obstetrics or anatomy, might well form the starting-point of what would become a course in the natural history of Homo Sapiens. Obviously greater familiarity with the organism in its healthy state, and with the characters which it has derived through the mysterious pathway of inheritance, must lead to a better understanding of the variable factor of susceptibility. Every aspect of a man must be known as accurately as possible before the diseases to which he is susceptible are fully understood. [...] The study of the individual dominate and surrounds the study of his affliction, and it is amazing how frequently the particular pathologic disturbance belongs irresistibly in the whole fabric of his personality. **(Ibid., pp. 22, 45)**

Significantly, Draper underlined that constitutional medicine was not an alternative approach to laboratory and experimental methods: "[...] the study of healthy children and adults by anthropologic methods provides an opportunity whereby another exact science may become a contributor to medicine." (Ibid., p. 31)

9.4 Conclusion

Even if the movement of "medical Darwinism" covers a period along which both medicine and Darwinism deeply changed their methods and theories, we can maintain a unique label because, apart from few exceptions, physicians who developed evolutionary theories of health and disease during these years shared some fundamental conceptions. In particular, we can affirm that medical Darwinism viewed disease as a negative variation in which natural selection had no place. Most of these physicians, in fact, viewed disease as a "dissolution," that is, the inverse process of evolution. The fundamental starting point was that organisms were variable and that this variation was mainly produced by heredity. The heredity of variations was the cornerstone of Darwinian theory, and we have seen that Darwin used many examples from medicine to demonstrate that point, showing how diseases could be transmitted through inheritance. Diseases were conceived, in their fundamental biological meaning, such as organic variations, spontaneous or acquired, of the same nature of those variations which were used by natural selection to shape adaptive characteristics. In his *Malignancy and Evolution*, the British physician Morley Roberts stated that "[...] any disease was itself variation" (Roberts, 1926). From this equivalence, a fundamental dichotomy was established. All variations could be divided into negative and positive ones, the former representing diseases and liabilities to diseases, and the latter representing potential gradations of new adaptations. Negative variations, that is, diseases, were the other side of the coin of natural variability which was the fundamental prerequisite of biological evolution. In other words, they were the inevitable price of natural variability and natural selection could do nothing to eliminate them in a definite way. At best, diseases were an instrument of natural selection for eliminating the unfit, but surely natural selection had no role in determining any characteristic related to diseases. The British surgeon Pitfield Mitchell arrived to deny the hereditary nature of almost all diseases, because otherwise natural selection would have already eliminated all of them (Mitchell, 1888).

The "dawn" of current Darwinian medicine occurred on a different theoretical basis. Randolph Nesse, describing how he founded the new approach with his friend and colleague George Williams (1926–2010), used the following words:

> We began our work by arguing for months about why selection shaped breast cancer, gastric ulcers, atherosclerosis, and the appendix. Finally... finally we recognized our error – selection does not shape diseases so looking for evolutionary explanations for diseases is like a walk in a bog. Selection does, however, shape traits that are surprisingly vulnerable to failure. The appendix and narrow birth canal are obvious, but inability to better regulate cell division and inflammation are responsible for huge amounts of disease. Such traits do have evolutionary explanations. It was this small change in the target of explanation, from diseases to traits that leave us vulnerable to diseases that kick started the field. Not that people had not asked such questions before. But we encouraged people to take them very seriously indeed, always comparing the body as it is, to the body as it could have been. Over and over again, we discovered that we were mistaken in our quick assumptions about how the body could be improved. However, we also found plenty of examples where path dependence or chance historical events left bodies with traits, such as our inside out eyeballs, that no engineer would countenance for a minute! Anyone who thinks evolutionary medicine sees the body as perfect is not paying attention. A deep evolutionary view reveals the bodies of all organisms as bundles of tradeoffs, compromises and limitations that ensure that perfection is impossible. (http://www.randolphnesse.com/articles/darwinian-medicine).

The shift from disease to vulnerability to disease is what mainly differentiates medical Darwinism from Darwinian medicine and characterizes current evolutionary approaches in medicine. Thanks to this new theoretical basis, new questions and research programs have arisen which are still in full development. In the period from 1950–1980, evolutionary biology achieved new results which proved to be fundamental for an evolutionary understanding of human diseases, including that evolution works on differential reproduction of genes, rather than individuals, and that evolution shapes adaptation that is optimal, not perfect (Zampieri, 2006). Moreover, current evolutionary biology stands on a scientific basis leaving no more room for racial theory, which represented one of the main dangers for a distorted version and use of evolutionary explanation in medicine.

References

Ackerknecht, E. H. (1982). Diathesis: The word and the concept in medical history. *Bulletin of the History of Medicine*, 56: 317–25.

Adami, G. (1907). Inheritance and disease. In Osler, W. (ed.) *Modern Medicine. Its Theory and Practice in Original Contributions by American and Foreign Authors*, vol. 1: (pp. 17–50). Philadelphia and New York: Lea Brothers & Co.

Adami, G. (1918). *Medical Contribution to the Theory of Evolution*. London: Duckworth and Co.

Adams, J. (1814). *A treatise on the supposed hereditary properties of diseases containing remarks on the unfounded terrors and ill-judged cautions consequents on such erroneous opinions, with notes illustrative on the subject, particularly in madness and scrofula*. London: J. Callow.

Aitken, W. (1885). Darwin's doctrine of evolution in explanation of the coming into being of some diseases. *The Glasgow Medical Journal*, 24: 98–107; 161–72; 241–53; 354–98; 431–46.

Allen, F. J. (1903). Darwinism and the increase of cancer. *The British Medical Journal*, 1: 1527.

Antolin, M. (2011). Evolution, medicine, and the Darwin family. *Evolution: Education and Outreach*, 4: 613–23.

Barker, L. F. (1922). The relation of the endocrine glands to heredity and development. *Science*, 55: 685–90.

Barrett, P. H. (1974). Early writings of Charles Darwin. In H. E. Gruber, (ed.) *Darwin on Man. A Psychological Study of Scientific Creativity; Together with Darwin's Early and Unpublished Notebooks. Transecribed and annotated by Paul H. Barrett, commentary by Howard E. Gruber. Foreword by Jean Piaget*, vol. II, (pp. 260–479). London: Wildwood House.

Bauer, J. (1941). *Constitution and Disease. Applied Constitutional Pathology*, (p. 4). New York: Grune & Stratton.

Bearn, A. G. (1993). *Archibald Garrod and the Individuality of Man*, (pp. 27–8. Oxford: Clarendon Press.

Beck, A. H. (2004). The Flexner Report and the standardization of American medical education. *JAMA*, 291: 2139–40.

Belon, P. (1555). *Histoire de la nature des Oyseaux avec leur description et naïf portraits retirés au naturel* (pp. 42–3). Paris: Chez Guillaume Cauellat.

Bendyshe, T. (1865), *The Anthropological Treatises of Johann Friedrich Blumenbach*. London: The Anthropological Society.

Bénichou, C. (1996). Dégénération, dégénérescence. Degeneration. In Tort, P. (ed.) *Dictionnaire du Darwinisme et de l'Evolution*, vol. I, (pp. 1151–57). Paris: Presses Universitaire de France.

Betta, E. (2015). Pende, Nicola. In *Dizionario biografico degli italiani*, vol. 82, (pp. 301–3). , Roma: Treccani.

Bland-Sutton, J. (1890). *Evolution and Disease*. London: Walter Scott.

Blumembach, J. F. (1798). *Über die natürlichen Verschiedenheiten im Menschengeschlechte*. Leipzig: Breitkopf und Härtel.

Brüne, M. (2007). On human self-domestication, psychiatry, and eugenics. *Philosophy, Ethics, and Humanities in Medicine*, 2: 1–9.

Bynum, W. (1983). Darwin and the doctors: Evolution, diathesis and germs in 19th-century Britain. *Gesnerus*, 1/2, 43–53.

Cameron, H. C. (1914). Diathesis in infancy: A plea for its closer study. *The British Medical Journal*, 2: 53–6.

Cantlie, J. (1885). *Degeneration Amongst Londoners*. London: Field & Tuer.

Castaldi, L. (1928). *Accrescimento Corporeo e Costituzioni dell'uomo*. Firenze: Ditta Editrice Luigi Nicolai.

Castiglioni, A. (1931). *L'orientamento neoippocratico del pensiero medico contemporaneo*. Torino: Tipografia editrice "Minerva".

Cirilli, E. (1984). Medicina costituzionalistica e fascismo. In Battimelli, G. De Maria, M. Rossi, A. (eds.) *La Ristrutturazione Delle Scienze tra le due Guerre Mondiali*, vol. 1, (pp. 347–54). Rome: La Goliardica.

Collins, W. J. (1888). Specificy and evolution in disease. *Public Health*, 2: 170–3.

Crichton-Browne, J. (1897). Ethics and individualism in medicine. *The Lancet*, 150: 835–42.

Darwin, C. (1868). *The Variation of Animals and Plants under Domestication*. London: John Murray.

Darwin, E. (1794–1796). *Zoonomia; or the Laws of Organic Life* vol. 1, (pp. 392–4). 2 vols. St. Paul's Church-Yard, London: Printed for J. Johnson.

Darwin, F. (1887). *The Life and Letters of Charles Darwin, Including an Autobiographical Chapter*, vol. 1, (p. 38). 3 vols. London: John Murray.

De Giovanni, A. (1898). *Morfologia del Corpo Umano*. Milano: Hoepli.

De Giovanni, A. (1904). *Commentarii di Clinica Medica Desunti dalla Morfologia del Corpo Umano* (pp. 59–60). Milano: Hoepli.

Draper, G. (1924). *Human Constitution. A Consideration of Its Relationship to Disease* (pp. 18, 21, 22, 45). Philadelphia and London: W. B. Saunders Company.

Duckworth, D. (1908). An address on the diatheses: The personal factor in disease. *The Lancet*, 171: 699–702.

Emery, A. E. (1989). Joseph Adams (1756–1818). *Journal of Medical Genetics*, 26: 116–18.

Freud, S. (1921). *Zeitgemäßes über Krieg und Tod*. Verlag, Leipzig: Internationaler Psychoanalytischer.

Garrod, A. (1902). The incidence of alkaptonuria: A study in chemical individuality. *The Lancet*, 160: 1616–20.

Garrod, A. (1909). *Inborn Errors of Metabolism*. H. Frowde, Hodder & Stoughton, London: Oxford University Press.

Garrod, A. (1927). The Huxley lecture on diathesis. Delivered at the Charing Cross Hospital, November 24th, 1927. *The British Medical Journal*, 2: 697–71.

Gould, S. J. (2002). *I Have Landed: The End of a Beginning in Natural History*. New York: Harmony Books.

Gregory, J. W. (1931). *Race as a Political Factor*. London: Watts & Co.

Haeckel, E. (1866). *Generelle Morphologie der Organismen: Allgemeine Grundzüge der Organischen Formen-Wissenschaft, Mechanisch Begründet Durch die von Charles Darwin Reformierte Deszendenz-Theorie*. Berlin: G. Reimer.

Haller, A. (1753). De partibus corporis humani sensilibus et irritabilibus. *Commentarii Societatis Regiae Scientiarum Gottingensis*, 2: (pp, 114–58).

Haller, J. (1981). *American Medicine in Transition 1840–1910*, (pp. 280–1). Chicago, London: University of Illinois Press, Urbana.

Haldane, J. B. S. (1949). Disease and evolution. *Ric Sci Suppl A*,;19: 68–76.

Heron, D. (1907). *First Study of the Statistics of Insanity and the Inheritance of the Insane Diathesis*. London: Dulau & Co.

Hufeland, C. H. (1800). *System der practischen Heilkunde; ein Handbuch für akademische Vorlesungen und für den practischen Gebrauch*. Leipzig: F. Frommann.

Hughlings Jackson, J. (1884). The Croonian lectures on evolution and dissolution of the nervous system. Delivered at the Royal College of Physicians, March 1884. *The British Medical Journal*, 1: 591–93, 660–3, 703–7.

Hunter, J. (1861). *Essay and Observations on Natural History, Anatomy, Physiology, Psychology, and Geology*, vol. 1, (p. 1). 2 vols. Paternoster Row, London: John Van Voorst.

Hurst, A. F. (1927). An address on the constitutional factor in disease. Delivered before the Ulster Medical Society on November 17th, 1926. *The British Medical Journal*, 1: 823–27.

Hutchinson, J. (1884). *The Pedigree of Disease; Being Six Lectures on Temperament, Idiosyncrasy and Diathesis, Delivered in the Theatre of the Royal College of Surgeons in the Session of 1881* (pp. 1, 127). London: J. & A. Churchill.

Keith, A. (1930). The inexorability of the law of evolution as manifested in modern medicine. *The British Medical Journal*, 1: 893–95.

Jordan, D .S. (1906). *The Blood of the Nation. A Study of the Decay of Races through the Survive of the Unfit*. Boston: American Unitarian Association.

Laennec, R. (1819). *De L'auscultation Médiate ou Traité du Diagnostic des Maladies des Poumons et du Cœur Fondé Principalement sur ce Nouveau Moyen D'exploration* (p. xxi). Paris: Brosson & Chaudé.

Lawrence, W. (1819), *Lectures on Physiology, Zoology, and the Natural History of Man*. London: J. Callow.

Lederberg J. B. S. Haldane (1949). On infectious disease and evolution. *Genetics*,153: 1–3.

Lindsay, J. A. (1909). The Bradshaw Lecture on Darwinism and Medicine. Delivered at the Royal College of Physicians, London, on November 2nd, 1909. *The British Medical Journal*, 2: 1325–32.

López-Beltrán, C. (2003). Heredity old and new; French physicians and l'hérédité naturelle in early 19th century. In Muller-Wille, S. (ed.) *A Cultural History of Heredity: 18th and 19th centuries*, (7–19). Berlin: Max Planck-Institute for the History of Science.

Lucas, P. (1847–1850). *Traité philosophique et physiologique de l'hérédité naturelle dans les états de santé et de maladie du système nerveux, avec l'application méthodique des lois de la procréation au traitement général des affections dont elle est le principe. Ouvrage où la question est considérée dans ses rapports avec les lois primordiales, les théories de la génération, les causes déterminantes de la sexualité, les modifications acquises de la nature originelle des êtres, et les diverses formes de névropathie et d'aliénation mentale*. 2 vols. J.-B. Paris: Baillière.

Lydson, G. F. (1902). The evolutionary aspect of infectious diseases, with especial references to the local venereal diseases. *JAMA*, 38: 1290–92.

Mcfarlane Burnet, F. (1940). *Biological Aspects of Infectious Disease*. Cambridge: Cambridge University Press.

Mendelsohn, J. A. (2001). Medicine and the making of bodily inequality in twentieth-century Europe. In Guadillière J. P. Lowy I. (eds.) *Heredity and Infection. The History of Disease Transmission*, (21–79). London and New York: Routledge.

Metchnikoff, E. (1892). *Leçons sur la Pathologie Comparée de L'inflammation*. Paris: G. Masson.

Millican, K. W. (1883). *The Evolution of Morbid Germs: A Contribution to Trascendental Pathology*. London: H. K Lewis.

Mitchell, C. P. (1888). *Dissolution and Evolution and the Science of Medicine*. London: Longmans, Green.

Morgagni, G. B. (1761). *De sedibus et causis morborum per anatomen indagatis*. Venetiis: Ex Typographia Remondiniana.

Morselli, E. & Tamburini, A. (1875). Contributo allo studio sperimentale delle degenerazioni fisiche e morali dell'uomo. *Rivista Sperimentale di Freniatria*, I: 49–61.

Morton, D. J. (1926). The relation of evolution to medicine. *Science*, 64: 394–96.

Nash, J. T. C. (1915). *Evolution and Disease*. ,Bristol: Wright.

Nicolle, C. (1937). *Le Destin des Maladies Infectieuses*. Paris: Félix Alcan.

Paget, J. (1882). The Bradshawe Lecture on some rare and new diseases. *The British Medical Journal*, 2: 1189–93.

Pancaldi, G. (1983). *Darwin in Italia*. Bologna: Il Mulino.

Pariset, J. & Villeneuve, L. (1814). Diathèse. In *Dictionnaire des Science Médicales*, Vol. IX, (pp. 247–65). Paris: Panckoucke

Pende, N. (1916). *Endocrinologia. Patologia e Clinica Degli Organi a Secrezione Interna*. Milano: Casa Editrice Dottor Francesco Vallardi.

Pende, N. (1927). *Trattato Sintetico di Patologia e Clinica Medica*. Manfredi Principato, Messina.

Pende, N. (1933). *Bonifica Umana Razionale e Biologia Politica*. Bologna: Cappelli.

Pinel, P. (1797). *Nosographie Philosophique, ou la Méthode de L'analyse Appliquée à la Médecine* (pp. xxiij–xxiv). 2 vols. Paris: Chez Maradan.

Porter, R. (1993). Disease of civilization. In Bynum W. Porter R. (eds.) *The Companion Encyclopedia of the History of Medicine*, vol. 1, (pp. 585–99). London & New York: Routledge.

Prichard, J. C. (1813), *Researches into the Physical History of Man*. London: John and Arthur Arch.

Proctor, R. N. (1988). *Racial Hygiene: Medicine under the Nazis*. Cambridge, MA: Harvard University Press.

Rainard, J. (1840). *Traité de pathologie et de thérapeutique générales vétérinaires* (p. 147). 2 vols. Paris: Bouchard-Huzard.

Ribber, H (1918). *Heredity, Disease and Human Evolution*. New York: Critic & Guide Co.

Roberts, M. (1926). *Malignancy and Evolution*, (p. 303). London: Nash & Grayson.

Ross, J. (1872), *The Graft Theory of Disease, being an Application of Mr Darwin's Hypothesis of Pangenesis to the Explanation of the Phenomena of the Zymotic Diseases*. London: Churchill.

Ryle, A. J. (1933). Medicine and eugenics. *Eugenics Review* 30: 9–19.

Scartabellati, A. (2008). Il dovere dei medici italiani nell'ora presente. Biopolitica, seduzione bellica e battaglie culturali nelle scienze umane durante il primo conflitto mondiale. *Medicina & Storia*, 14: 65–94.

Singleton, M. M. (2014). The "science" of eugenics: America's moral detour. *Journal of American Physicians and Surgeons*, 19: 122–5.

Smith, G. M. C. (1908). The inheritance of the insane diathesis. *The British Medical Journal*, 2: 1457.

Spencer, H. G. & Paul, D. B. (1998). The failure of a scientific critique: David Heron, Karl Pearson and Mendelian eugenics. *The British Journal for the History of Science*, 31: 441–52.

Stewart, G. T. (1968). Limitations of the germ theory. *The Lancet*, 291: 1077–81.

Taccari, E. (1978). Castaldi, Luigi. In *Dizionario biografico degli italiani*, vol. 21, (pp. 301–3). Roma: Treccani.

Tait, L. (1869). *Has the Law of Natural Selection by Survival of the Fittest Failed in the Case of Man?* Dublin: John Falconer.

Tait, R. L. (1966). Lawson Tait (1845–1899), Scottish gynecologist. *JAMA*, 198: 1363.

Weindling, P. (1993). *Health, Race, and German Politics Between National Unification and Nazism, 1870–1945*. Cambridge: Cambridge University Press.

Wells, K. D. (1973). William Charles Wells and the Races of Man. *Isis*, 64: 215–25.

Wells, W. C. (1818). *Two Essays: One Upon Single Vision with Two Eyes; the other on dew. A Letter to the Right Hon. Lloyd, Lord Kenyon and an Account of a Female of the White Race of Mankind, Part of whose Skin Resembles that of a Negro; with some Observations on the Causes of the Differences in Colour and Form Between the White and the Negro Races of Men*. London: Printed for Archibald Constable and Co.

Williams, G. C. & Nesse, R. M. (1991). The dawn of Darwinian medicine. *The Quarterly Review of Biology*, 66: 1–22.

Worboys, M. (2000). *Spreading Germs: Diseases Theories and Medical Practice in Britain 1865–1900*. Cambridge: Cambridge University Press.

York, G. K. & Steinberg, D. A. (2006). *An Introduction to the Life and Work of John Hughlings Jackson with a Catalogue Raisonné of his Writings*. London: The Welcome Centre for the History of Medicine at UCL.

Zampieri, F. (2006). *Storia e Origini Della Medicina Darwiniana* (pp. 136–68). Parma: Mattioli.

Zampieri. F. (2009). Medicine, evolution and natural selection: An historical overview. *The Quarterly Review of Biology*, 84(4): 333–55.

Zampieri, F. (2012). Early Darwinism and Italian psychiatry: The case of Cesare Lombroso. *Medicina & Storia*, 1–2: 111–38.

Zampieri, F. (2016). *Il Metodo Anatomo-Clinico Fra Meccanicismo ed Empirismo. Marcello Malpighi, Antonio Maria Valsalva e Giovanni Battista Morgagni*. L'Erma di Bretschneider.

CHAPTER 10

Tinbergean Approach to Clinical Medicine

B. Natterson-Horowitz

10.1 Nobel Prize

In 1973, two ornithologists and a bee biologist were awarded the Nobel Prize in Physiology or Medicine. The decision to award a trio of animal behaviorists with this prize served not only as a ratification of the deep and essential connection between human and non-human animals; it also suggested that universal principles, rooted in shared evolutionary heritage, were at play in all behavior, health and disease.

It is also notable that the Nobel Prize in physiology or medicine was granted to three men, only one of whom, Konrad Lorenz, was a trained physician (Burkhardt, 2005). Lorenz practiced medicine only briefly during World War Two before abandoning the field for his study of the instincts of geese and other birds. Karl von Fritsch, a co-recipient, is best known for his work characterizing the waggle dance of the honeybee (Kruuk, 2004). Of the Laureates, it was Nikolaas Tinbergen, a Dutch ornithologist, whose theoretical framework would be most foundational to the emergence of the field of evolutionary medicine decades later (Nobel Prize, 2018).

A prolific writer and researcher, Tinbergen is credited with bringing much-needed experimental rigor to the nascent field of animal behavior which had historically focused on observation and for encouraging a synthesis of experimental and observational approaches. The scientific partnership that developed between Tinbergen and Konrad Lorenz was strengthened by the contrast between Tinbergen's experimental design and implementation and Lorenz's observational and intuitive approach. While some of their experimental findings were later challenged, their systematic efforts to identify the biological and environmental forces shaping behavior elevated the study of animals from a primarily descriptive to unequivocally scientific enterprise (Schleidt, Shalter & Moura-Neto, 2011). Still, the importance of this work to fields beyond animal behavior was not something even they recognized during the early days of collaboration.

Tinbergen himself described his authentic surprise that his work on bird and fish behavior had been deemed worthy of the Nobel Prize in physiology or medicine. In an autobiographical essay, Tinbergen reports responding to the caller from Swedish Embassy with an incredulous, "Are you sure?" (Dewsbury, 1985). He reports having considered whether Lorenz's work on imprinting with its impact on emerging psychiatric theories might have risen to that level but he was truly shocked that his work had also been selected. Of course, others recognized that the early ethologists were providing not only a foundation for the future discoveries in human psychology and psychopath-

Horowitz, B. N., *Tinbergean Approach to Clinical Medicine* In: *Integrating Evolutionary Biology into Medical Education—for maternal and child healthcare students, clinicians, and scientists*. Edited by: Jay Schulkin and Michael L. Power, Oxford University Press (2020). © Oxford University Press. DOI: 10.1093/oso/9780198814153.003.0011

ology but an expanded, ecologically-informed approach to all biologic phenomenon.

Tinbergen's most impactful and enduring contributions came in the form of a heuristic structures for conceiving causation in biology. Building on the work of Julian Huxley and Ernst Mayr, Tinbergen asserted that a holistic understanding of any biologic phenomenon required a consideration of causation from four discrete perspectives: the proximate, developmental (ontogenetic), phylogenetic (comparative) perspective, and functional/adaptive ((Huxley, 1914; Mayr, 1961; Burkhardt, 2005). Although elements of this structure appear in his earliest work, it was the 1963 publication of *On the Aims and Methods of Ecology* that the framework was presented in its entirety (Tinbergen, 1963).

The awarding of the Nobel Prize for the first time for "non-reductionistic behavioral research" sent a strong message to scientific communities around the world that for any biological phenomenon to be fully understood, its context was crucial (Kirk, 2009). An essay by Marler and Griffen published in *Science* in 1973 supported the broad future application of their work: "(they) showed how an ecological decision made in phylogeny can reverberate through many aspects of the Biology(*sic*) of a species" (Marler, 1973; Dewsbury, 1985).

Over the past decade, interest in leveraging evolutionary biology to better understand human health and pathology has led to a "rediscovery" of Tinbergen's powerful framework for looking at theories of causation. Published works such as Hans Kruuk's *Niko's Nature: The Life of Niko Tinbergen* and Joan Straussmann's "Tribute to Tinbergen: The Place of Animal Behavior in Biology," both provide a pronounced analysis of his works and contributions to the framework in which we use to better frame our understanding of animal behavior and scientific investigations (Kruuk, 2004; Straussmann, 2014).

What is the evidence that evolutionarily-informed theories of causation advance biomedical thinking and innovation? Until quite recently, the explicit use of evolutionary ideas in biomedical research was limited (Nesse, 2001). Increasingly, evolutionary biologists, ecologists, and clinicians are creating collaborations to better understand health and disease. Important books offering a framework for this approach include Stephen Stearns and Ruslan Medzhitov's *Evolutionary Medicine*, Gluckman, Beedle, Buklijas, and Hanson's *Principles of Evolutionary Medicine*, Michael Power and Jay Schulkin's *The Evolution of Obesity*, and Randolph Nesse's *Why We Get Sick*, among others (Stearns & Medzhitov, 2015; Gluckman & Beedle, 2016; Buklijas & Hanson, 2016; and Nesse & Williams, 1996). A growing community of scholars are contributing to the conceptualization and translational aspects of the field. Professional societies and academic programs are emerging to support growing interests in this interdisciplinary approach to medicine.

Tinbergen's "Four Questions" have provided a unifying foundation upon which multiple disciplines can collaborate to advance evolutionary and ecological thinking about human medical disorders. In fact, an interdisciplinary lens is embedded in Tinbergen's "Four Questions" of causation. Tinbergen's "Four Questions" are highly salient for physicians because they address a central focus of medicine: understanding disease causation: 1) What are the mechanistic explanations? 2) What are the developmental explanations? 3) What are the phylogenetic perspectives? 4) What are the adaptive explanations? Questions 1) and 2) can be broadly conceived of as providing proximate causal explanations while questions 3) and 4) express a more evolutionarily-informed perspective.

The importance of the Tinbergean approach for physicians may be best appreciated through the lens of these four questions because they address a central focus of medical education and practice. It is widely appreciated among clinical practitioners that multiple contributing factors are typically responsible for the emergence of pathology. This concept of "multifactoriality" is embedded within diagnostic thought. Typically, a multifactorial explanation for disease takes into account factors including genetic, environmental effects, social and psychological influences and others. Tinbergen's four questions expand the window through which causation might be understood, adding to factors which must be considered the phylogenetic and adaptive perspectives.

In modern medical practice, diseases are typically understood and explained through proximate (mechanistic and developmental) explanations. The addition of comparative and adaptive considerations widens the aperture through which biological phenomenon can be understood. It is important to consider that Tinbergen's theoretical work was focused exclusively on understanding the cause of animal behavior. Tinbergen spent the later decades of his life applying his heuristic framework to human behavioral disturbances, primarily autism. However, he never attempted to apply this theoretical scaffolding to advancing the understanding of the etiology or natural history of somatic medical illnesses (Tinbergen, 1963).

While Tinbergen himself might not have foreseen how foundational his work would be to the field of medicine, his work has provided a structure for considering evolutionary contributions to medical and psychiatric illnesses. Milestones in the history of modern medicine over the past 400 years reflect a focus on, even preoccupation with, identifying reductionistic and proximate explanations for disease. From van Leeuwenhoek's discovery of bacterium in 1683, to Koch and Pasteur's identification of germ theory as a cause of disease, to Wilhelm Roentgen's 1895 discovery of the X-ray, to Watson and Crick's DNA analyses, investigation at a reductionistic level has dominated conceptualization of and hypotheses for disease causation. Emerging from this approach have been many extremely clarifying mechanistic and developmental explanations which have sparked many therapeutic approaches. Consider, for example, how these proximate (mechanistic and developmental) perspectives dominate common clinical explanations for myocardial infarction, cerebrovascular accident, melanoma, or lung carcinoma.

Physicians regularly inform their patients that the "causes" of their myocardial infarctions and strokes are a build-up of atherosclerotic plaque or the development of a blood clot (Virmani et al., 2006). A patient's skin or lung cancer is attributed to spontaneous-, inherited- or exposure-related (sun or cigarette) mutations (Kennedy et al., 2003; Hecht, 1999). Developmental explanations often accompany these mechanistic explanations. Atherosclerotic plaques that lead to myocardial infarctions often begin decades before the cardiac event, patients are told. Similarly, melanoma or lung-cancer causing mutations are explained through retrospective identification of teenaged sunburn or decades of smoking, respectively (Kennedy et al., 2003; Hecht, 1999). Developmental explanations, while drawing from a broader time frame than hyper-mechanistic ones, are still proximate in nature.

While the mechanistic and developmental explanations offered by physicians for thousands of conditions and disorders are not incorrect, they are also not complete. They offer little insight into the contributions of ecological and evolutionary effects in the pathogenesis and natural history of health and disease. Bringing evolutionarily-informed considerations into discussions of causation spark new directions of investigations and generate novel hypotheses. But what is an evolutionary approach to disease causation and how can it lead to novel testable hypotheses?

An evolutionary approach to understanding disease causation requires asking two questions not typically considered by physicians. First, a phylogenetic perspective question: what non-human animals might be vulnerable to these diseases? Secondly, an adaptive perspective question: whether there are fitness-enhancing elements embedded within the vulnerability. Both of these questions are not traditionally considered in medical practice or investigation. However, each offers scientific insights that the traditional mechanistic and developmental explanations cannot offer.

Phylogenetic analyses facilitate the identification of patterns between and across taxa and the correlations between life history and a species' vulnerability to pathology. Recent work comparing rates of cancer across mammalian species led to the identification of a possible evolved cancer suppression mechanism (Sulak et al., 2016; Abegglen et al., 2015; Aktipis et al., 2015). Moreover, looking between and across animal taxa is a powerful hypothesis generator. This approach can facilitate the identification of natural animal models of disease vulnerability, resistance or the recognition of shared environmental factors in the genesis of diseases across multiple species (Singer, 2007).

Adaptive perspectives identify possible evolutionary explanations for why an organism evolved or retained susceptibility to a disease or disorder. Why does vulnerability to a particular disease exist and why does it persist? What are the potential elements of vulnerability to this disease that might have fitness benefit for the individual? Adaptive hypotheses emerge from recognition that while a particular behavior or trait may itself appear *non-adaptive*, embedded within the trait is some *adaptive* component. The adaptive frame recognizes the ancient basis of vulnerability to disease and asks if aspects of this vulnerability may have conferred adaptive benefits to our animal ancestors. For example, the acute formation of blood clots (thrombi) leading to the occlusion of arteries in the brain and cardiovascular systems are responsible for many strokes and heart attacks. While these events are highly maladaptive, vulnerability to rapid blood clotting can be understood through an evolutionary and adaptive perspective: having rapid emergency hemostasis systems is important for animals in which trauma and potential exsanguination have been important, existential threats (Furie & Furie, 1992).

Still, evolutionary thought has propelled investigation and important advances into some of the most high-impact pathologies in humans. The emergence of resistance to therapies, for example, in both infectious pathogens and cancer cells, is being increasingly understood as a real-time consequence of selection, a primary evolutionary process. Indeed, some of the most efficacious therapeutic strategies that are emerging to counter resistance are informed by the evolutionary processes (Pepper et al., 2009; Thomas et al., 2013).

Cancer biologists are reconsidering the "scorched earth" chemotherapeutic approaches which may select for the most aggressive cell lines. Cancer-related mortality is most commonly the result of the metastatic spread of the disease and not from the localized effects of malignancies (Fidler, 1978). Biologists traditionally believed metastasis occurred later in the natural history of a malignant tumor and standard therapy and treatments involved the wiping out of the local, cancerous tissues and cells before it could spread (DeVita & Chu, 2008). However, cancer researchers are now borrowing concepts from our understanding of infectious agents to devise new methods for cancer therapy, given the less-than-optimal patient outcomes from the "scorched earth" approach (Boddy et al., 2015; DeVita & Chu, 2008: Casas-Selves & DeGregori, 2011). These therapies may, in fact, inadvertently facilitate the spread of metastatic tumors and promote resistance to chemotherapeutic agents. Cancer biologists have illustrated this point utilizing models from infectious disease studies to demonstrate in-host competition and its effects on altering the fitness costs and benefits of resistance in cancer (Read, Day, & Huijben, 2011). Principles of microbial survival and evolution are now being utilized to address the limitations and reduce the probability of resistance to drug treatments.

One unsung "triumph" of evolutionary perspectives in modern medicine is the markedly reduced mortality due to advanced congestive heart failure (Lechat, 2006). Until the late 1980s, congestive heart failure was conceived of as primarily an issue of dysfunction of the ventricular myocardial—the cardiac "pump." Therapies were guided by this model and included a heavy reliance on inotropic agents derived from digitalis. If the problem was a weak pump, as it was widely believed, then the therapy should address focus on making the pump work harder. Inotropes, drugs which can increase the vigor of cardiac contraction, were widely prescribed as they can, at least transiently, improve the pumping function of a weak ventricular muscle. Unfortunately, inotropes were not effective in reducing mortality. In past decades, a diagnosis of congestive heart failure carried a prognosis worse than some malignancies (Schocken et al., 1992).

The emergence of a new model for heart failure in the 1980's would change therapy and outcomes for heart failure. The pathological focus shifted from the failing heart muscle itself, to what was described as the body's "maladaptive response "(Ferrari et al., 1996). Injury to the ventricle was activating neuroendocrine systems. which themselves were contributing to patient morbidity and mortality.

The use of the term maladaptive signaled a recognition that the same physiologic responses which evolved as functional adaptations in our animal ancestors may be deleterious in modern human bodies. For example, activation of neuroendocrine systems leading to volume retention and vasoconstriction is adaptive in settings of hypovolemia due dehydration or bleeding, existential threats in the lives of most vertebral species. However, volume retention and vasoconstriction underlie much of the morbidity and mortality in human heart failure cases. This shifting of the model from "pump dysfunction" to maladaptive responses led to more effective therapies targeting maladaptation (Yusuf, 1993; Sacks, Jarcho, & Curfman, 2014).

Examples of other evolutionarily-informed hypotheses leading to improvements in patient care are slowly emerging. Still, evolution remains greatly unleveraged as an approach to improving our understanding of our species' most pressing medical and psychiatric concerns.

Tinbergen's questions, particularly his phylogenetic and adaptive queries, serve as a clarifying guide for this effort. Encouraging physicians to consider the non-uniqueness of medical disease to humans can substantially improve the quality of hypotheses and investigation. As physicians become aware of the spontaneous occurrence of the most common and deadly so-called "human" diseases—atherosclerosis, breast cancer, melanoma—in non-human animals—it reframes their understanding of disease causation. This comparative view expands the window through which they understand the genesis and progression of these disorders.

The 2015 publication of a study by pediatric oncologist Joshua Schiffman and his research colleagues at University of Utah provides an excellent example of how a Tinbergean approach can spark a series of new questions from the collaborative work of clinicians and scientists. (Abegglen et al., 2015). An observation that elephants have less cancer than predicted if the number of cancer cells, mutations, and animal size were simply or clearly linked, led to an investigation of the elephant genome and the identification of an amplification of TP53, a cancer suppression gene, in proboscideans, both living and extinct. A comparison of how lymphocytes responded to radiation demonstrated markedly augmented apotosis—a cancer suppression mechanism—in elephants as compared with healthy humans.

This suggests that as proboscideans increased in size about 30 million years ago, cancer may have presented a constraint and this lineage may have co-evolved cancer suppression mechanisms (Sulak et al., 2016). Efforts to use this finding in elephants to protect humans with increased cancer susceptibility are underway. While the results of these investigations are not yet known, what is certain is that this research path would not exist had a comparative and phylogenetic observation not been made (Abegglen et al., 2015).

An equally powerful and also underleveraged evolutionary tool which can be applied to biomedical investigation is derived from Tinbergen's fourth question regarding adaptation. Diseases themselves are rarely, if ever, adaptive. Yet, embedded within the vulnerability to many diseases may be elements of physiology and anatomy which have been adaptive in some way for our animal ancestors. Consider the heart failure example previously described: having decompensated heart failure with marked fluid retention and an elevated systemic vascular resistance is clearly *maladaptive.* However, embedded within the vulnerability to this process is a dynamically responsive neuroendocrine system responding with volume retention and powerful vasoconstriction to decreased perfusion pressure—a highly *adaptive* response to dehydration and exsanguination.

The Tinbergean approach can be applied to any question in medicine and sparks novel, testable hypotheses. It starts with questioning why human patients are vulnerable to pathology and considers a range of mechanistic and developmental explanations. However, if the question remains unanswered, a phylogenetic analysis is performed to reveal the non-human animals in which the disorder spontaneously occurs. Based on these findings, a specific, testable adaptive hypothesis—rooted in ecological principles and considerations of life history—can be developed. Examples of the application of this framework to highlight high-impact human diseases follow.

10.2 Hematologic

Why are humans vulnerable to the rapid formation thrombus in the vasculature–a phenomenon that contributes to myocardial infarction and stroke morbidity and mortality? (Furie & Furie, 1992).

1) In which non-human animals do clotting cascades lead to rapidly-forming thrombus?
2) What are the life histories of animals with slower or faster throbogenesis?
3) What environmental factors accelerate or slow the clotting cascade in these species?
4) Correlating life histories with coagulation profiles of these species, what adaptive benefits of thrombogenesis can be identified?

10.2.1 Testable Adaptive Hypothesis

Rapid thrombogenesis leading to vascular occlusion is maladaptive in modern humans. However, across a wide range of vertebral taxa, robust emergency hemostasis offers fitness benefits for animals in which exsanguination related to trauma is a leading, existential threat.

10.3 Cardiovascular 1

Why are humans vulnerable to atherosclerosis, leading to heart attacks and strokes?

1) In which non-human animals does atherosclerosis spontaneously develop?
2) What are the life histories of species in which spontaneously-occurring atherosclerosis have been identified?
3) What are the environmental factors that may be contributing to this vulnerability to atherosclerosis?
4) Correlating life histories with specific endothelial characteristics, what are potential adaptive benefits to vulnerability to atherosclerosis?

10.3.1 Testable Adaptive Hypothesis

Atherosclerosis leading to myocardial infarction and stroke is maladaptive in modern humans. However, across a wide range of vertebral taxa, vascular (endothelial) vulnerability atherosclerosis may be linked to pre-reproductive protection from bactermias, fungemias and viremias.

10.4 Cardiovascular 2

Why are children and young adults susceptible to fear-induced syncope (acute bradycardia and vasodilation)?

1) In which non-human animals do young individuals exhibit a slowing of the heart in response to fear?
2) What are the life histories of species in which acute slowing of the heart occurs in response to fear?
3) What are the environmental factors that increase or decrease vulnerability to fear-induced bradycardia?
4) Correlating life histories with species' cardiovascular response to fear, what adaptive benefits can be identified for acute bradycardia and vasodilation?

10.4.1 Testable Adaptive Hypothesis

Fear-induced bradycardia and vasodilation leading to syncope is maladaptive in modern humans. However, across a wide range of vertebral taxa, the tonic immobility resulting in these autonomic responses offers anti-predation benefit, especially for younger individuals.

For every disorder, syndrome, or diagnosis seen in humans, consideration of phylogenetic and adaptive perspectives can generate novel testable hypotheses which could not emerge without this approach. This is true for the biggest killers of our species: heart disease, cancer, and other disorders. It is also true for the psychopathologies which cause so much morbidity and mortality for our species. It is worth noting that Darwin recognized emotions and behavior in non-human animals and humans as part of an evolutionary inheritance. Yet, according to Tinbergen, "his behavior work…long remained underestimated, almost ignored" (Dewsbury, 1985). In fact, a serious consideration of mental illness as emerging through natural selection and related evolutionary processes has, until very recently, been absent in mainstream modern psychiatric thought. Tinbergen developed a keen interest in early childhood autism later in his career connecting ethology to evolution in behavior (Tinbergen & Tinbergen, 1972; Tinbergen et al., 1985). While the application of evolutionary thought to understanding mental illness is widely perceived as a modern approach, an appreciation of the comparative approach to understanding emotion was recognized in the nineteenth century. Charles Darwin's last book, *On the Expression of Emotions in Man and Animal*, revealed a recognition that emotion, along with all other aspects of human anatomy and physiology, emerged through natural selection (Darwin, 1872). This early recognition presaged the emerging modern recognition that a fundamental understanding of emotional responses—normal or pathological—requires an evolutionary perspective. A growing number of investigators are beginning to use the tools of evolutionary biology to understand the range of emotional and behavioral responses including Wendy Trevathan on the evolution of women's health; Robert Perlman, Jay Schulkin on corticotropin releasing factor (CRF) signaling; Paul Keedwell on the evolution of depression; Martin Brune on psychosomatic medicine; and Dan Blumstein on fear (Trevathan, 2010; Perlman, 2013; Schulkin, 2017; Keedwell & Barker, 2008; Brune, 2015; Stankowich & Blumstein, 2005). Again, applying a Tinbergen lens, a phylogenetic survey of "human" psychopathologies in non-human animals reveals a range of disorders which, at least phenomenologically, closely resemble each other.

Phylogenetic surveys identifying non-human animals presenting with biobehavioral disorders including separation anxiety, post-traumatic stress disorder (PTSD), compulsions, abnormal post-partum behavior, and self-injury, can spark novel testable hypotheses about these conditions. This is true interspecifically and between species. For example: Why are some horse breeds more vulnerable to postpartum foal rejection and harm? Why are some dog breed prone to anxiety and compulsions? Whya re some breeds so prone to overeating? (Natterson-Horowitz & Bowers, 2012).

10.5 Psychiatric

Why are humans vulnerable to compulsions and stereotypies?

1) Which non-human animals present with compulsions and stereotypies?
2) What are the life histories traits associated with species vulnerable to compulsions and stereotypies?
3) What are the environmental factors which increase or decrease this vulnerability?
4) Correlating life histories to species' responses to anxiety, what are possible adaptive benefits to compulsions and stereotypies?

10.5.1 Testable

Compulsions and stereotypies are maladaptive for modern humans. However, vulnerability to both compulsions and stereotypies across a wide range of group-living vertebral taxa suggests components of this vulnerability offer conflict-resolving and stress-reducing adaptive benefit.

10.6 A Multifactorial Approach

There is a trend in medicine away from strictly reductionistic approaches to those which consider a broader context. In this environment, there may be unique receptivity to the inherently multifactorial and contextual Tinbergean approach to understanding disease causation. How can physicians and investigators be encouraged to consider evolutionary insights in their work? The Tinbergean framework offers a structured rubric which breaks down evolutionary thinking, encouraging broadly comparative insights and adaptive considerations. This approach can be framed as an extension of the multifactorial models of disease causation with which physicians already have familiarity.

The Tinbergean approach expands how physicians conceive of multifactoriality to include phylogenetic perspectives and, having recognized shared susceptibility to disease, consider potential adaptive value associated with this vulnerability. The specific steps to developing a Tinbergean analysis are clear and can readily be put into the hands of clinicians, investigators and medical educators.

Given the power of a Tinbergean approach to strengthen our understanding of health and disease and recognizing how underleveraged it currently is in medical practice, efforts should focus on evolutionary content to medical students. Leading physicians and evolutionary biologists have called for increased collaboration and engagement between their fields (NESCent, 2018). The aim is to place principles of evolutionary biology into the hands of physicians and investigators, to spark novel hypotheses, inspire innovative thinking, and improve both clinical practice and outcomes. Increasingly, examples of the value of evolutionarily-informed thinking and approaches are emerging.

Despite the growing belief that learning and adopting Tinbergean and evolutionary perspectives could bring much to medicine, the effort to bring evolutionary content into the many phases of medical education has been limited. An inadequate understanding of the benefits of the evolutionary perspective, a belief that there is "not enough room" in the standard medical education, and the current lack of evidence that this effort will improve educational, practice and investigational outcomes underlies the limited presence of this content in medical schools.

A group of physicians, educators, and scientists came together from 2009–2014 to develop strategies for infusing evolutionary thinking into medical curriculum (NESCent, 2018). The group worked to identify opportunities within curricula at all levels of medical education and training. Through outreach to US medical school dean's offices, premedical curricula development personnel, current and past medical students, and leaders in evolutionary biology and medical education, the group identified numerous courses and training experiences in which evolutionary content could be introduced. Furthermore, analysis of the goals and expectations of these educational activities suggests that the evolutionary content would aid in the achievement of these goals while significantly enhancing the students' understanding of the presented material.

This working group identified multiple opportunities, distributed across the course of a physician's training and development, where exposure to evolutionary concepts, frameworks, and modes of reasoning are most likely to be usefully absorbed.

There is some evidence that these efforts are accelerating this content into premedical and medical curriculum. In 2015, the Medical College Admissions Test (MCAT) began including specific questions on topics in evolutionary biology, providing further incentive for instructors of premedical education to incorporate explicit treatment of evolutionary concepts. There are many other opportunities like these, including within clinical settings and academic medical centers, to leverage opportunities to further infuse evolutionary thought and analysis into the biomedical research and the everyday practices of delivering patient care.

It has been 45 years since the Nobel Prize in physiology or medicine was awarded to a trio of animal behaviorists. The foresight of the Nobel Committee in bestowing this honor on scientists whose work was not directly or obviously connected to the study of disease and pathophysiology is hard to overstate. It represents one of the most forward-thinking moments in the committee's history. The recognition and prestige of the prize fueled interest, support for and growth of the fields of, first, ethology, later, behavioral ecology, and currently, evolutionary medicine. Modern physicians and their patients owe a great dept of gratitude to the scientists who developed a framework for understanding health and disease from an ecological and evolutionary context.

Again, we ask: How can medical education be improved by applying a Tinbergean lens to somatic and biobehavioral disorders? One important contribution is the inclusion of comparative perspectives of "human" pathology. Tinbergen advocated for phylogenetic information to be included in theories of causation. Too often in medicine, the blindfold of human exceptionalism has prevented physicians from recognizing that non-human animals share our vulnerability to many diseases. The spontaneous occurrence in non-human animals of high-impact human diseases including cancer, heart disease, autoimmune disorders, and biobehavioral conditions is crucial information for physicians and investigators working to understand these diseases.

Both the research and educational benefits of comparative medical information have been greatly underleveraged. Medical education should include questions about human disease in non-human animals. Phylogenies of disease vulnerability and resistance should be developed. Awareness of shared vulnerability between animal taxa can accelerate the development of help novel hypotheses and research based on Tinbergen's central question about causation: Why do we humans and these animals share this vulnerability? These comparative questions encourage the kind of broad thinking that can generate testable adaptive hypotheses.

Because, as Theodor Dobzhansky noted, "nothing in biology makes sense outside the context of evolution," the inclusion of phylogenetic and adaptive perspectives is crucial for physicians and medical students. Even at the earliest stages of their careers, the Tinbergean perspective can push students to see beyond the human exceptionalism which has narrowed the kinds of questions asked about the nature and origin of disease. Beyond this, it encourages a broad, species-spanning approach to medicine which recognizes that human pathology arises from hundreds of millions of years of animal health and disease.

References

Abegglen, L. M., Caulin, A. F., Chan, A., Lee, K., Robinson, R., Campbell, M. S., Kiso, W. K., Schmitt, D. L., Waddell, P. J., Bhaskara, S., Jensen, S. T., Maley, C. C. & Schiffman, J. D. (2015). Potential mechanisms for cancer resistance in elephants and comparative cellular response to DNA damage in humans. *JAMA*, 314(17): 1850–60.

Aktipis, C. A., Boddy, A. M., Jansen, G., Hibner, U., Hochberg, M. E., Maley, C. C., & Wilkinson, G. S. (2015). Cancer across the tree of life: Cooperation and cheating in multicellularity. *Phil Trans R Soc Lond B. Biol Sci*, 370(1673): pii: 20140219. doi: 10.1098/rstb.2014.0219.

Boddy A. M., Kokko, H., Breden, F., Wilkinson, G. S., Aktipis, C. A. (2015). Cancer susceptibility and reproductive trade-offs: A model of the evolution of cancer defences. *Philos Trans R Soc Lond B Biol Sci*, 370(1673): pii: 20140220. doi: 10.1098/rstb.2014.0220.

Brune, M. (2015). *Textbook of Evolutionary Psychiatry and Psychosomatic Medicine: The Origins of Psychopathology*. Oxford: Oxford University Press.

Burkhardt, R. W. (2005). *Patterns of Behavior: Konrad Lorenz, Niko Tinbergen, and the Founding of Ethology* (pp. 133, 431). Chicago: University of Chicago Press.

Casás-Selves, M. & DeGregori, J. (2011). How cancer shapes evolution and how evolution shapes cancer. *Evolution: Education and Outreach*, 4(4), 624–34.

Clark, Colin W., & Ekman, Jan. (1995). Dominant and Subordinate Fattening Strategies: A Dynamic Game. *Oikos*, 72(2): 205–212.

Darwin, C. (1872). *The Expression of the Emotions in Man and Animals*. London: John Murray. http://dx.doi.org/10.1037/10001-000

DeVita, Jr V. & Chu, E. (2008). A history of cancer chemotherapy. *Cancer Research*, 68(21): 8643–53.

Dewsbury, D. A. (1985). *Leaders in the Study of Animal Behavior: Autobiographical Perspectives* (p. 455). Lewsburg: Bucknell University Press.

Ferrari, R., Ceconi, C., Curello, S., Ferrari, F., Confortini, R., Pepi, P., & Visioli, O. (1996). Activation of the neuroendocrine response in heart failure: Adaptive or maladaptive process? *Cardiovascular Drugs and Therapy*, 10(2), 623–29.

Fidler, I. J. (1978). Tumor heterogeneity and the biology of cancer invasion and metastasis. *Cancer Research*, 38(9), 2651–60.

Furie, B. & Furie, B. C. (1992). Molecular and cellular biology of blood coagulation. *N Eng J Med*, 326(12): 800–6.

Gluckman, P., Beedle, A., Buklijas, T., Low, F., & Hanson, M. (2016). *Principles in Evolutionary Medicine* (2nd Edition). Oxford: Oxford University Press.

Griffin, P. C., Griffin, S. C., Waroquiers, C., & Mills, L. S. (2005). Mortality by moonlight: Predation risk and the snowshoe hare. *Behavioral Ecology* 16(5): 938–44.

Hecht, S. S. (1999). Tobacco smoke carcinogens and lung cancer. *Journal of the National Cancer Institute*, 91(14): 1194–210.

Huxley, J. S. (1914). The courtship-habits of the great crested grebe (Podiceps cristatus); with an addition to the theory of sexual selection. In: *Proceedings of the Zoological Society of London*, (491–562).

Keedwell, P. & Barker, P. (2008). *How Sadness Survived: The Evolutionary Basis of Depression*. Baco Raton: CRC Press.

Kennedy, C., Willemze, R., de Gruijl, F. R., Bouwes Bavinck, J. N., & Bajdik, C. D. (2003). The influence of painful sunburns and lifetime sun exposure on the risk of actinic kerotoses, seborrheic warts, melanocytic nevi, atypical nevi and skin cancer. *Journal of Investigative Dermatology*, 120(6): 1087–93.

Kirk, R. G. W. (2009). Between the clinical and the laboratory: Ethology and pharmacology in the work of

Michael Robin Alexander chance, c.1946–1964. *Medical History*, 53(4): 513–36.

Kruuk, H. (2004). *Niko's Nature* (pp. 85, 125). Oxford: Oxford University Press.

Lechat, P. (2006). The evolution of heart failure management over recent decades: From CONSENSUS to CIBIS. *European Heart Journal Supplements*, 8(C), C5–C12.

Levathan, W. (2010). *Ancient Bodies, Modern Lives: How Evolution has Shaped Women's Health*. New York: Oxford University Press.

Marler, P. (1973). The 1973 Nobel Prize for physiology or medicine, *Science*, 182(4111): 464–66.

Mayr, E. (1961). Cause and effect in biology: Kinds of causes, predictability, and teleology are viewed by a practicing biologist. *Science*, 134(3489): 1501–6.

Natterson-Horowitz, B. & Bowers, K. (2012). *Zoobiquity: The Astonishing Connection Between Human and Animal Health*. New York: Vintage.

NESCent (National Evolutionary Synthesis Center) https://www.nescent.org/about/index.html. Accessed on June 5, 2018.

Nesse, R. M. (2001). How is Darwinian medicine useful? *Journal of Medicine*, 174(5): 358–60.

Nesse, R. M. & Schiffman, J. D. (2003). Evolutionary Biology in the Medical School Curriculum. *BioScience*, 53(6): 585–87.

Nesse, R. M. & Williams, G. C. (1996). *Why We Get Sick*. New York: Vintage.

NobelPrize.Org. Press Release, Karolinska Institutet, The Nobel Prize in Physiology or Medicine 1973. Accessed on March 10, 2018 at https://www.nobelprize.org/nobel_prizes/medicine/laureates/1973/press.html.

Pepper, J. W., Findlay, C. S., Kassen, R., Spencer, S. L., & Maley, C. C. (2009). Cancer research meets evolutionary biology. *Evolutionary Applications*, 2(1): 62–70.

Perlman, R. (2013). *Evolution and Medicine*. New York: Oxford University Press.

Powers, M. L. & Schulkin, J. (2009). *The Evolution of Obesity*. Baltimore: Johns Hopkins University Press.

Read, A.F., Day, T., & Huijben, S. (2011). *The evolution of drug resistance and the curious orthodoxy of aggressive chemotherapy*. Retrieved from http://www.pnas.org/content/108/Supplement_2/10871.full

Sacks, C. A., Jarcho, J. A., & Curfman, G. D. (2014). Paradign Shifts in Heart-Failure Therapy—A Timeline. *The New England Journal of Medicine* 371: 989–91.

Schleidt, Wolfgang, Shalter, Michael D., Moura-Neto, Humberto. (2011). The hawk/goose story: The classical ethological experiments of Lorenz and Tinbergen, revisited. *Journal of Comparative Psychology*, 125(2): 121–33.

Schocken, D. D., Arrieta, M. I., Leaverton, P. E., & Ross, E .A. (1992). Prevalence and mortality rate of congestive heart failure in the United States. *Journal of the American College of Cardiology*, 20(2): 301–6.

Schulkin, J. (2017). *The CRF Signal*. Oxford: Oxford University Press.

Singer, M. A. (2007). *Comparative Physiology, Natural Animal Models and Clinical Medicine: Insights into Clinical Medicine from Animal Adaptation*. London: Imperial College Press.

Stankowich, T. & Blumstein, D. T. (2005). Fear in animals: A meta-analysis and review of risk assessment. *Proceedings of the Royal Society* B 272(1581): 2627–34.

Stearns, S. C. & Medzhitov, R. (2015). *Evolutionary Medicine*. Sunderland: Sinauer Associates.

Straussmann, J. E. (2014). Tribute to Tinbergen: The Place of Animal Behavior in Biology. *Ethology*, 120(2): 123–6.

Sulak, M., Fong, L., Mika, K., Chigurupati, S., Yon, L., Mongan, N. P., Emes, R. D., & Lynch, V. J. (2016). TP53 copy number expansion is associated with the evolution of increased body size and an enahnced DNA damage response in elephants. *eLife* 5: e11994.

Thomas, F., Fisher, D., Fort, P., Marie, J. P., Daoust, S., Roche, B., Grunau, C., Cosseau, C., Mitta, G., Baghdiguian, S., Rousset, F., Lassus, P., Assenat, E., Gregoire, D., Misse, D., Lorz, A., Billy, F., Vainchenker, W. Delhommeau, F., Koscielny, S., Itzykson, R., Tang, R., Fava, F., Ballesta, A., Leoutre, T., Krasinska, L., Dulic, V., Raynaud, P., Blache, P., Quittau-Prevostel, C., Vignal, E., Trauchessec, H., Perthame, B., Clairambault, J., Volpert, Solary, E., Hibner, U., & Hochberg, M. E. (2013). Applying ecology and evolutionary theory to cancer: A long and winding road. *Evolutionary Applications*, 6: 1–10.

Trevathan, Wenda (2010). *Ancient Bodies, Modern Lives: How Evolution Has Shaped Women's Health*. Oxford, UK: Oxford University Press.

Tinbergen, E.A. & Tinbergen, N. (1972). *Early Childhood Autism: An Ethological Approach*. Milton Park, Abingdon-on-Thames, Oxfordshire United Kingdom: Taylor and Francis.

Tinbergen, N. (1963) On aims and methods of Ethology. *Zeitschrift fur Tierpsychologie* 20(4): 410–33. This journal was renamed *Ethology*, in 1986.

Tinbergen, N., Tinbergen, E. A., Welch, M., & Zappella, M. (1985). *Autistic Children: New Hope for a Cure*. Crows' Nest, Australia: George Allen & Unwin.

Vajaranant, T. S., Nayak, S., Wilensky, J. T., & Joslin, C. E. (2010). Gender and glaucoma: What we know and what we need to know. *Curr Opin Opthalmol*, 21(2): 91–9.

van der Horst, F. C. P. (2011). *John Bowlby—From Psychoanalysis to Ethology: Unraveling the Roots of Attachment Theory*, (pp. 75–102). Hoboken: Wiley-Blackwell.

Virmani, R., Burke, A. P., Farb, A., & Kolodgie, F. D. (2006). Pathology of the Vulnerable Plaque. *Journal of the American College of Cardiology*, 47(8) Supplement: C13–18.

Yusuf, S. (1993). Reduced mortality and morbidity with the use of angiotensin-coverting enzyme inhibitors in patients with left ventricular dysfunction and congestive heart failure. *Herz*, Supplement 1: 444–8.

CHAPTER 11

The Role of the Immune System from an Evolutionary Perspective

Carsten Schradin and Rainer H. Straub

11.1 Introduction

Evolutionary medicine emphasizes that to better understand modern human health and disease, the underlying evolutionary history and evolutionary consequences must also be understood. Thus, evolutionary medicine will contribute significantly to our understanding of the fitness consequences and the phylogeny (evolutionary history) of diseases and disease related traits, explaining why disease exist even though it reduces fitness (Gluckman et al., 2009; Perlman, 2013; Stearns et al., 2016).

For example, when trying to understand why chronic inflammatory diseases occur, evolutionary biologists pointed out that it is crucial to consider the effect of chronic inflammatory diseases on lifetime reproductive success. This success depends on the species-specific life history traits such as age at first reproduction, reproductive lifespan, and life expectancy (Straub et al., 2016). However, evolutionary medicine faces the challenge and difficulty to integrate physiological mechanisms with evolutionary function (Gluckman et al., 2009; Lozano, 2010; Nesse et al., 2012). This challenge is not unique to evolutionary medicine, but is one of the main topics in organismal research in biology (Mayr, 1961), and has been especially addressed in the research on animal behavior (Tinbergen, 1963).

Biological research has traditionally been divided into the study of proximate (physiological or pathophysiological) mechanisms on the one hand and ultimate (evolutionary) causes on the other hand (Mayr, 1961). Most researchers take only one of these two approaches into account (Drickamer, 1998). For example, immunological research is mainly proximate, focusing on the pathophysiological mechanisms underlying the most prevalent infectious diseases (malaria as an example), allergic diseases, and chronic inflammatory diseases. Studies in behavioral ecology, on the other hand, are primarily ultimate, focusing on the fitness consequences of behavior, which means whether individuals showing a specific behavior have more offspring than other individuals (Krebs et al., 1993; Wilson, 1975). Behavioral ecologists have realized that studying physiological mechanisms as evolved traits that increase fitness provides an integrative understanding of animal behavior (Wikelski et al., 2001; Wingfield, 2008).

For example, the major histocompatibility complex (MHC), a set of proteins that plays a significant role in the immune response (a proximate factor), is known to affect behavior, enabling females to choose mates with complementary MHC proteins (the set of genes is called human leukocyte antigen (HLA) gene complex), which increases fitness of common offspring (Kamiya et al., 2014; Lerner et al., 1991). Thus, behavioral ecologists came to a more comprehensive understanding of behavior by

Schradin, C. and Straub, R., *The Role of the Immune System from an Evolutionary Perspective* In: *Integrating Evolutionary Biology into Medical Education—for maternal and child healthcare students, clinicians, and scientists.* Edited by: Jay Schulkin and Michael L. Power, Oxford University Press (2020). © Oxford University Press.
DOI: 10.1093/oso/9780198814153.003.0012

taking into account the fitness consequences of its underlying physiological mechanisms.

At the beginning of this chapter, we will highlight why inflammation and the immune response is a key category for evolutionary medicine (in section 11.2.2). Then, based on the four questions of Tinbergen that he applied to animal behavior (Tinbergen, 1963), we want to tackle the role of the immune system from an evolutionary perspective (section 11.3—11.4). With this knowledge, the special situation of chronic inflammatory diseases will be addressed using the Tinbergen approach (section 11.5). Finally, medical applications are presented, which can be relevant for medical education and medical decision making (section 11.6).

11.2 Inflammation and immune response—Why is it a key category of evolutionary medicine?

When we think of the major challenges during our evolutionary history, the following threats, the immediate responses, and the response systems come into mind (see bullet point list). Concerning response systems, the sympathetic nervous system (SNS), the hypothalamic-pituitary-adrenal axis (HPA axis), and the renin-angiotensin-aldosterone system (RAA system) belong to the hormonal stress axes that are controlled by the brain.

- *Predators* (response: fight-or-flight; response system: brain with SNS and HPA axis)
- *Intra-specific competition for resources, including mates* (fight-or-flight; brain with SNS and HPA axis)
- *Wounding and hemorrhage* (blood pressure stabilization and blood coagulation; brain with SNS, HPA axis, and SNS-dependent RAA system, and standby system of blood coagulation)
- *Food shortage* (foraging; brain with SNS and HPA axis)
- *Thirst* (water seeking; brain with SNS, HPA axis, and SNS-dependent RAA system)
- *Cold* (warming-up and warm seeking; brain with SNS and HPA axis)
- *Heat* (sweating/cooling and cold seeking; brain with SNS and HPA axis)
- *Infection* (inflammation and immune response; immune system)
- *Wounding and infection* (inflammation and immune response; immune system)

From this list, it is evident that only two major response systems are liable to protect the body from harmful threats: the brain and the immune system. The brain stimulates hormonal and neuronal stress axes and the locomotor system so that the brain controls activity of muscles. Brain and muscles are tightly coupled to build the psychomotor system.

Since the psychomotor system and the immune system require enormous amounts of energy (Box 11.1), they possess separate selfish rules to demand energy-rich fuels from energy stores by inducing insulin resistance (Straub, 2014; Straub, 2017). These energy stores are the fat tissue (fatty acids, glycerol) and muscles (mainly amino acids; e.g., for glucose generation). If a psychomotor or immune process takes very long, e.g., during chronic illness (psychiatric diseases and chronic inflammatory diseases as examples), selfishness of the two systems becomes more and more obvious. The separate rules of either the brain or the immune system inhibit the respective other system (Straub, 2017). This can lead to brain-induced immunosuppression or immune-mediated fatigue or depression.

Box 11.1 Energy expenditure of the major organs in the body.

An individual with a sedentary lifestyle who is 1.80 m tall and weighs 85 kg requires 10000 kJ/d (7500 kJ/d of which represents the resting metabolic rate) (Blaxter, 1989; Speakman et al., 2010). The total amount of energy required by the immune system in an inactivated state is 1600 kJ/d, and this level can increase by 25% during mild activation to 2,000 kJ/d (Straub et al., 2010). During sepsis, the immune system needs 30–60% more energy (Blaxter, 1989;, Torine et al., 2007). In contrast, the brain needs approximately 2200 kJ/day and it does not need much more when activated. The increased amount of energy expenditure through psychomotor activity depends on active muscles that need 2500 up to 30000 kJ/d (the latter number is relevant for a Tour de France bicyclist) (Straub et al., 2010).

It is important to understand selfishness of brain and immune systems as a fundamental principle because this explains why inflammation and immune response can be key categories of evolutionary medicine. Concerning the immune system and chronic inflammatory diseases, we discussed the evolutionary aspects in several publications (Straub et al., 2003; Straub et al., 2010; Straub, 2014; Straub, 2015; Straub et al., 2016; Straub 2017). Similarly, the brain and the psychomotor system have a central role in evolutionary medicine. In the late 1970s and early 1980s, several authors started to discuss evolutionary principles in the context of psychiatric diseases. They focused on maladaptive behaviors that can lead to chronic illness (Chagnon et al., 1979; McGuire et al., 1981; Nesse, 1984). By inclusion of evolutionary medicine, energy regulation, and neuroendocrine–immune crosstalk, we understand many maladaptive responses in chronic inflammatory diseases, in psychiatric diseases, and during ageing (Straub, 2017).

Long-term demand of energy-rich fuels by the brain or the immune system also blocks voluntary physical and mental activity (Straub, 2017). The long-term consequences of reduced physical and mental activity include cardiovascular diseases (Alves et al., 2016), metabolic diseases and obesity, cognitive dysfunction and pain (Chang et al., 2017; Paley et al., 2016), neurodegenerative disease (Phillips et al., 2014; Phillips et al., 2015), psychiatric diseases (Hallgren et al., 2016), frailty (McPhee et al., 2016), and other chronic diseases such as cancer (Kyu et al., 2016). These conditions all constitute risk factors for early death. An overactive immune system has a central role for these risk factors, and, thus, it is a key category of evolutionary medicine.

11.2.1 The four questions of Tinbergen

Tinbergen formulated four questions in 1963 (Tinbergen, 1963), which are the basis for an integrative approach to understand biological traits (Krebs et al., 1993). Each question is of equal importance. The four questions are (see also Bateson et al., 2013; Nesse, 2013):

- What is the *Function*, which means the fitness value of the trait?
- What is the *Phylogeny*, which means the evolutionary history of the trait?
- How did the trait develop in individuals over their lifetime (*Ontogeny*)?
- What physiological machinery causes the observable trait (*Causation*)?

The questions concerning the function and phylogeny address ultimate factors, whereas questions concerning ontogeny and causation address proximate factors. Evolutionary medicine focuses on the two ultimate factors. The question about the function concerns the fitness consequences of a trait, i.e., why did they evolve? In order to evolve through the process of natural selection, a specific trait must have fitness benefits for the individual, which means increase the number of off-spring produced during the individual's lifetime. The question regarding the phylogeny of a trait concerns the evolutionary history. Tinbergen regarded the responding central nervous system as the machinery causing animal behavior. He defined ontogeny of behavior as the "change of behavioral machinery during development." Thus, the question addressing causation is the question about the behavioral machinery leading to the observed behavioral pattern, and the question about ontogeny asks how this machinery develops until the behavior is shown (Tinbergen, 1963). In the context of an immune response, this machinery is called the "immune machinery."

11.2.2 The four questions of Tinbergen and the immune response

11.2.2.1 Function—fitness value of the trait

Reactions to acute inflammation serve the body to overcome short-lived infectious diseases and other threats such as exposure to allergic antigens. In both situations, the body tries to get rid of the stimulus. The mechanisms are thus adaptive, increasing individual fitness. Typically, inflammatory responses are highly adaptive, enabling individuals to overcome infections, to combat allergic threats, and to heal wounds. Thus, alleles of genes enabling this adaptive response got fixed within the gene pool.

Thanks to the last two decades of genome-wide association studies, we now recognize that hundreds of genetic loci associated with chronic inflammatory diseases overlap with genes relevant in infectious and allergic diseases (Brinkworth et al., 2014). While in earlier time (before 2000), chronic inflammatory diseases were thought to have one unique immune disease pathways, it is now very clear that this is incorrect. In chronic inflammatory diseases inflammation is typically regulated by a network of genes that are implicated in multiple diseases (Cotsapas et al., 2011). The unique disease pathway is known in monogenic autoinflammatory diseases, which also employ immune pathways relevant to infections and allergies.

11.2.2.2 Phylogeny—evolutionary history of the trait

The evolutionary history of human immunological pathways is shared with sharks, birds, and rodents, spanning more than 420 million years (Boehm, 2011). The best known association between infections and genes has been established for the HLA allele system, which is an important element of antigen presentation in the MHC molecules. Polymorphic HLA alleles have been retained because they helped to overcome various infections (Mangalam et al., 2013), leading to positive selection. For example, the genetic factor HLA-DR4 (DRB1*04) protects from dengue hemorrhagic fever (LaFleur et al., 2002). Furthermore, clearance of virus infections is much better for hepatitis B virus in the presence of HLA-DR13 (Wang et al., 2016) and for human immunodeficiency virus in the presence of HLA-B27 (Chakrabarti et al., 2010). There are many more examples also outside the HLA allele system, presentation of which would go beyond the scope of this chapter. The reader is referred to important reviews (Brinkworth et al., 2014).

11.2.2.3 Ontogeny—development of the trait in individuals over a lifetime

The environment of an individual during development and adult life together with the genotype determine how the inflammatory response system develops and will later respond. The immune system is equipped with a system to memorize antigens, and this influences its later responsiveness to the same antigens. We call it immunological memory (Murphy et al., 2011), which is very personal depending on the exposure of the individual to infectious agents during the lifetime. For example, most people in Europe have been exposed to smallpox virus and are thus immune to it. This was completely different for native Americans at a time point of arrival of European seafarers in the sixteenth century, leading to mass extinction of local inhabitants (Patterson et al., 2002).

Another example in the neuroendocrine immune field further demonstrates the impact of ontogeny. Neonatal exposure of a new-born rat to lipopolysaccharide changes the hypothalamic-pituitary adrenal axis markedly (Shanks et al., 1995). As adults, neonatally endotoxin-treated animals exhibited significantly greater adrenocorticotrophic hormone and corticosterone responses to restraint stress than controls. Neonatal endotoxin treatment reduced glucocorticoid receptor density across a wide range of brain regions (Shanks et al., 1995). Exposure to gram-negative lipopolysaccharide in early life can alter the development of neural systems, which govern endocrine responses to stress and may thereby predispose individuals to stress-related pathology. Indeed, it has been demonstrated that neonatal endotoxin treatment of rats protects animals from experimental arthritis later in life (Shanks et al., 2000). This long-term imprinting of central nervous changes in the early phase of ontogeny can also change the behavior of the immune system throughout life.

11.2.2.4 Causation—the "immune machinery"

Inflammation is caused by specific antigens, including pathogens or cell material of damaged tissue (wounds), to which the immune system responds (Murphy et al., 2011). There are several supportive reactions of the neuronal and endocrine systems that foster the immune response in acute infectious diseases (Straub, 2015), altogether representing the causation of inflammation and immune response. In line with Tinbergen's terminology (i.e., behavioral machinery), this might be called the immune machinery. The immune machinery consists of immune cells (neutrophils, lymphocytes, macrophages, dendritic cells, natural killer cells, and many others) with specific surface receptors to activate these cells, causing the release of inflammatory mediators (Murphy et al., 2011).

The immune machinery is highly dependent on the availability of energy-rich substrates in the body because immune cells demand enormous amounts of energy (Gaber et al., 2017, Straub et al., 2010). We may call the immune system selfish, because during infection it is the major dominating factor operating at the highest hierarchical level to acquire energy-rich substrates (Straub, 2014). This selfishness has been positively selected for infectious diseases, and it is an important determining aspect of the immune machinery (causation).

11.2.2.5 Integrative approach—the immune system in evolutionary medicine

The immune machinery has been positively selected for short-lived infectious diseases (causation), which leads to measurable immune responses and inflammation (diagnostic, red box in Figure 11.1), protecting the individual from infections/helping during wound healing. This increases the chance of survival and reproduction at a later stage, increasing evolutionary fitness (Schmid-Hempel, 2011). Genes improving the immune response increased in frequency in the gene pool (phylogeny). The individual genotypes together with the individual specific environment lead to an individual specific development of the immune system (ontogeny). A classical ontogenetic factor of the immune response is immunological memory of the adaptive immune system towards pathogen-specific antigens (Murphy et al., 2011). If a microbial antigen is known to the immune machinery (causation), then it is rapidly activated, i.e., an immediate response towards an infectious agent is generated. The immune machinery with memory consists of educated specific T or B cells (physiological response)

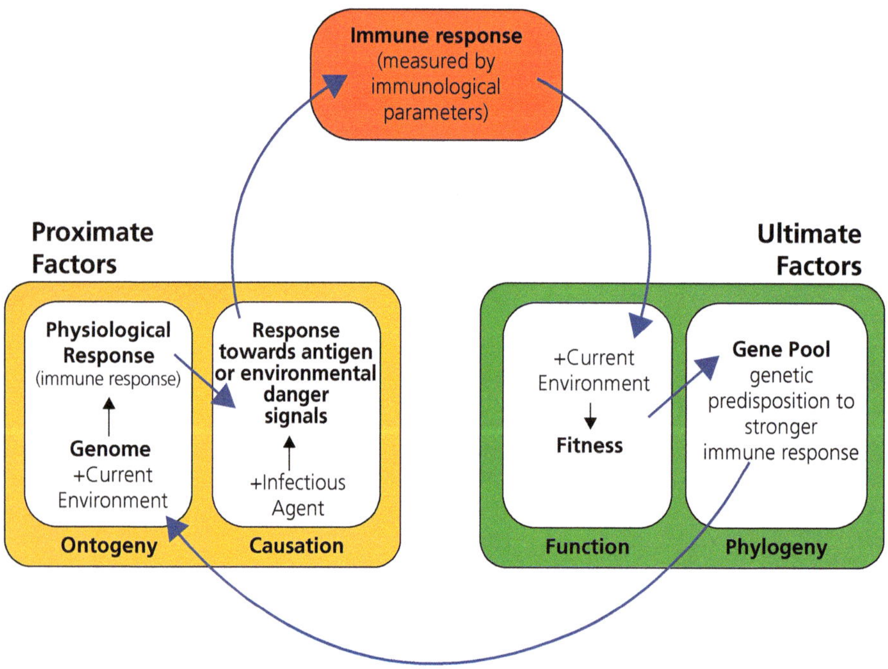

Figure 11.1 Integration of the four questions of Tinbergen to come to a comprehensive understanding of the immune system. The individual's immune system is stimulated by an infectious episode induced by an infectious agent (Causation). Inflammation starts, which can be measured as a change in immunological parameters (e.g., increase of antibodies, secretion of cytokines in circulation, proliferation of antigen-specific T cells etc.). This influences the evolutionary fitness (Function) of the individual, influencing the gene-pool (Phylogeny) of future generations. This can lead to a stronger and faster immune response. Offspring with the inherited genetic predisposition install a stronger immune response during their development (Ontogeny) leading to a better recognition and presentation of antigens in an encounter between antigen-presenting cells and T/B lymphocytes, secretion of the correct cocktail of cytokines and efficient phagocytosis (Causation). In adaptive short-term inflammation with the correct response towards a microbial antigen, the immune machinery enables the individual to survive critical infections and, thus, to reproduce later and increase individual fitness

that can be observed as a phenotypic trait, causing a stronger and faster immune reaction towards the antigen (causation).

11.3 The four questions of Tinbergen and chronic inflammatory systemic diseases

Chronic inflammatory diseases are characterized by a continuous immune response towards harmless autoantigens (reaction against innocuous self) or harmless microbes on the surface of the body (reaction against innocuous foreign) (Firestein et al., 2013). Examples for chronic inflammatory diseases are rheumatoid arthritis, multiple sclerosis, ankylosing spondylitis, systemic lupus erythematosus, Sjögren syndrome, pemphigus vulgaris, and many others. While an immune response is usually confined to a small space, it becomes chronic and widespread in chronic inflammatory diseases (Hochberg et al., 2015; Straub, 2015). A chronic inflammatory disease leads to new energy requirements in an affected person, which supports many unwanted disease-related signs and symptoms recently demonstrated (Straub et al., 2016).

11.3.1 Ultimate factors—function and phylogeny

While reactions to acute inflammation serve the body to overcome short-lived infectious diseases, a long-term chronic inflammatory disease is detrimental. In chronic inflammatory diseases, the positively selected immune mechanisms, which usually support fitness and function (Figure 11.1), are continuously activated by harmless antigens, and the response becomes maladaptive (Straub et al., 2016). Importantly, a chronic inflammatory disease—while costly—will not have a significant influence on fitness (function), because it typically happens after the reproductive phase (Figure 11.2).

Chronic inflammatory diseases have a multifactorial genetic background. The HLA allele system

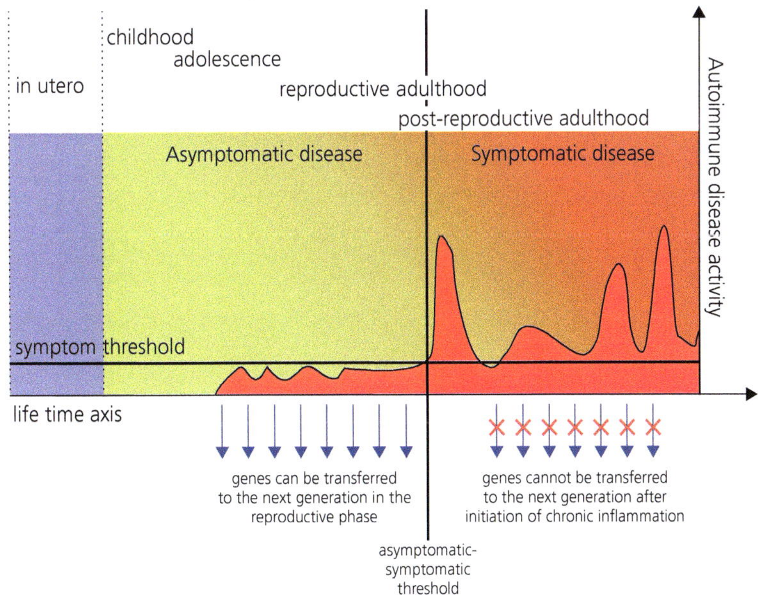

Figure 11.2 The influence of chronic inflammatory diseases on fitness. There is a critical difference between typical inflammation and chronic inflammatory systemic diseases, which separates the asymptomatic phase from the symptomatic phase of a chronic inflammatory systemic disease, the asymptomatic-symptomatic-threshold (a-s-threshold). Importantly from an evolutionary point of view, reproduction is only impeded during the symptomatic, but not during the asymptomatic phase. Genes enabling an adaptive inflammatory response, allowing the organism to overcome an infection, will thus increase survival probability and the potential for future reproduction and as such evolutionary fitness in young age. After initiation of chronic systemic inflammation, however, reproduction is inhibited, and this for prolonged periods, often for years and until death, causing fitness costs. However, as these costs typically occur only at the end of the reproductive life-history stage or in post-reproductive age, these fitness costs are lower than the fitness benefits in early life leading to an overall increase in Darwinian fitness.

plays an important role in chronic inflammatory diseases. For some HLA alleles, the risk to develop a chronic inflammatory disease is ninety times increased (Brewerton et al., 1973; Schlosstein et al., 1973). However, negative selection against specific HLA alleles was most probably weak, because in most individuals carrying these alleles the disease did not develop due to the lack of additional risk factors (Straub et al., 2016) that are important during ontogeny. HLA alleles were positively selected due to their role in infection (see section 11.3.2). We described the link between HLA-DR4 (DRB1*04) and dengue hemorrhagic fever, but the same HLA allele is known to be positively associated with rheumatoid arthritis, type 1 diabetes mellitus, and other chronic inflammatory diseases.

We described HLA-B27 as an important factor to overcome human immunodeficiency virus infection (see section 11.3.2), but the same allele is linked to ankylosing spondylitis (Brewerton et al., 1973; Schlosstein et al., 1973). Thus, many traits have an evolutionary history (phylogeny) developed in the context of infectious disease, but these traits are used again in another somatic environment in later life (Williams, 1957). This phenomenon is called antagonistic pleiotropy (Williams, 1957), which means fitness benefits at one life history stage being higher than fitness costs at another life history stage. Here, they can support the development of a chronic inflammatory disease.

11.3.2 Ontogeny—development of the trait in individuals over a lifetime

We learned that the immune system has a memory for antigens. However, cross-reactivity towards harmless autoantigens (reaction against innocuous self) or harmless microbes on the surface of the body (reaction against innocuous foreign) as in chronic inflammatory diseases can also occur. Now, the immune memory is directed against the harmless antigens, which leads to a chronic immune attack in chronic inflammatory diseases. The new memory is an autoimmune memory.

Sometimes autoantigens are post-translationally processed, which can be supported by factors that appear during individual development. For example, smoking may change self-peptides leading to an exchange of arginine with citrulline (Quirke et al., 2011). This post-translational modification with the new citrulline is foreign to the immune system and may be attacked as an antigen. Thus, smoking is one environmental factor during development known to increase the risk that the immune system later attacks harmless autoantigens fostering chronic inflammatory diseases (Klareskog et al., 2011; Mathews et al., 1973). Another environmental risk factor for some chronic inflammatory diseases is silica exposure, for example during construction work, which might overstimulate the immune system and trigger the autoimmune process leading to chronic inflammation (Pollard, 2016).

Another environmental risk factor is psychological stress. We already saw that neonatal endotoxin exposure can change the adult immune system response, and it seems that the childhood immune system is more vulnerable to stressors as compared to the adult immune system (see 11.3.3). In several studies, it was demonstrated that childhood trauma (death of a parent, sexual abuse, violence in the family etc.) was positively linked to later appearance of chronic inflammatory diseases (Dube et al., 2009, Neufeld et al., 2013, Spitzer et al., 2013). These examples are indicative of how the individual development influences the immune system and the appearance of overt chronic inflammatory diseases in the future.

11.3.3 Causation—the immune machinery

In chronic inflammatory diseases, the role of the immune machinery was described for more than 100 years, reaching enormous detail since the start of molecular biology and functional genomics in the 1980s and 1990s. Today, we realize that many important pathways have been discovered that cause and support chronic inflammatory diseases (Firestein et al., 2013; Hochberg et al., 2015). These novel factors led to the development of biological therapy such as anti-TNF strategies (Feldmann et al., 2010). Nevertheless, much detail needs to be learned, particularly, with respect to human chronic inflammatory diseases. Importantly, the immune machinery for chronic inflammatory diseases is the same as the immune machinery for infectious diseases and allergy.

On top of the specific immune machinery, which was called selfish (see 11.3.4), a general bodily reaction supports the immune system (Straub et al., 2013). The immune system is selfish because it makes use of other systems (nervous and endocrine) to redistribute energy-rich fuels (Straub, 2014). While this has been positively selected for short-lived infectious diseases, the long-term application of these adaptive programs leads to many unfavorable signs and symptoms in chronic inflammatory diseases (Straub et al., 2016). These are daytime fatigue, low mood, sleeping alterations, anorexia, inflammation type of anemia, muscle loss (sarcopenia), bone loss, insulin resistance, decreased fertility and libido, high blood pressure and volume overload, and increased blood coagulation (Straub et al., 2016). These phenomena were positively selected as a machinery outside the immune system that supports the immune machinery. However, if these pathways are used too long, many problems appear that need specific extra treatment besides immunosuppression. Long-term use of these programs is maladaptive (Straub et al., 2016; Straub, 2017).

11.3.4 Integrative approach to understand CIDS in evolutionary medicine

Ultimate factors have been positively selected for short-lived infectious diseases and not for chronic inflammatory diseases (Straub et al., 2016). Net fitness is increased because the fitness benefits in early life are much higher than fitness costs in post-reproductive life, so that favorable genes stay in the gene pool (function and phylogeny). The individual genotypes together with the individual specific environment lead to an individual specific development of the immune system (ontogeny). Here, environmental factors such as smoking, silica exposure, and childhood stress can prepare the immune machinery for the unfortunate development of a chronic inflammatory disease. When several factors come together (genetic and environmental), the individual can develop an immune response towards harmless self and harmless foreign antigens. Then, the response is imprinted in the immune memory, which leads to measurable immune responses such as autoantibodies towards citrullinated peptides (diagnostic, red box in Figure 11.3).

11.4 Medical applications and conclusions

Many chronic diseases might be studied and understood using the same framework as outlined in Figure 11.3. The evolutionary approach here considers three points:

1. One must take into account the fitness consequences of the trait leading to disease.
2. To be done at the different life history stages (pre-reproduction, reproductive age, and post-reproductive; Figure 11.2).
3. To determine life-time reproductive success.

This will enable us to understand the evolutionary history of the disease, especially why alleles favoring the disease have been established in the gene pool and have not been out-selected. On the basis of the "antagonistic pleiotropy theory" of Williams (Williams, 1957), we start to understand that disease alleles found related to chronic diseases often play a beneficial role in pre-reproductive and reproductive life history stage. In later life, the same alleles and related mechanisms can play a harmful role by stimulating a chronic disease in a different life history stage. Clinicians do not know this phenomenon and often think that disease-associated alleles are specific for a certain chronic disease without understanding their overall adaptive function. In other words, the idea is completely wrong that specific alleles have been positively selected to support a chronic disease. The example of HLA-DR4 was given previously.

Our approach might thus also benefit clinicians: it is often less relevant to know which genes and alleles favor the expression of a specific disease (there are often many little influences with small effects), but it is more important to know the interplay with environmental factors, which trigger the disease. This is because we cannot easily change the genotype of individuals, but we can interfere with environmental threats in an individual life. The example of smoking was given.

Another example is described as the "Barker phenomenon" according to David Barker (Barker et al., 1989; Hales et al., 1992). The hypothesis says that

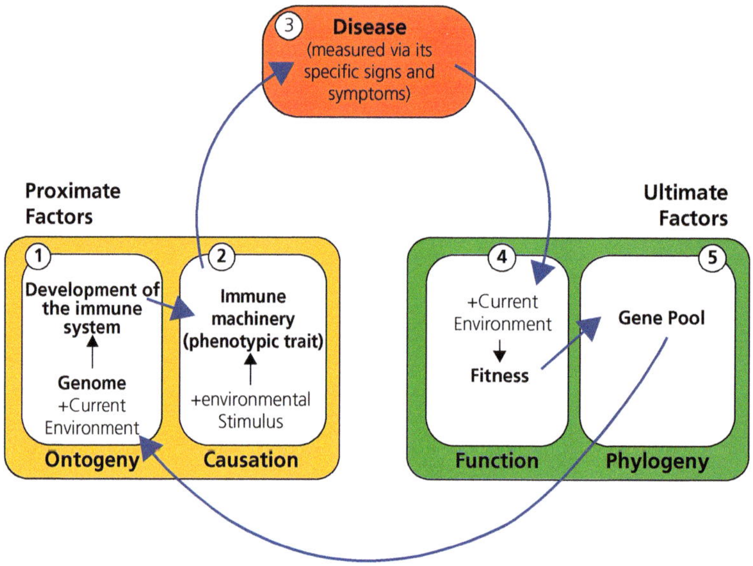

Figure 11.3 Integration of the four questions of Tinbergen to come to a comprehensive understanding of the development of chronic inflammatory systemic diseases

1. The immune machinery depends on the development of the immune system in an individual that can be influenced by different stimuli in the vulnerable phase of fetal, perinatal, and postnatal life or childhood (ontogeny), for example smoking or the exposure to silica dust.
2. In CIDs, the immune machinery is switched on by an immune response towards an autoantigen to induce autoimmunity or by monogenic defects in immune pathways to induce autoinflammation (causation).
3. The disease can be measured through specific signs and symptoms, one of which are autoantibodies.
4. Usually, the disease influences evolutionary fitness (function), but the influence is often very low, because CIDs appear late in or after the reproductive phase (Figure 11.2).
5. Thus, we do not expect a marked change in the gene pool that would support or prevent an autoimmune disease. The gene pool was positively selected for infectious diseases but not for CIDs (Figure 11.1).

reduced fetal growth by fetal constraints like undernutrition of the pregnant mother is strongly associated with a number of chronic conditions later in life in humans but also in experimental animals like rats. These chronic problems include coronary heart disease, stroke, diabetes, and hypertension. Since the phenomenon is stimulated during fetal life, it must have a strong environmental component. Typically, poor nutritional conditions were mentioned as the major trigger but other forms of fetal constraint play similar roles.

A third major element that is relevant to clinicians is the understanding that there is major crosstalk between different systems in the body. The most important systems are the psychomotor (brain and muscles) and immune systems due to their selfish role in combating threats. Since both systems use independent measures to achieve their goals, understanding of these interrelated pathways helps to explain many problems in chronic diseases and aging (Straub, 2017). These disease sequelae frequently seen in patients with chronic inflammatory diseases are fatigue, depressed mood, sleep alterations, loss of appetite, muscle wasting, cachectic obesity, bone loss, hypertension, and many others. The new understanding is based on the principles of evolutionary medicine, energy regulation, and neuroendocrine-immune crosstalk (Straub, 2017).

Fourth, understanding the evolutionary function of the immune machinery leading to CIDS could be of important psychological consequence. When clinicians can explain to patients how overall function of this system in earlier life is beneficial, this could reduce psychological and emotional suffering. The clinician can explain to the patient that nothing is genetically wrong and that the disease is rather a combination of unfortunate circumstances.

It helps the patient to understand that there is no personal weakness that causes the disease.

Finally, the absence of physiological mechanism acting against a chronic disease is no indicator that such a mechanism does not exist or is impossible; it only indicates that such a mechanism would not necessarily increase fitness. However, our human aim is to reduce suffering and costs of public health, not to increase evolutionary fitness. This leads to the conclusion that biomedical research might be most successful in diseases that have no or low lifetime fitness cost, but high costs for public health and a high load of individual suffering.

Last but not least, current research topics in evolutionary medicine where our graphical model might fail could nevertheless benefit by taking the four questions of Tinbergen into account. Examples could be maladaptation, environmental mismatch, and evolved traits that are not adaptive in the current modern environment. By applying the model and then identifying where it fails, the reason as well as the consequences why the trait is not adaptive or why environmental mismatch occurs could be identified. Thus, not all disease can be explained and described by Tinbergen`s four question, but identifying why and when this concept fails will nevertheless improve our understanding of why our body is constructed in a way that sometimes leads to disease.

Funding and conflict of interest

The authors have no conflict of interest. CS was supported by the CNRS, RS by the University of Regensburg.

References

Alves, A. J., Viana, J. L., Cavalcante, S. L., Oliveira, N. L., Duarte, J. A., Mota, J., Oliveira, J. & Ribeiro, F. (2016). Physical activity in primary and secondary prevention of cardiovascular disease: Overview updated. *World J Cardiol*, 8(10): 575–83.

Barker, D .J., Winter, P. D., Osmond, C., Margetts, B., & Simmonds, S. J. (1989). Weight in infancy and death from ischaemic heart disease. *Lancet*, 2(8663): 577–80.

Bateson, P. & Laland, K. N. (2013). Tinbergen's four questions: An appreciation and an update. *Trends Ecol Evol*, 28(12): 712–8.

Blaxter, K. (1989). *Energy Metabolism in Animals and Man*. Cambridge: Cambridge University Press.

Boehm, T. (2011). Design principles of adaptive immune systems. *Nat Rev Immunol*, 11(5): 307–17.

Brewerton, D. A., Hart, F. D., Nicholls, A., Caffrey, M., James, D. C., & Sturrock, R. D. (1973). Ankylosing spondylitis and HL-A 27. *Lancet.*, 1(7809): 904–7.

Brinkworth, J. F. & Barreiro, L. B. (2014). The contribution of natural selection to present-day susceptibility to chronic inflammatory and autoimmune disease. *Curr Opin Immunol*, 31: 66–78.

Chagnon, N. A. & Irons, W. (1979). *Evolutionary Biology and Human Social Behavior: An Anthropological Perspective*. North Scituate, MA: Duxbury Press.

Chakrabarti, L. A. &Simon, V. (2010). Immune mechanisms of HIV control. *Curr Opin Immunol*, 22(4): 488–96.

Chang, Y. K., Chu, C. H., Chen, F. T., Hung, T. M., & Etnier, J. L. (2017). Combined effects of physical activity and obesity on cognitive function: Independent, overlapping, moderator, and mediator models. *Sports Med*, 47(3): 449–68.

Cotsapas, C., Voight, B. F., Rossin, E., Lage, K., Neale, B. M., Wallace, C., Abecasis, G. R., Barrett, J. C., Behrens, T., Cho, J., De Jager, P. L., Elder, J. T., Graham, R. R., Gregersen, P., Klareskog, L., Siminovitch, K. A., van Heel, D. A., Wijmenga, C., Worthington, J., Todd, J. A., Hafler, D. A., Rich, S. S., Daly, M. J., & FOCiS Network of Consortia (2011). Pervasive sharing of genetic effects in autoimmune disease. *PLoS Genet*, 7(8): e1002254.

Drickamer, L. C. (1998). Vertebrate behavior: Integration of proximate and ultimate causation. *American Zoologist*, 38(1): 39–42.

Dube, S. R., Fairweather, D., Pearson, W. S., Felitti, V. J., Anda, R. F., & Croft, J. B. (2009). Cumulative childhood stress and autoimmune diseases in adults. *Psychosom Med*, 71(2): 243–50.

Feldmann, M., Williams, R. O., & Paleolog, E. (2010). What have we learnt from targeted anti-TNF therapy? *Ann. Rheum Dis*, 69 (Suppl 1): i97–i99.

Firestein, G. S., Kelley, W. N., Budd, R. C., Gabriel, S. E., McInnes, I. B., & O'Dell, J. R. (2013). *Kelley's Textbook of Rheumatology*. Philadelphia, PA: Elsevier/Saunders.

Gaber, T., Strehl, C., & Buttgereit, F. (2017). Metabolic regulation of inflammation. *Nat Rev Rheumatol*, 13(5): 267–79.

Gluckman, P., Beedle, A. & Hanson, M. (2009). *Principles of Evolutionary Medicine*. New York: Oxford University Press.

Hales, C. N. & Barker, D. J. (1992). Type 2 (non-insulin-dependent) diabetes mellitus: The thrifty phenotype hypothesis. *Diabetologia.*, 35(7): 595–601.

Hallgren, M., Herring, M. P., Owen, N., Dunstan, D., Ekblom, O., Helgadottir, B., Nakitanda, O. A., & Forsell, Y. (2016). Exercise, physical activity, and sedentary behavior in the treatment of depression: Broadening the scientific perspectives and clinical opportunities. *Front Psychiatry*, 7: 36.

Hochberg, M. C., Silman, A. J., Smolen, J. S., Weinblatt, M. E., & Weisman, M. H. (2015). *Rheumatology*. Philadelphia: Elsevier Mosby.

Kamiya, T., O'Dwyer, K., Westerdahl, H., Senior, A., & Nakagawa, S. (2014). A quantitative review of MHC-based mating preference: The role of diversity and dissimilarity. *Mol Ecol*, 23(21): 5151–63.

Klareskog, L., Malmstrom, V., Lundberg, K., Padyukov, L., & Alfredsson, L. (2011). Smoking, citrullination and genetic variability in the immunopathogenesis of rheumatoid arthritis. *Semin Immunol*, 23(2): 92–8.

Krebs, J. R. & Davies, N. B. (1993). *An Introduction to Behavioural Ecology*. Oxford: Blackwell Science Ltd.

Kyu, H. H., Bachman, V. F., Alexander, L. T., Mumford, J. E., Afshin, A., Estep, K., Veerman, J. L., Delwiche, K., Iannarone, M. L., Moyer, M. L., Cercy, K., Vos, T., Murray, C. J., & Forouzanfar, M. H. (2016). Physical activity and risk of breast cancer, colon cancer, diabetes, ischemic heart disease, and ischemic stroke events: Systematic review and dose-response meta-analysis for the Global Burden of Disease Study 2013. *BMJ*, 354: i3857.

LaFleur, C., Granados, J., Vargas-Alarcon, G., Ruiz-Morales, J., Villarreal-Garza, C., Higuera, L., Hernandez-Pacheco, G., Cutino-Moguel, T., Rangel, H., Figueroa, R., Acosta, M., Lazcano, E., & Ramos, C. (2002). HLA-DR antigen frequencies in Mexican patients with dengue virus infection: HLA-DR4 as a possible genetic resistance factor for dengue hemorrhagic fever. *Hum Immunol*, 63(11): 1039–44.

Lerner, S. P. & Finch, C. E. (1991). The major histocompatibility complex and reproductive functions. *Endocr Rev*, 12(1): 78–90.

Lozano, G. A. (2010). Evolutionary explanations in medicine: How do they differ and how to benefit from them. *Med Hypotheses*, 74(4): 746–9.

Mangalam, A. K., Taneja, V., & David, C. S. (2013). HLA class II molecules influence susceptibility versus protection in inflammatory diseases by determining the cytokine profile. *Journal of Immunology*, 190(2): 513–8.

Mathews, J. D., Whittingham, S., Hooper, B. M., Mackay, I. R., & Stenhouse, N. S. (1973). Association of autoantibodies with smoking, cardiovascular morbidity, and death in the Busselton population. *Lancet*, 2(7832): 754–8.

Mayr, E. (1961). Cause and effect in biology. *Science*, 134(3489): 1501–6.

McGuire, M. T. & Essock-Vitale, S. M. (1981). Psychiatric disorders in the context of evolutionary biology. A functional classification of behavior. *J Nerv Ment Dis*, 169(11): 672–86.

McPhee, J. S., French, D. P., Jackson, D., Nazroo, J., Pendleton, N., & Degens, H. (2016). Physical activity in older age: Perspectives for healthy ageing and frailty. *Biogerontology*, 17(3): 567–80.

Murphy, K. M., Travers, P., & Walport, M. (2011). *Janeway's Immunobiology*. Oxford: Taylor & Francis.

Nesse, R. M. (1984). An evolutionary perspective on psychiatry. *Compr Psychiatry*, 25(6): 575–80.

Nesse, R. M. (2013). Tinbergen's four questions, organized: A response to Bateson and Laland. *Trends in Ecology & Evolution*, 28(12): 681–2.

Nesse, R. M., Ganten, D., Gregory, T. R., & Omenn, G. S. (2012). Evolutionary molecular medicine. *J Mol Med (Berl)*, 90(5): 509–22.

Neufeld, K. M., Karunanayake, C. P., Maenz, L. Y., & Rosenberg, A. M. (2013). Stressful life events antedating chronic childhood arthritis. *J Rheumatol*, 40(10): 1756–65.

Paley, C. A. & Johnson, M. I. (2016). Physical activity to reduce systemic inflammation associated with chronic pain and obesity: A narrative review. *Clin. J Pain*, 32(4): 365–70.

Patterson, K. B. & Runge, T. (2002). Smallpox and the Native American. *Am J Med Sci*, 323(4): 216–22.

Perlman, R. L. (2013). *Evolution & Medicine*. Oxford: Oxford University Press.

Phillips, C., Baktir, M. A., Das, D., Lin, B., & Salehi, A. (2015). The link between physical activity and cognitive dysfunction in Alzheimer disease. *Phys Ther*, 95(7): 1046–60.

Phillips, C., Baktir, M. A., Srivatsan, M., & Salehi, A. (2014). Neuroprotective effects of physical activity on the brain: A closer look at trophic factor signaling. *Front Cell Neurosci*, 8: 170.

Pollard, K. M. (2016). Silica, silicosis, and autoimmunity. *Front Immunol*, 7: 97.

Quirke, A. M., Fisher, B. A., Kinloch, A. J., &Venables, P. J. (2011). Citrullination of autoantigens: Upstream of TNFalpha in the pathogenesis of rheumatoid arthritis. *FEBS Lett*, 585(23): 3681–8.

Schlosstein, L., Terasaki, P. I., Bluestone, R., & Pearson, C. M. (1973). High association of an HL-A antigen, W27, with ankylosing spondylitis. *N Engl J Med*, 288(14): 704–6.

Schmid-Hempel, P. (2011). *Evolutionary Parasitology*. Oxford: Oxford University Press.

Shanks, N., Larocque, S., & Meaney, M. J. (1995). Neonatal endotoxin exposure alters the development of the hypothalamic-pituitary-adrenal axis: Early illness and later responsivity to stress. *J Neurosci*, 15(1 Pt 1): 376– 84.

Shanks, N., Windle, R. J., Perks, P. A., Harbuz, M. S., Jessop, D. S., Ingram, C. D. & Lightman, S. L. (2000). Early-life exposure to endotoxin alters hypothalamic-pituitary-adrenal function and predisposition to inflammation. *Proc Natl Acad Sci U. S. A*, 97(10): 5645–50.

Speakman, J. R. & Westerterp, K. R. (2010). Associations between energy demands, physical activity, and body composition in adult humans between 18 and 96 years of age. *Am. J Clin Nutr*, 92(4): 826–34.

Spitzer, C., Wegert, S., Wollenhaupt, J., Wingenfeld, K., Barnow, S., & Grabe, H. J. (2013). Gender-specific association between childhood trauma and rheumatoid arthritis: A case-control study. *J Psychosom Res*, 74(4): 296–300.

Stearns, S. C. & Medzhitov, R. (2016). *Evolutionary Medicine*. Sunderland, MA: Sinauer Associates, Inc.

Straub, R. H. (2014). Insulin resistance, selfish brain, and selfish immune system: An evolutionarily positively selected program used in chronic inflammatory diseases. *Arthritis Res Ther*, 16(Suppl 2): S4, 1–15.

Straub, R. H. (2015). *The Origin of Chronic Inflammatory Systemic Diseases and their Sequelae*. San Diego: Academic Press.

Straub, R. H. (2017). The brain and immune system prompt energy shortage in chronic inflammation and ageing. *Nat Rev Rheumatol*, 13(12): 743-751.

Straub, R. H. & Besedovsky, H. O. (2003). Integrated evolutionary, immunological, and neuroendocrine framework for the pathogenesis of chronic disabling inflammatory diseases. *FASEB J*, 17(15): 2176–83.

Straub, R. H., Bijlsma, J. W., Masi, A., & Cutolo, M. (2013). Role of neuroendocrine and neuroimmune mechanisms in chronic inflammatory rheumatic diseases—The 10-year update. *Semin Arthritis Rheum*, 43(3): 392–404.

Straub, R. H., Cutolo, M., Buttgereit, F. & Pongratz, G. (2010). Energy regulation and neuroendocrine-immune control in chronic inflammatory diseases. *J Intern Med*, 267(6): 543–60.

Straub, R. H .& Schradin, C. (2016). Chronic inflammatory systemic diseases: An evolutionary trade-off between acutely beneficial but chronically harmful programs. *Evol Med Public Health*, 2016(1): 37–51.

Tinbergen, N. (1963). On aims and methods of ethology. *Zeitschrift für Tierpsychologie*, 20(4): 410–33.

Torine, I. J., Denne, S. C., Wright-Coltart, S., & Leitch, C. (2007). Effect of late-onset sepsis on energy expenditure in extremely premature infants. *Pediatr Res*, 61(5 Pt 1): 600–3.

Wang, L., Zou, Z. Q., & Wang, K. (2016). Clinical Relevance of HLA Gene Variants in HBV Infection. *J Immunol Res*, 2016:9069375.

Wikelski, M. & Ricklefs, R. E. (2001). The physiology of life histories. *Trends in Ecology & Evolution*, 16(9): 479–81.

Williams, G. C. (1957). Pleiotropy, natural selection, and the evolution of senescence. *Evolution*, 11(4): 398–411.

Wilson, E. O. (1975). *Sociobiology: The new Synthesis*. Cambridge, MA: Harvard University Press,.

Wingfield, J. C. (2008). Comparative endocrinology, environment and global change. *Gen. Comp Endocrinol.*, 157(3): 207–16.

CHAPTER 12

Evolution, Genomics, and the New Genetic Technologies

Louise Wilkins-Haug

12.1 Introduction

With advancing technologies, the concept of 'genetics' is undergoing change as well. Genetics, the study of alterations in DNA sequences, is now broadened to "genomics"—an interdisciplinary approach incorporating gene-gene interactions, controllers of DNA transcription beyond sequence alterations, and a broader view of the process from gene to protein to eventual impact on the individual. Genomics is a driver behind precision medicine and remaining knowledgeable about these fast-paced discoveries is a daunting task for health care providers and patients. Providing an infrastructure for incorporating genomic advances rather than focusing on a specific technology benefits the medical community as well as the public. Evolutionary medicine with principles encompassing both the mechanistic (proximate) cause of the disease and the ultimate cause (the evolutionary nature) provides such a foundation (Nesse et al., 2010; Gluckman & Bergstrom, 2011)

This chapter covers three important areas to help the reader appreciate the emerging role of genomic technologies within the context of evolutionary medicine. In the first part, basic concepts of evolution are reviewed to develop an infrastructure for application of more advanced genomic testing. Secondly, a review of the past and current options for genomic testing affords a perspective on the information that is now readily available to physicians and the public. This section emphasizes the revolutionary change in genomic testing over the past 15 years and how tremendous amounts of DNA information are retrievable from an individual relatively quickly and at low cost. Lastly, understanding these concepts within a framework of evolutionary medicine provides a structure for incorporation of future genomic discoveries. This last goal is reliant upon education around evolutionary medicine as well as a broader reach to all individuals about genomic concepts.

12.1.1 Evolution

Nothing in Biology Makes Sense Except in the Light of Evolution **Theodosius Dobzhansky**

Our understanding of evolution incorporates several broad principles which are important to understand as one considers the impact of the new molecular technologies. Darwin's theory of evolution as originally put forth in 1859, in *On the Origin of Species by Means of Natural Selection, or the Preservation of Favoured Races in the Struggle for Life* has stood the test of time.

Core concepts of this theory include:

1. More individuals are produced each generation than can survive.
2. Phenotypic variation exists among individuals and the variation is heritable.

Wilkins-Haug, L., *Evolution, Genomics, and the New Genetic Technologies* In: *Integrating Evolutionary Biology into Medical Education—for maternal and child healthcare students, clinicians, and scientists*. Edited by: Jay Schulkin and Michael L. Power, Oxford University Press (2020). © Oxford University Press.
DOI: 10.1093/oso/9780198814153.003.0013

Table 12.1 Central concepts and common misconceptions of evolution theory.

Central concepts	Common misconceptions
Evolution occurs in populations, not individuals	All traits of organisms are adaptations
Evolution is a change in a population's allele frequency	All members of a population develop new traits at the same time
Genetic variation is required for natural selection and genetic drift	"Fitness" refers to strength, size of speed of individual
Genetic variation arises from mutation and gene flow	Evolution strives for perfection
Evolution can occur through random chance	All mutations are harmful
	Evolution is slow and gradual

Modified from Lee, (2017). (Springer open access) Teresa W. Lee, Kathleen E. Grogan & Justine S. Liepkalns. Making evolution stick. In *Evolution: Education and Outreach* (2017) 10:11. https://doi.org/10.1186/s12052-017-0074-2

3. Those individuals with heritable traits better suited to the environment will survive.
4. When reproductive isolation occurs, new species will form.

These core tenets are the basis for concepts such genetic (heritable) variability, natural selection, and reproductive fitness (Table 12.1). They remain central to the past and ongoing evolution of mankind. As an example, *Homo sapiens* are thought to arise from a common ancestor, *Homo erectus*—upright man in Latin. Dispersion and refinement of the evolving mankind has been aided by migratory evolution. Currently, application of advanced genetic technologies support an "out of Africa" progression of *Homo sapiens* as opposed to "multiregional" origins theory. Findings such as extensive variation at the genetic level, tracking of the maternally inherited DNA in the mitochondria ("mitochondrial Eve") and analysis of unique gene sequences on the Y chromosome all support an African origin. Natural selection, reproductive fitness and intermingling with closely related populations led to further variation and refinement of the human genome across the globe. Two notable examples of genetic contribution from closely related populations are thought to have occurred with *Homo neanderthalensis*, originating in Europe and migrating to Asia and from the *Denisovans* populations, with likely origins in Russian regions (Paabo, 2015) (Figure 12.1).

A closer look at heritable variation is now possible with the newer genomic technologies and warranted when examining Darwin's theory of evolution. With regard to human variation, after 200,000 years, the actual differences between the genetic sequences of two humans is relatively small, occurring as infrequently as at 1 in 1000 base pairs. Interestingly, less variation in the human genome is noted when compared to the rates in other species. This may reflect both the relatively recent evolution of humans and the probable small size of the original population (as few as 10,000 individuals). Many other species have existed longer and began with a substantially larger breeding population. Greater variation within a species affords a greater ability to respond to a changing environment.

Heritable variability is key to maintaining a range of responses to changing environments. Two drivers of variability are genetic selection and genetic drift. Over generations, genetic selection operates as a response to an environment through reproductive advantage to optimize a species' survival. Reproductive fitness and its impact on the continuation of a genetic change, rather than longevity, is the goal of natural selection. Divergent lines within a species develop as genetic variability favors the reproduction of the successful phenotypes (Figure 12.2). This is perhaps best exemplified by Darwin's observations in the Galapagos Finch species with different beak profiles optimized for the food source specific to their microenvironment. The genetic determination of the variations in Finch beak size and contour was later shown to be directed by altered timing of *Bmp*[4] protein (Abzhanov et al., 2004).

A second driver of variation is genetic drift; the ongoing, random change of DNA sequences in a population. Genetic drift also leads to alterations in gene frequency. Genetic drift is altered by bottleneck effect (events which occur to drastically reduce a population) and founder effect (loss of the allelic variation that accompanies a new smaller population). Bottleneck effect occurs with random loss of gene variation through various environmental constraints to population size (physical, social, ideo-

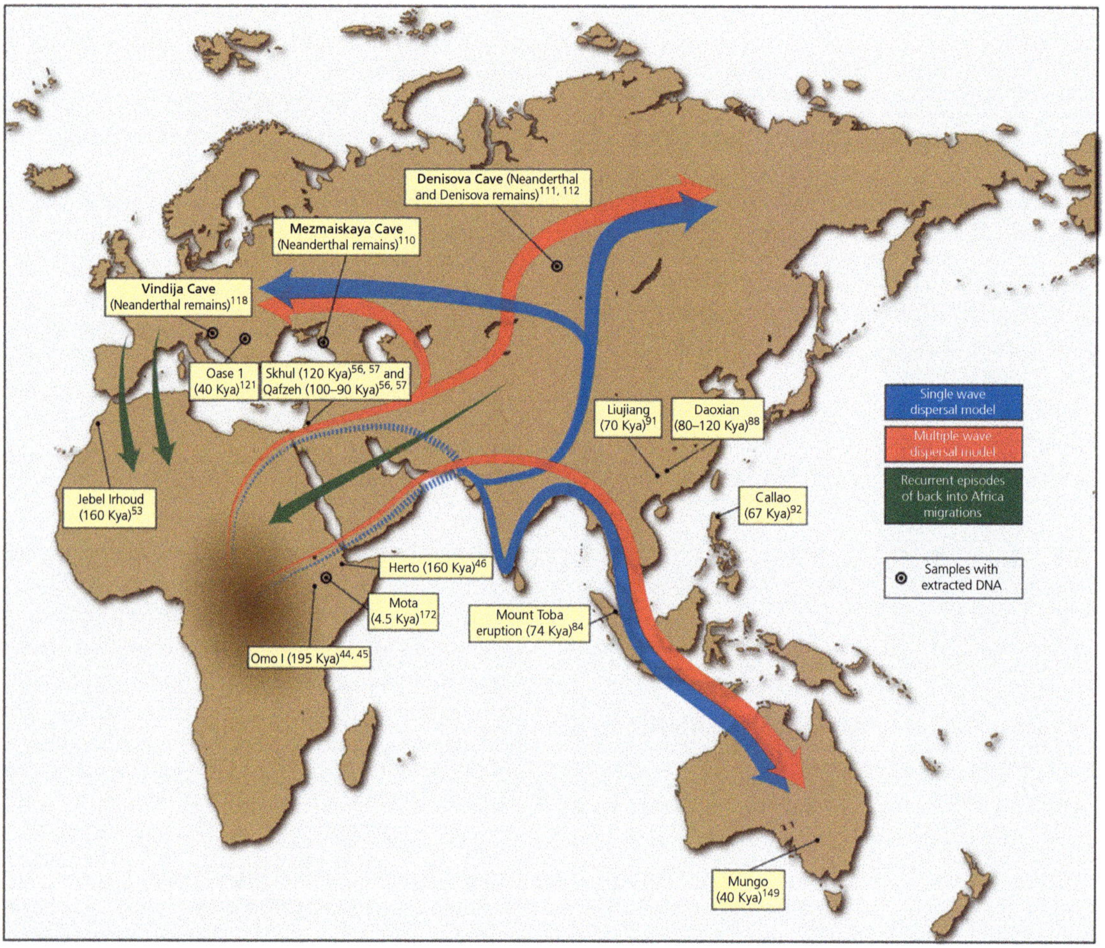

Figure 12.1 Putative migration waves out of Africa and back migrations into the continent, as well as the locations of major ancient human remains and archeological sites

López, S., van Dorp, L., & Hellenthal, G. (2015). Human dispersal out of Africa: A lasting debate. *Evolutionary Bioinformatics Online*, 11(Suppl 2), 57–68. http://doi.org/10.4137/EBO.S33489 https://www.ncbi.nlm.nih.gov/pmc/articles/PMC4844272/, CC BY 3.0, https://commons.wikimedia.org/w/index.php?curid=50508700

logical, etc.). Founder effect relates to the decreased variability in the genes of a smaller population which has removed itself from the main group. (Table 12.2) For example, as a result of targeted hunting, the Northern Elephant Seal population was reduced to as few as twenty seals by the end of the nineteenth century. With changes in hunting practices over the next century, over 30,000 elephant seals now exist. Among these Northern Elephant Seal populations, a more limited range of allelic variation exists. By comparison, the Southern Elephant Seals have greater genetic variation never having had their populations diminished in size by targeted hunting practices (Hoelzel et al., 1993).

Human examples exist for both bottleneck and founder effect reflecting diminished genetic variation. Huntington's disease (HD), an autosomal dominant condition of later-life neurologic disease, is more frequent among the South African Afrikaner population. The dominant gene for HD was introduced by the Dutch settlers (bottleneck) and the maintenance of a higher frequency among

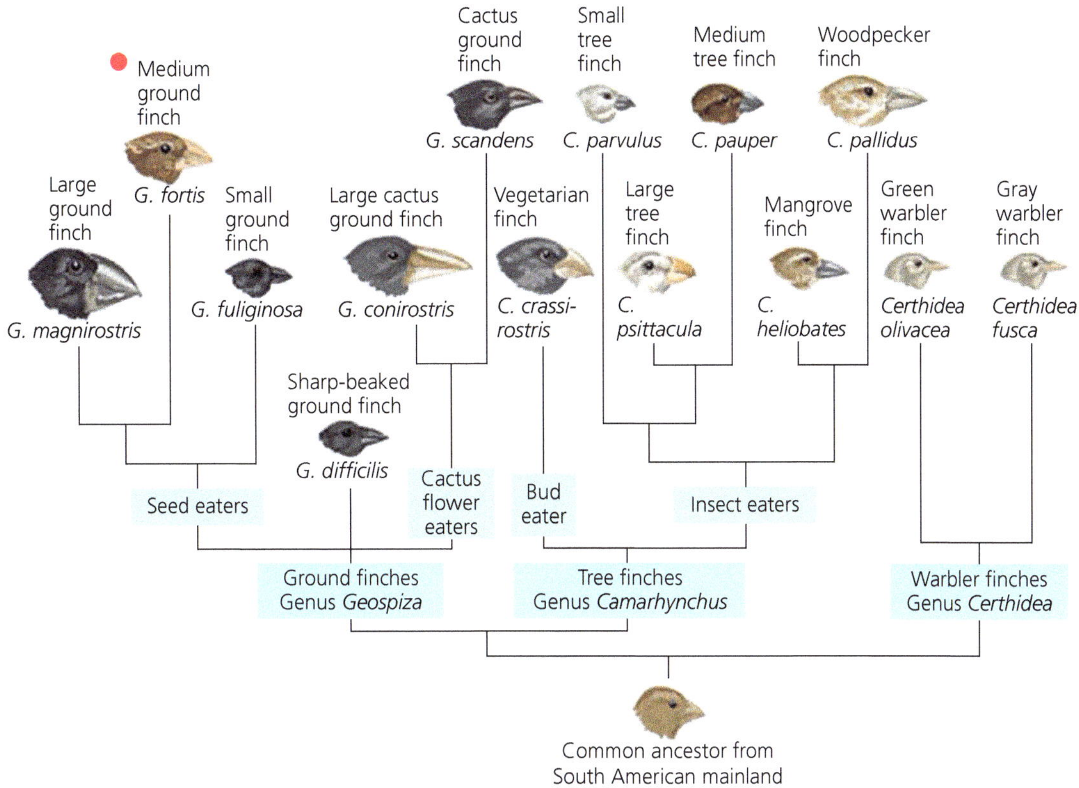

Figure 12.2 Evolution of Darwin's finches from South American species, those adapted to various food supplies in the Galapagos Islands Copyright at Pearson Education Inc, publishing as Bemjamin Cummings.

Afrikaners is a founder effect with social and geographic constraints decreasing the variation within this population (O'Donald, 1967).

Genetic variability, however, does not always directly translate to phenotypic variability. "Concealed variability" (Charlesworth & Charlesworth, 2018) represents the unseen genetic variation between individuals which is key to their response to the environment. As an example, among autosomal recessive genetic disorders, mutant alleles can be maintained within a population by carriage in an unaffected individual (a heterozygote carrier). Preferential survival advantages for the heterozygote carrier within their environment perpetuate the gene's presence. Such advantages are exemplified by improved survival of sickle cell carriers when exposed to malaria.(Williams et al., 2005) The higher heterozygotic carriage of the delta508 gene for cystic fibrosis among Northern Europeans likely provided altered sodium channels theorized to decrease dehydration during bubonic plague. Concealed variability is more often proposed than actually documented, however.

12.1.2 New genetic technologies

Only time and money stand between us and knowing the composition of every gene of the human genome

Francis Crick, 1986

12.1.2.1 Gene sequencing

The desire to understand the exact sequence of bases in a segment of DNA and its relationship to a specific disease has existed for decades. Genetic diseases were initially approached from an epidemio-

Table 12.2 Summary of mechanisms of evolutionary change.

Mechanism	*Description*	*Effect on gene alleles*
Gene flow	Transfer of alleles between populations	Introduces new variant allele
Genetic drift	Random fluctuations due to change	Allele frequencies vary
Natural selection	Individuals with trait have higher reproductive fitness	Nonrandom allele selection and transmission to next generation
Bottleneck	Any event which severely reduces population size	Allele frequencies in surviving population differ or may be missing
Founder effect	When a fraction of a population separates to from a secondary population	Allele frequency in subsequent population differ or may be missing

Modified from Lee, (2017). (Springer open access) Teresa W. Lee, Kathleen E. Grogan & Justine S. Liepkalns. Making evolution stick. In Evolution: Education and Outreach (2017) 10:11 https://doi.org/10.1186/s12052-017-0074-2

logic basis when similar findings segregated within families. For many genetic conditions, the now familiar transmission patterns such as autosomal dominant, autosomal recessive, and X-linked recessive were discerned by study of large kindreds. In the 1970s, with the introduction of Sanger DNA sequencing, genetics moved from a mostly statistical and epidemiologic approach into the world of molecular technologies. Sanger sequencing through sequential incorporation of labeled, complimentary bases along a single segment of DNA elicits the ordering of the bases and identifies when altered nucleic acid base sequences occur. As these nucleic acid base sequences are then responsible for production of a protein, Sanger sequencing opened extensive avenues for examination of genetic disease at a molecular level. Altered base sequences were found to be responsible for a protein not being produced, formation of an abnormally shaped protein, or protein production in insufficient quantities (Heather & Chain, 2016). Through these early days of gene discovery, sickle cell anemia represents an historic example for several reasons. As early as 1949, Linus Pauling related the altered shape of the sickle hemoglobin to its diminished oxygen carrying capacity, a first ever association of an altered protein as the source of a disease. It was another decade before the identification by Ingram and Hunt that the altered hemoglobin protein was the result of a single nucleic acid base change resulting in the substitution of one amino acid for another (Ingram, 2004).

Sanger sequencing was a revolutionary concept at its introduction in the 1970s providing base by base determination, a "reading of the gene." However, the Sanger process is time consuming and laborious. Only a single strand of DNA can be deciphered at a time by examining the labeled nucleic bases added one at a time. Initial reports of sequencing a fragment of a yeast gene required years to accomplish, a few bases a year. However, a proof of principle was established—changes or alterations in the amino acid base order of a DNA segment could alter the protein produced and downstream, result in the physical characteristics associated with a specific genetic disease. Using Sanger sequencing, our ability to understand base sequence alterations, the resultant gene changes and the associated diseases progressed but at only a moderate pace. In the human, this was most notably the detection of base sequence changes which correlated to recognized genetic disorders, such as Tay Sach (TS) and cystic fibrosis. While a remarkable stride forward in our understanding of genetic disorders, Sanger sequencing, even with modifications, is a time consuming and tedious process which could only evaluate a single gene at a time.

Despite its limitations, Sanger sequencing with modifications and the resultant information attained was utilized for several decades. Such sequencing could provide confirmation of a clinical diagnosis, examine a specific gene when symptoms were clustered within a family, and provide a tool for identification of carriers of genetic conditions. In many cases, this information supplemented prior work on the particular disorder gained from biochemical and metabolic studies. Through the 1980s and until 2004, hundreds of thousands of DNA sequence changes were added to the Genbank, an open repository for identifying the impact of gene sequence variations (http://www.ncbi.nlm.nih.gov/genbank).

One genetic disorder which was extensively studied during this time was TS. Due to a disorder of the enzyme hexoaminidase A, affected individuals had abnormal sphingomyelin stored in brain and neural tissue. In the most severe form, affected infants lost their attained developmental milestones around 6 months and eventually died during childhood. Carriers had approximately 50% levels of this important enzyme. However, this enzyme test was difficult to implement as the results could be influenced by several variables including whether the mother was on oral contraception pills. The introduction of Sanger sequencing, and identification of the gene mutations responsible forTS, provided a tool for confirmation of diagnosis as well as reliable carrier detection and pregnancy evaluation. Limited by the labor and time constraints of Sanger sequencing, genetic carrier screening for TS disease was reasonably applied to individuals of Ashkenazi heritage, a population with a higher gene frequency to utilize the technology most efficiently (Slatkin, 2004). As importantly, Sanger sequencing revealed that among carriers and affected individuals, the abnormal enzyme could be produced by several different sequence alterations of the gene. This knowledge advanced our understanding of the complex nature of genetic disease. Previously held paradigms of one mutation producing one disease, such as sickle cell anemia, turned out to be the exception.

As work continued with Sanger sequencing and disease correlation, advances in our understanding of gene mutations, and especially genetic variability emerged. For example, the classic presentation of cystic fibrosis was associated initially with 3 and then 23 mutations within the Cystic Fibrosis Transmembrane Regulator (CFTR) gene. Soon, however, more than 1000 genetic changes in this gene were reported. (Figure 12.3, Table 12.3) As

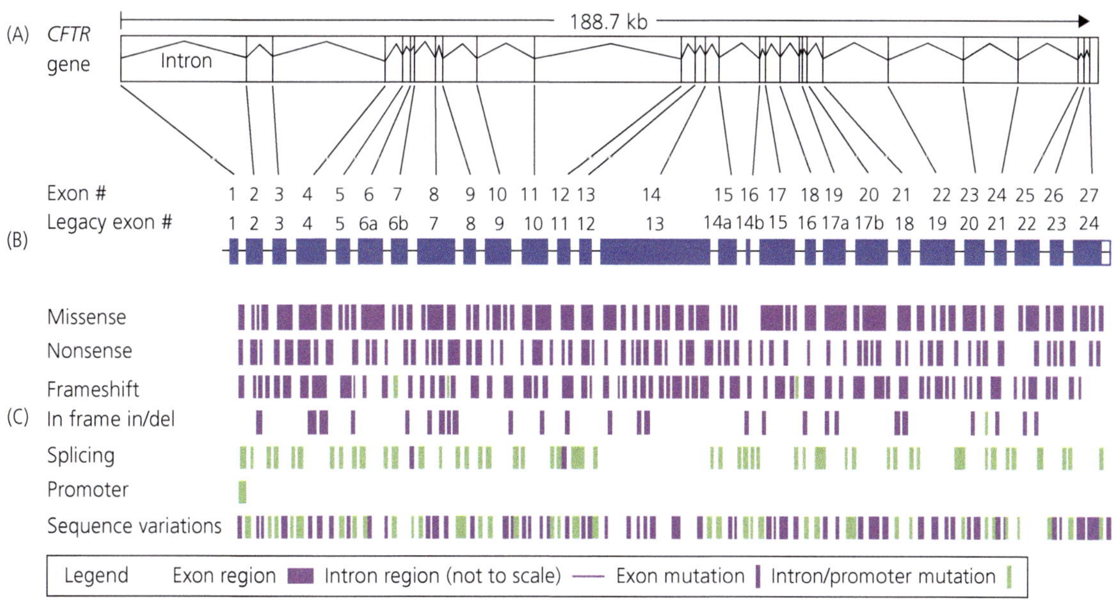

Figure 12.3 Diversity of sequence alterations identified in Cystic Fibrosis Transreceptor (CFTR) gene
Tsui, L. C. & R. Dorfman (2013). The cystic fibrosis gene: a molecular genetic perspective. *Cold Spring Harb Perspect Med*, 3(2): a009472. doi: 10.1101/cshperspect.a009472. In John R. Riordan, Richard C. Boucher, & Paul M. Quinton (eds.) *Additional Perspectives on Cystic Fibrosis*, available at www.perspectivesinmedicine.org

Table 12.3 Diversity of CFTR mutation types by Jan. 2011.

Mutation type	Count	Frequency (%)
Missense	765	40.41
Frameshift	303	16.01
Splicing	225	11.89
Nonsense	160	8.45
In frame in /del	37	1.95
Large in / del	49	2.59
Promoter	15	0.79
Sequence variation	269	14.21
Unknown	70	3.7
TOTAL	1893	

Tsui, L. C. and R. Dorfman (2013). The cystic fibrosis gene: a molecular genetic perspective. *Cold Spring Harb Perspect Med*, 3(2): a009472. doi: 10.1101/cshperspect.a009472.

genotype–phenotype correlations emerged, the findings reflected not only a wide range of mutations among the CFTR gene but also a spectrum of clinical findings with aberrant calcium transport as a central feature (Tsui & Dorfman, 2013). This included classic cystic fibrosis (CF), CF with and without the associated pancreatic insufficiency, adult respiratory issues to bilateral absence of the vas deferens in males. The identification of each of these newly recognized changes in the CFTR gene sequence, a large gene, required not only time but also considerable investment.

In 2003, after 13 years and $2.7 billion dollars, sequencing of the entire human genome was accomplished based on the DNA from several volunteers. While providing important information about human DNA genes, the continued sequencing limitations of time and cost prohibited immediate translation to clinical care. However, from the human genome project, emerged technologies which fueled significant advances in gene sequencing. In the first decade of the twenty-first century, faster and lower-cost sequencing tools revolutionized delineation of human genetic disease as well as variation. First generation sequencing (Sanger sequencing) was supplanted by next generation sequencing (NGS). NGS accomplished the sequencing of its first human genome (that of J. D. Watsun, Noble Prize winner) in two months and at markedly lower cost.

NGS collectively refers to the advances made on "first generation" sequencing—Sanger sequencing. NGS, which can sequence fragments rapidly and simultaneously across numerous DNA fragments, disrupted the field of genetic analysis with shorter time to results and lower costs. This technology is far reaching with applications for both quantitative and qualitative DNA analysis. From a quantitative aspect, early in pregnancy a sample of the mother's blood can be assessed by NGS to determine if too many copies of a particular gene (s) are present. For instance, when there is an abnormally higher presence of genes from chromosomes 21, a high risk of trisomy 21 (Down syndrome) is suspected. In a qualitative fashion, NGS can efficiently, quickly, and at lower cost, examine a multitude of genes at the same time for sequence changes related to specific diseases. This forms the basis for current genetic testing panels crafted to specific disease states such as cancers.

For NGS, the rate limiting step is not production of the raw data but the compilation and interpretation of the nucleic acid base sequences produced. For any one individual, if their genome sequence was printed in 12-point font, it would stretch from New York to California. The curation of the identified base sequences is known as the bioinformatics. Numerous online sites are instrumental in converting the sequencer raw base reads to clinically useful information. This has resulted in an explosion of NGS applications including assessing DNA for base sequence changes leading to genetic diseases at all life stages, variation in response to drug treatments, and identification of risk for later onset disease such as cancer, and Alzheimer's disease.

NGS can be targeted to a specific gene, or more broadly reaching, the entire human genome through whole genomic screening (WGS). To provide efficiency, the protein coding segments of the DNA, the exomes, are targeted with whole exome screening (WES). The exome represents only 1–2% of the total DNA but accounts for over 85% of known inherited disease. Restriction to WES as opposed to WGS does lose the ability to discern changes outside the exome such as those mutations which alter start and stop signals. Currently the cost of sequencing a particular gene or, in fact, an entire whole exome ranges from $500 to $2,000 driven by the intensity of the bioinformatics involved. Rapid progression of these technologies has raised a myriad of unanswered questions concerning privacy of genetic information, commercialization, ownership of the DNA sequences, and possible uses for harm such as bioterrorism.

NGS has also contributed to a greater understanding of variation at the genetic level between species, within species and within the same genetic disorder. Between species, most DNA sequences are remarkably similar—95% of the DNA content is the same between higher primates, chimpanzee (*Pan troglodytes*) and humans. Within species, much of an individual's DNA (98%) is remarkably like that of other humans. Within genetic diseases, however knowledge of greater variability gained from NGS can be seen at multiple levels:

1. In persons with identical presenting diseases but DNA changes in different genes.
2. Multiple sites of sequence change within one gene producing the same disease.
3. Different sequence changes within one gene producing clinically different disorders.
4. Variation not previously reported in gene gsequences.

With increased technical resolution the identification of DNA sequence variation unique to an individual is increasing. The ability to differentiate which DNA sequence change is a benign variation and which is likely to be disease producing is complex. Jointly, the American College of Medical Genetics and Genomics, Association of Molecular Pathology, and the American College of Pathology have identified five classes of variants. A wide range of criteria within four domains is applied to make these variant determinations (Table 12.4). For example, for sequence variations not previously identified, the impact is based on a compilation of familial segregation, "in silico" computational modeling of the amino acid changes produced by the new DNA sequence, and functional studies. Laboratories reporting DNA sequence results should conform to Clinical Laboratory Improvement Amendments (CLIA) oversight with interpretation by a board-certified clinical molecular geneticist, molecular genetic pathologist, or the equivalent (Richards et al., 2015).

Given the fast pace at which variants are being identified, reassessment and reclassification has occurred in some of the earlier databases. In initial databases, the finding that reanalysis revealed only 34% of sequence variations were interpreted the same instead of correctly by two independent laboratories is unsettling(Amendola et al., 2016). A National Institute of Health (NIH) sponsored partnership between CLINVAR and ClinGEN, is focused on improving these interpretations. As NGS studies increase, the question of what is disease causing and what is benign variation remains one of the most challenging aspects, especially when a variant is paired with a known pathogenic allele.

Understanding the principle of variation at the DNA level informs NGS interpretation whether it is applied to genetic disorders of childhood, response to pharmacologic agents, or the detection of adult cancers. The recognition that everyone has a unique DNA profile with disease causing variants as well as benign variation is a relatively new concept. Appreciating the uniqueness of everyone's DNA profile is key to the wider use of NGS now within reach for active clinical implementation.

12.1.2.2 Gene editing

The transition from identification of gene sequence alterations to correction of the errant genes once seemed unreachable. In the last 5 years it has become a reality. The ability to alter DNA sequences at the molecular level was introduced to the scientific literature just recently—in the Doudna lab at

Table 12.4 Guidelines for determining level of pathogenicity and confidence in assessment from the American College of Genetics and Genomics Guidelines.

Pathogenic
Likely pathogenic
Benign
Likely benign
Uncertain significance

Richards, S., N. Aziz, S. Bale, D. Bick, S. Das, J. Gastier-Foster, W. W. Grody, M. Hegde, E. Lyon, E. Spector, K. Voelkerding, H. L. Rehm, & Acmg Laboratory Quality Assurance Committee (2015). Standards and guidelines for the interpretation of sequence variants: a joint consensus recommendation of the American College of Medical Genetics and Genomics and the Association for Molecular Pathology. *Genet Med* 17(5): 405–24. doi: 10.1038/gim.2015.30.

University of California, Berkley in 2012 and the lab of Fang Zhang at the Broad, Harvard, in 2013 (Wiedenheft, Sternberg, & Doudna, 2012; Cong et al., 2013). Their approaches use a natural system in bacteria which identifies and "cuts out" viral DNA to which previous exposure has occurred. This is accomplished by ongoing incorporation of small segments of viral DNA within the bacterial genome to act as early warning signals. Known as CASPR-cas9, (Clustered Regularly Interspaced Short Palindromic Repeats), the ability to directly target and modify specific genes represents a sea change technology in genetics, medicine, and basic biology research (Figure 12.4a, b). To gene edit using a CRISPR-cas 9 system, the desired DNA is targeted using a specific RNA, excised with the attached Cas 9 enzyme and the DNA of concern removed. Lastly, a normal sequence of DNA is inserted or the sequence allowed to heal itself using the normal allele as a template. With additional refinements, this system can control reading of a targeted DNA sequence (activation and suppression of transcription), fuse proteins, modulate the epigenetic state, and provide a site for tagging for image analysis of biologic systems.

Applications of CRISPR technology have emerged in all avenues of science and medicine. Among the plant world, the concept of genetic modification, inserting beneficial or deleterious genes (usually tagged to a viral DNA) has existed for over 20 years as "genetic modification." However, longer term concerns continue to plaque this approach as the inserted genes are placed at random (potentially harming adjacent genes) and the carrier, foreign DNA, remains. Despite these concerns, since 1996, such genetic modifications of numerous field crops (soybeans, corn, cotton) are commercially available. In most cases, the genetic modification is an insertion of bacterial DNA to increase resistance to herbicides. Herbicidal treatment would then effectively destroy only the non-DNA engineered, nondesired growth, such as weeds. Federal regulations on food for humans required a notification that the item was "partially produced with genetic engineering." While proving to be effective in increasing crop production, the method of randomly inserting a foreign sequence of DNA continued to raise concerns about as yet undetected effects on the crop and the consumer (Abdallah, Prakash, & McHughen, 2015). Clear harm from the inserted, foreign DNA, which typically remains in the altered agricultural product, has not been demonstrated although a span of 20 years is small from an evolutionary perspective.

With the emergence of CRISPR, however, its ease of use combined with targeted gene editing has swept the agricultural world. The relative simplicity of CRISPR, the ability to remove deleterious traits as well as add beneficial characteristics and the insertion at a known location has contributed to an explosion of CRISPR gene edited agricultural products in only a few short years. Increased crop production, improved stress tolerance, greater disease and pest resistance, lower cost, and increased nutritional value are all highlighted as moving agricultural production from genetic modification to gene editing with CRISPR (Abdallah, Prakash, & McHughen, 2015).

Among insects and animals, CRISPR-cas9 also has been adopted enthusiastically. Further potential uses include 1) development of animal models to understand biologic processes, and 2) gene drives. Decades of animal studies have contributed a greater understanding of basic biologic pathways to the scientific literature. Previously, however, the creation of animal models whether through selective breeding or gene knock-out was costly and time intense. With gene editing through CRISPR, investigators can now succinctly remove the gene sequence of interest and either substitute a different gene sequence or allow the removed DNA sequence to be filled by incorporation of nucleic acids from the homologue sequence (Figure 12.4 a, b). Different approaches are used to accomplish different goals. For example, in ants, their social clustering and coordinated work has been linked to a unique pheromone. Removal through gene editing with CRISPR-cas9 of the specific pheromone receptors renders the ant colony unable to communicate and coordinated work ceases (Kronauer & Levine, 2017). While such experiments were possible with the prior gene knock-out methods, CRISPR-cas9 has proved more effective, less costly, and provides greater specificity for the intended sequence of interest.

Secondly, gene drive applications of CRISP-cas9 are emerging in agricultural reports and increas-

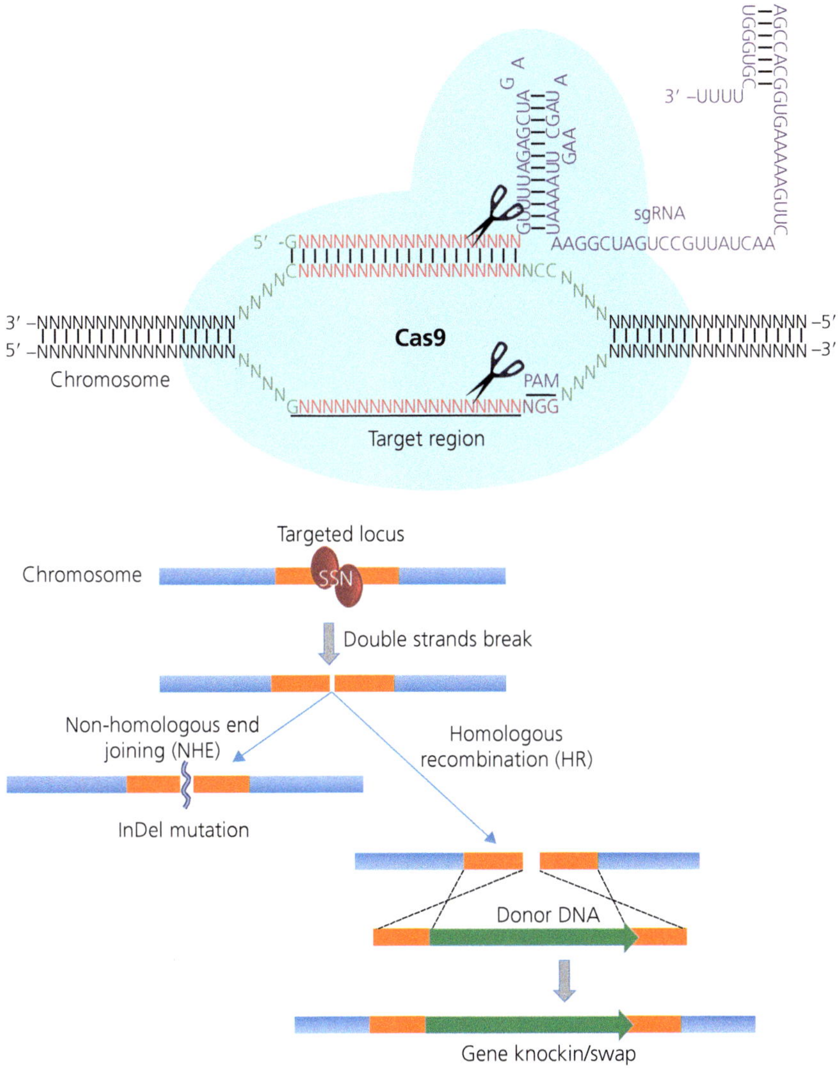

Figure 12.4 a,b) Use of CRISPR derived RNA to target a specific DNA segment for subsequent cas9 enzymatic cutting of the double stranded DNA. b) Possible repair mechanisms producing a deletion/ insertion alteration or alternatively a knock-in alteration
Abdallah, N. A., C. S. Prakash, & A. G. McHughen (2015). Genome editing for crop improvement: Challenges and opportunities. *GM Crops Food*, 6 (4): 183–205. doi: 10.1080/21645698.2015.1129937.

ingly in insect/animal work. Gene drives refer to the insertion of a deleterious gene which is then maintained in future generations though naturally occurring DNA replication (Figure 12.5). While gene drives can occur in nature when the transmission of an allele reaches greater than 50%, among sexually reproducing species, introduction and then selective transmission of a desired gene through CRISPR-cas9 is quickly becoming possible (Figure 12.6).

Examples include:

1. Introduction of a gene into a laboratory animal line which will then be propagated quickly to subsequent generations and serve as a source for examining biologic process.
2. Applications in wild populations such as extinguishing ("crashing") a species of insects which spread disease. Selective gene drives have been proposed to introduce sterility into mosquito

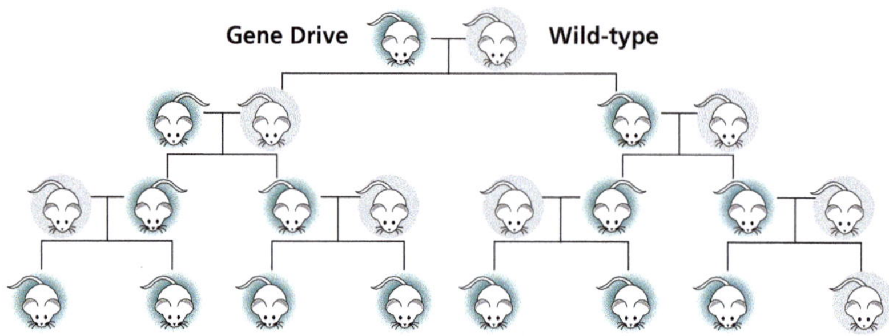

Figure 12.5 Schematic illustrating in mice how in a few generations a population could be eliminated through gene editing incorporating a sterility gene
Esvelt, K. M. & Gemmell, N. J. (2017). Conservation demands safe gene drive. *PLoS Biol*, 15(11): e2003850. https://doi.org/10.1371/journal.pbio.2003850

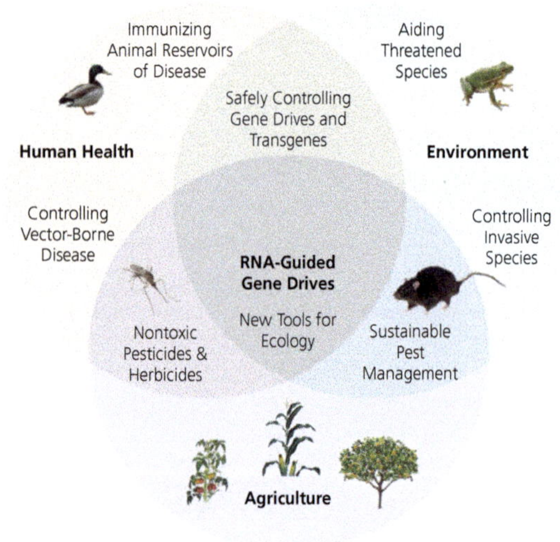

Figure 12.6 Potential applications of RNA-guided gene drives. Clockwise from left: genetically engineered malarial mosquitos resistant to pathogen acquisition or to be eliminated with a suppression drive, protective gene drives which spread engineered genes in soon-to-be extinct species, genetically modified for demise species which are invasive, sensitizing drives to improve safety of pesticide and herbicides, gene drives to test ecologic hypotheses
Modified from Esvelt, (2014). *eLife.* doi:10.7554/eLife.03401

populations which transmit malaria, dengue, and Zika. Initial steps for crashing the mosquito population transmitting Zika in Florida received Food and Drug Administration (FDA) approval in the summer of 2016. Other examples include "crashing" a specific invasive species to protect a native species. With the introduction to the New Zealand island nation of small mammals by man, the native aviary population has markedly decreased as the mammals continue their consumption of the eggs of the native birds. A goal of *2050 Predator Free New Zealand* is to protect the endogenous species such as parrots and kiwi from extinction. In addition to trapping and creating sanctuaries, a CRISPR-cas9 gene drive of introduced sterility is proposed in New Zealand. A CRISPR-Case9 project is underway to eradicate 8 predator mammalian species (rats, weasels, possums) for New Zealand's *2050 Predator Free* Initiative.

Further advances are expected in this sphere of research with both independent donors (Bill and Melinda Gates Foundation) and governmental agencies (Defense Advanced Research Projects Agency (DARPA), and the World Health Organization (WHO)) investing hundreds of millions of dollars. There continues to be enthusiasm for reaching select gene drives (mice, mosquitoes) before 2020. There remains concern however around fully understanding the evolutionary impact of gene drives prior to their introduction.

In humans, gene editing with CRISPR-cas9 is also emerging. These applications are taking one of two approaches:

1. Somatic cell editing—removal and replacement of a DNA sequence in individuals with an inherited disease.
2. Germline editing—alterations in embryos to prevent genetic diseases.

Somatic cell editing is currently being pursued in tissue culture and animal models. Disorders cur-

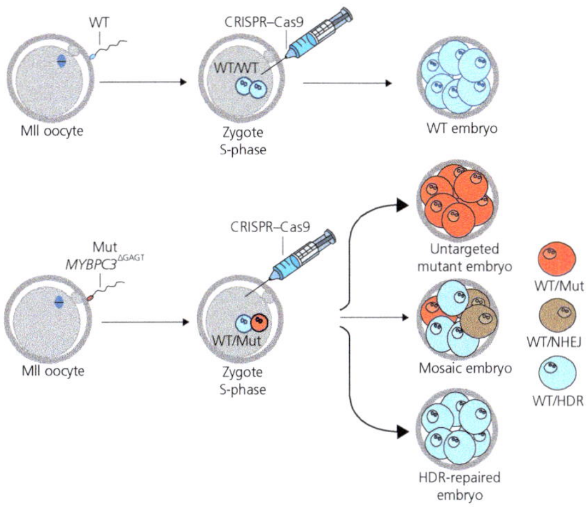

Figure 12.7 Germline editing through gene correction in S-phase injected human embryos. Fertilization with wild type (non mutated) sperm, and subsequent CRISPR-Cas9 injection results in a wild type embryo. However when the mutant gene is present in the sperm, subsequent CRISPR-Cas9 gene editing removes the altered sequence, allowing repair to insert a normal sequence. The embryos derived are repaired, unrepaired, and mosaic
Ma, H., N. Marti-Gutierrez, et al. (2017). Correction of a pathogenic gene mutation in human embryos. *Nature*, 548(7668): 413–19. doi: 10.1038/nature23305.

rently targeted are those with known, often single base or small sequence changes such as Duchenne Muscular Dystrophy and sickle cell anemia. Ultimately, genetic disorders which confer risk such as cancer could be candidates. Similarly, disorders not resulting from defects of the gene sequence but due to the mutations of associated enhancer genes are also being examined. CRISPR success with identification and promotion of enhancer genes opens the possibility for a wider range of approaches to genetic disease. Progress in these areas is being reported and research initiatives are likely to move to higher primates before proceeding to humans.

Germline, primarily embryonic gene editing, has also emerged. The first reports of targeting and specifically correcting a designated gene in an embryo occurred in 2017 (Tang et al., 2017). Among nonviable embryos, CRISPR-cas9 successfully removed and then repaired the defect in the gene responsible for beta thalassemia. While exemplifying proof of principle, embryonic gene editing faces several challenges including high unintentional changes in the non-targeted genes and successful gene editing in some but not all cells (mosaicism). Alternative approaches to germline gene editing are insertion of the CRISPR-cas9 into the egg at the time of fertilization to address the mosaicism difficulties (Ma et al., 2017) (see, Figure 12.7). Gene editing in the embryo will lead to an individual spared the consequences of the gene mutation. Additionally, as the mutation is corrected early in the embryo, the germ cells of the individual are also spared the mutation essentially removing it from the population. The conclusions of this initial report have been challenged with alternative explanations possible for the initially promising results. Nonetheless, the potential for embryonic gene editing is recognized and led to the National Academy of Science February 2017 report supporting investigations restricted to serious genetic disease but not enhancements (height, intelligence, etc.). Subsequently, joint societal position statements emerged (Table 12.5).

12.1.2.3 Gene writing

As science moves from gene reading to gene editing, the potential for gene writing is being discussed. Gene writing refers to building a complete genome from scratch. More detailed than simply

Table 12.5 American Society of Human Genetics Position Statement on Human Germline Genome Editing.

1) inappropriate to perform germline gene editing that culminates in human pregnancy
   a. no reason to prohibit in vitro germline genome editing on human embryos and gametes
   b. appropriate oversight and consent from donors, to facilitate research on the possible future clinical applications of gene editing
2) no prohibition on making public funds available to support this research.
3) Future clinical application of human germline genome editing should not proceed unless, at a minimum, there is:
   a. compelling medical rationale
   b. evidence base that supports its clinical use
   c. ethical justification
   d. transparent public process to solicit and incorporate stakeholder input.

Summarized from Ormond, K. E., Mortlock D. P, et al. (2017). Human germline genome editing. American Journal of Human Genetics, 101 (2, 3)167–6 https://doi.org/10.1016/j.ajhg.2017.06.012

aligning the correct order of amino acid bases, the construct of an entire genome will require incorporation of integral wrapping of the DNA around histones, the control by the epigenome, and other forces driving gene function. With reference to the Human Genome Project—the first Human Genome Project Read (HG-R), and the second Human Genome Project- Write (HG-W) are being pursued (Boeke et al., 2016). Francis Collins described the HG-R as "the first glimpse of our own instruction book" and set the stage for the next phase of HG-writing. However, current abilities to write a genome are confined to short segments and smaller genomes from bacteria. Compilation of the 3 billion nucelic acid letters of the human genome will require considerable effort. The range of impact, however, would be tremendous and again revolutionize our understanding of basic biologic processes, treatment of disease, and the wider life sciences.

The launch of HG-write was targeted for 2016 to coincide with the twenty-fifth anniversary of HGP-read. An independent, nonprofit Center for Excellence for Engineering Biology serves as a coordinating center for this new initiative. A 10-year plan starts with genome writing on a smaller scale (bacteria, yeast), moving to human genomes at year 5 (2021) and completing with far-reaching expansions to medicine, the environment and population sustainability (Boeke et al., 2016).

12.1.3. Evolutionary impact

Natural selection involves no plan, no goal, and no direction—just genes increasing and decreasing in frequency depending on whether individuals with those genes have, relative to other individuals, greater or lesser reproductive success.

Randolph M. Nesse

NGS and gene editing indeed have value to the individual. With NGS, to discern if an individual is a carrier of a rare autosomal mutation, assess the genetic makeup of an early pregnancy for a deleterious condition, or target specific therapies to their cancer, all hold tremendous "personalized" value. The prospect of identifying and then altering the gene for sickle cell anemia such that the individual no longer faces a lifetime of painful sickle crises is tempting on a personal basis. Likewise, individually targeted cancer treatments are revolutionizing survival rates. And much of healthcare is driving in this direction to individualized treatment plans with initiatives such as President Obama's Precision Medicine.

Despite the strides these technologies afford on an individual level, the overall population impact of these rapidly advancing genomic techniques has received less attention. The new technologies are establishing further that everyone possesses a unique DNA profile with deleterious, risk-altering genes as well as normal variation. How gene frequencies will change when natural evolution drivers are replaced, at least in part, by decisions made on an individual basis, remains unknown. Or agricultural gene drives are initiated for intended good but have unintended, unpredictable effects on plant, insect, and small mammalian evolution. Certainly, history would suggest that maintaining high levels of variation within a population provides the best position for the ability to adapt to a changing environment. A further discussion of genetic variation and the major contributors to genetic variation, genetic selection and genetic drift may provide some answers.

12.1.3.1 Genetic variation and advances in molecular testing

NGS and gene editing place gene variation at risk. As noted previously, NGS provides less costly and faster access to the human genome and in particular the protein coding components in the exomes. Importantly, interpretation of the pathogenicity of previously unrecognized variation within gene sequences is currently elementary despite extensive efforts. This variation, while disconcerting when attempting to predict an expected function of the specific gene sequence, serves as the basis for evolution. Heterozygote advantage in today's environment has received scant investigation. Universal genetic carrier screening made possible by NGS increasingly identifies heterozygote carriers for over a hundred disorders. While statistically, removal of the homozygotes for these rare disorders will alter the gene frequency little, the possible longer-term impact on heterozygote frequency and any potential survival advantage they possess has received only preliminary discussion. Lastly, the potential removal from the population of genes with anticipated but unproven concerns for pathogenic effects may have unintended consequences on gene frequency and ultimately the heterozygotes themselves.

12.2 Evolutionary medicine

An understanding of evolutionary processes provides an essential perspective on what is health, what is disease, and why certain symptoms appear **Peter Gluckman**

Darwinian Medicine, also known as Evolutionary Medicine, incorporates these theories of evolution into the context of medicine and the traits which result in disease. This approach to medicine focuses on questions such as why natural selection has left us vulnerable to disease (Nesse et al., 2010). Evolutionary medicine shifts an understanding of disease processes from the classic "What has made this individual ill?" To: "How is this disease or condition a component of population health?" As all major species have now been sequenced, evolutionary medicine also now permits the comparison of a human gene sequence to that of other species. This has led to the understanding that genes which change little in a species over time (conserved) are more likely to be important to reproduction and survival.

From an evolutionary medicine viewpoint, the examination of man's traits including disease states should involve both proximate mechanistic questions as well as an evolutionary focus about origins and function. This has been referred to as examining traits for: "How they work" as well as: "How they got that way" (Nesse et al., 2010). An understanding that the body is not a machine to be repaired but evolutionarily derived with capacity for variation and natural selection will inform future disease research. Encompassed within this evolutionary framework is a closer look at disease characteristics which are troublesome but have an evolutionarily derived purpose (i.e., cough to clear lung secretions) (Nesse, 2001).

Key to understanding an evolutionary approach includes an understanding of natural selection as an operative on reproductive fitness and not overall survival. While a tenet of public health has been improving a population's ability to survive, application of evolutionary theory to public health focuses on interventions which alter reproductive fitness, which is often different than improved survival (Wells et al., 2017). An example lies in an unexpected but evolutionarily driven side effect of a public health initiative to supply water to an Ethiopian village via an easily accessed tap rather than individually carried from a distant source. However, the intention of improving women's health and nutritional status was not met. Easier water access improved their overall health initially resulting in increased fertility, a larger number of children, and resultant poorer endpoint nutrition and decreased reproductive survival (Gibson & Mace, 2006).

Also key to understanding an evolutionary medicine approach is consideration of the time for subsequent generations to be produced when responding to new, rapidly introduced environmental variation (Nesse, 2001; Stern & Dawes-Hoang, 2010) Introduction of an altered environment will be assimilated differently driven by the varied speeds of reproduction. A current example is embodied in the rapidly escalating prevalence of obesity. Relatively rapid changes over a generation or two in diet composition are outpacing the evolutionary ability to invoke change through on inherited variation, gene selection, gene drive, natural selection, and reproductive fitness. Finally, as genetic technologies advance, the understanding that while a few genetic diseases may be the result of a single gene mutation, for most traits and conditions, multiple genes and genetic changes are involved. This concept reflects the effects of selection based on variability and response to the environment over many generations (Stern & Dawes-Hoang, 2010). The understanding that common conditions in humans such as diabetes and hypertension are not due to a single genetic change but often multiple genetic variations has arisen from the extensive genetic-wide association studies performed over the last decade.

Expansion of genomic technologies has also furthered the concept of "developmental plasticity"– the ability to anticipate and change to one's environment (Gluckman et al., 2009) (Figure 12.8). Such plasticity to respond to one's environment can

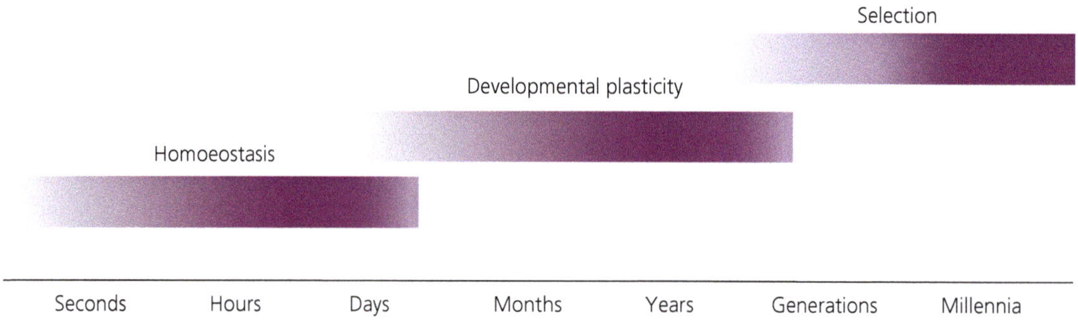

Figure 12.8 Modes of human adaptation with respect to time
Gluckman, P. D., M. A. Hanson, et al. (2009). Towards a new developmental synthesis: adaptive developmental plasticity and human disease. *Lancet* 373(9675): 1654–7. doi: 10.1016/S0140-6736(09)60234-8.

occur either at the time of environmental change or in an anticipatory sense as predictive developmental plasticity, more recently, refined as predictive anticipatory response (PAR) (Bateson, Gluckman, & Hanson, 2014). Epidemiologic support of predictive plasticity is reflected in Barker's original premise of Development Origins of Health and Disease (DOHD). Children born during times of food shortages, despite subsequent adequate nutrition, entered adult years with higher rates of obesity and metabolic syndrome associated conditions (cardiovascular disease, diabetes, hypertension) (Barker & Osmond, 1988; Barker et al., 1989). Fetal exposures to dietary restriction as well as excess are now considered responsible for later-life conditions and most remarkably, are considered heritable between generations (Gluckman & Hanson, 2004).

Such immediate response to the environment and its intergenerational inheritance relates to a unique aspect of genomic medicine, epigenetics. Epigenetics encompasses changes to DNA transcription and resultant protein production which occur without underlying sequence alterations. Epigenetic "marks" occur throughout the genome including methylation leading to silencing of gene transcription, or modifications to histones (Figure 12.9). Epigenetics provides an individual with anticipatory developmental plasticity to respond to their environment but also set the template for future generations to respond in a similar fashion. During fetal life, maternal exposures as well as nutritional excess or deficiencies drive fetal anticipatory developmental plasticity through epigenetic change. For example, maternity programs of supplemental nutrition to improve birthweight can lead to accelerated childhood adiposity shown to be associated with higher rates of diabetes in later life (Uauy & Kain, 2002). Establishment of epigenetic marks as the key functional elements of anticipatory development plasticity has emerged in numerous species including humans. Additionally, documentation of transgenerational passage of these epigenetic marks also exists (Duncan, Gluckman, & Dearden, 2014). Interim analysis of the Gusto study (Growing up in Singapore Towards Healthy Outcomes) provides support for alterations in specific sites of methylation, maternal contributors, and newborn through age four years parameters (Lin et al., 2017).

Acquiring an understanding of evolutionary medicine throughout education and especially in medical school would enable the physician to view a disease process as shaped by evolution rather than the consequence of biologic malfunction. The new genomic technologies will serve important roles in deciphering this information as well as hold the potential for the introduction of alterations, potentially without a full understanding of the evolutionary impact. As importantly, outside of medicine, individuals educated with an evolutionary medicine framework will be better equipped to understand the far reaching possibilities of the new and progressing genomic technologies. As these technologies advance, as this chapter has reviewed, the potential exists for altering any one of the basics tenets of evolution—genetic variation, natural

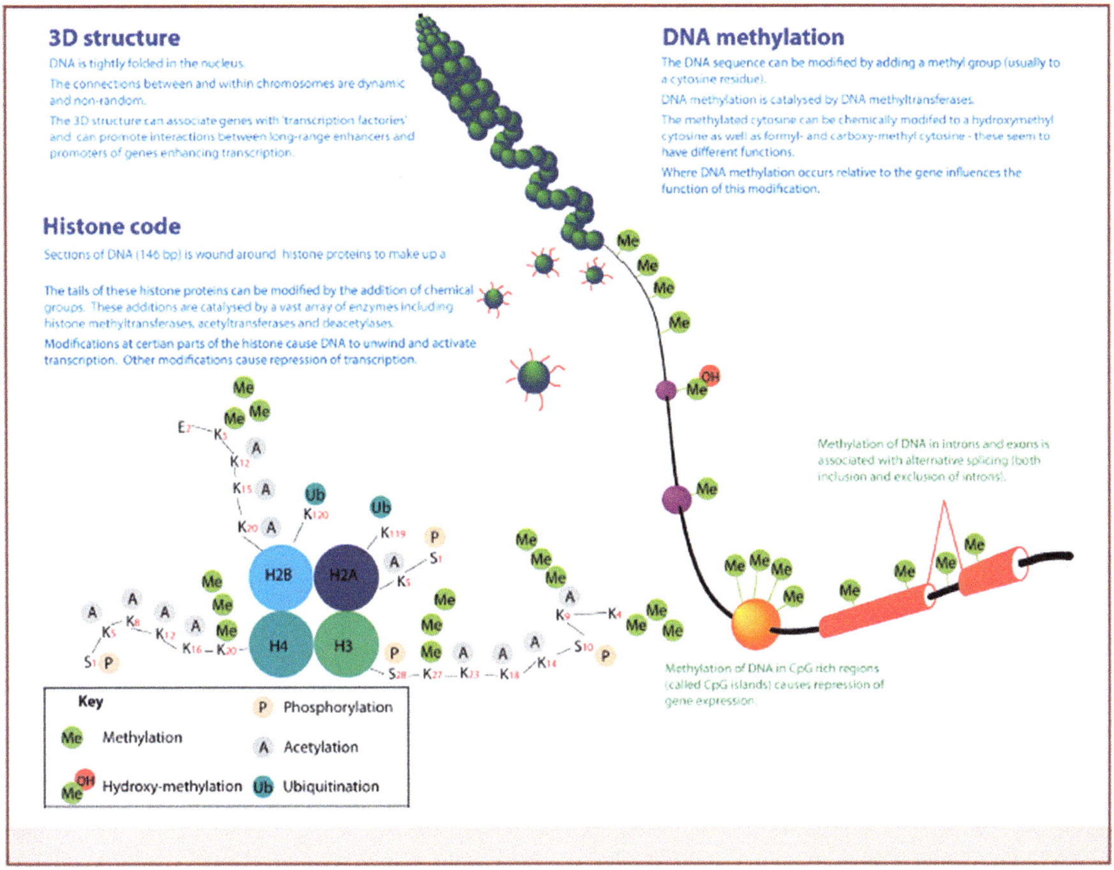

Figure 12.9 Types of epigenetic modifications and their potential effects on gene expression
Duncan, E. J., P. D. Gluckman, & P. K. Dearden (2014). Epigenetics, plasticity, and evolution: How do we link epigenetic change to phenotype? *J Exp Zool B Mol Dev Evol*, 322(4): 208–20. doi: 10.1002/jez.b.22571.

selection, or reproductive fitness. Understanding the slower pace of multigenerational evolutionary change through gene alteration as compared to the often single generation epigenetic responses driving developmental plasticity has far reaching implications for agriculture, medicine, zoonotic interactions between humans and animals, animal husbandry, and much more.

12.2.1 Educational approaches

Education is not the learning of facts, but the training of minds to think. **Albert Einstein**

Incorporation of evolutionary theory into the framework of rapidly advancing genomic testing will require substantial education for both the general population as well as physicians. Currently, in both groups, gaps exist in genetic literacy, baseline knowledge, and the ability to interpret and integrate genetic information to reach a decision. Furthermore, an understanding of genetic variation as a driver of evolution, and the benefits and potential harms of increasingly precise genetic technologies remain largely unexplored. Identifying and addressing these gaps will influence how personalized genetics is integrated into an individual patient's health care, the possible refinements of treatments and, as importantly, the understanding of the effects of genomic theories and evolution on the health of a population.

Among the general population, genetic literacy is challenged by limited education opportunities/resources as well as misperceptions. The difference between genes and chromosomes is not understood by half of individuals, and 75% believe that single genes determine behaviors (Molster et al., 2009; Christensen et al., 2010). Furthermore, genetic literacy is closely tied to numeracy, that ability to conceptualize quantitative alternatives to a decision. In adults with diabetes, 38% had inadequate genetic literacy to understand genetic testing based on familiarity scores testing (Vassy et al., 2012). Increasingly researchers are recognizing that for two out three people, lack of literacy and numeracy is such that even with genetics familiarity, incorporation of that knowledge into genetic decision-making is hindered. To address this need, contextualizing genetic literacy, and providing tools for research and enhancing genetic education has been proposed (Lea et al., 2011).

Concerted efforts for public education about genomics are emerging at a national level. Most notably, the National Institutes of Health, and specifically the National Human Genome Research Institute assembled a broad panel of stakeholders ranging from physicians, to geneticists, to commercial entities to understand and craft avenues for increasing the general population's genomics knowledge in 2017. Known as the Genetic Literacy, Education, and Empowerment (GLEE), this group is tasked with a national campaign to improve genetic literacy in three areas: K-16 students, the general public, and healthcare providers. While still in its initial stages, there is an ambitious rollout of 5 years to 1) provide information, tools, and resources, 2) address genetic literacy disparities for underserved populations, and 3) measure health benefits in order to minimize problems with genomics and family health history information (genome.gov/27568594/genomic-literacy-education-and-engagement-glee-initiative/). GLEE will work closely with the NIH supported precision medicine cohort (AllofUs) also to be implemented in the next 1–5 years. Crucial to the success of GLEE will be not only increasing genomic knowledge and access but also integrating the concept of genetic variation as an evolutionary force that can shape a population and ultimately a species.

For physicians, perhaps even more concerning gaps in genetic literacy exist (Metcalfe et al., 2002; Baars et al., 2005; Houwink et al., 2011). Recent work suggests that as genetic testing has evolved to genomic medicine, even less is understood by physicians (Ginsburg, 2008; Avard & Knoppers, 2009). In the area of pharmacogenetics, only a minority (29%) of physicians had received specific education relating to this topic, with only 10% feeling adequately informed for patient counseling (Stanek et al., 2012). Similar findings exist in the area of direct consumer genetic testing. Not unexpectedly, physicians also had less training and were less comfortable with the more recent advances in genomics such as genetic wide association studies (GWAS) (Selkirk et al., 2013). This same questionnaire study of a diverse range of physicians, including both primary and specialists, found higher levels of comfort with genetic testing and genomics among those further from their residency. This may, however, not reflect their actual knowledge of genetics (Wilkins-Haug, Erickson, et al., 2000; Wilkins-Haug, Hill, et al., 2000). Counter-intuitively, greater genetic knowledge may fuel greater discomfort, such as among more recently trained physicians who remain skeptical. Lack of time and knowledge were noted as significant barriers to incorporation of genomic medicine by the same group of physicians. (Selkirk et al., 2013).

How best to educate and continue the learning of genomics for physicians is unclear. Over half of current MDs (60%) indicate that their continuing medical education (CME) availabilities are inadequate to address these knowledge gaps. Greater use of digital support has been suggested incorporating ongoing email updates to MDs, or access to an institutionally supported education portal (Selkirk et al., 2013). Electronic medical record prompts and best practice links are also being considered (Guttmacher, Porteous, & McInerney, 2007). Recent studies suggest that innovative approaches such as partnering with a commercial company for genomic based care learning may be helpful (Ormond et al., 2011). Currently, however, the primary focus of these interventions is raising basic genetic knowledge, incorporating new testing paradigms for genetic technologies, and patient education/consent for a specific gene test. While these approaches may well increase current genetic test understanding, the ability to sustain ongoing assimilation of genomics over time is best served by broader exposure to medical evolutionary science as a companion to genetic principles.

12.2.1.2 Genetics education and evolutionary medicine

For current and future medical school students, introduction of genetic variation and evolutionary theory into the curriculum of genomics will need support. Increasing their knowledge not only of basic genetic concepts but also the role of genetic variation and its relationship to evolution should be a key goal. These future physicians will also be challenged by the increasing amounts of genomic information they will be asked to assimilate into their clinical practice (Nelson & McGuire, 2010). Medical students and their educators should work to develop an infrastructure upon which to integrate genetic technologies. This will require medical educators to reach beyond explanation of the basics of genetics to how this information will be used in virtually every medical specialty (Korf 2011; Korf, Irons, & Watson, 2011). Similarly, an appreciation for genomics should be incorporated not only into discussion of disease but as a crucial component of health. In all areas of health and disease, physicians play a critical role in educating their patients.

The Sackler Colloquium on Evolution in Health and Medicine convened almost 10 years ago (www.nasonline.org/Sackler_Evolution_Health_Medicine), and highlighted medical education as one of its goals. Bringing together multidisciplinary specialists across evolutionary theory, medical education, and genetics, the current gaps in medical education were examined and proposals put forward (Nesse et al., 2010). Similar to the lack of education around basic genetic concepts as described previously, lack of instruction on medical evolution as a framework for genetics was notably absent from most medical schools. An understanding of the core concepts of medical evolution including natural selection, mutation, genetic drift, and bottlenecks as drivers of genetic change were proposed as a framework upon which new genetic information could continually be incorporated (Nesse et al., 2010). These concepts were integrated into the newly developed 2009 American Association of Medical College and the Howard Hughes Medical Institute recommendations for competencies of individuals entering medical school. Recommendation E8, supported incorporation of medical evolution as a basic science, similar to chemistry, with knowledge needed before and during medical education.

The overall goals of this colloquium can be summarized as:

1. Education about evolution and its applications in medicine which benefit all levels of medical care providers.
2. Evolution as a basic science to be taught in medical school.
3. Current courses have insufficient evolutionary materials and approaches to increase are varied.
4. Evolution educators will need to recognize the already over-crowded medical school curriculum and consider innovative approaches to inclusion.
5. Evolutionary biology is a unifying framework for other medical knowledge.

Contemporaneously with the Sackler symposium, genetics in medical schools followed the recommendations of the Association of Professor of Human and Medical Genetics Curriculum (Plunkett-Rondeau, Hyland, & Dasgupta, 2015). While new technologies were being added (personalized medicine and directions to consumer testing), among 15% of programs on evolutionary genetics were being eliminated. Similarly, at the third and fourth year North American medical student level, only about a third (36%) felt they received genetics as part of their training and half felt they lacked basic competencies for applying genetic tools (Wolyniak, Bemis, & Prunuske, 2015).

Following from the Sackler Report, and of significance to the ongoing incorporation of genetics into medical education, was a 2105 survey of medical school deans. The range at which medical school deans could assimilate evolution into the curriculum was a broad 4–74% of schools, with limited resources and lack of faculty expertise major barriers. Concerns regarding controversy surrounding teaching evolutionary theory were anticipated by

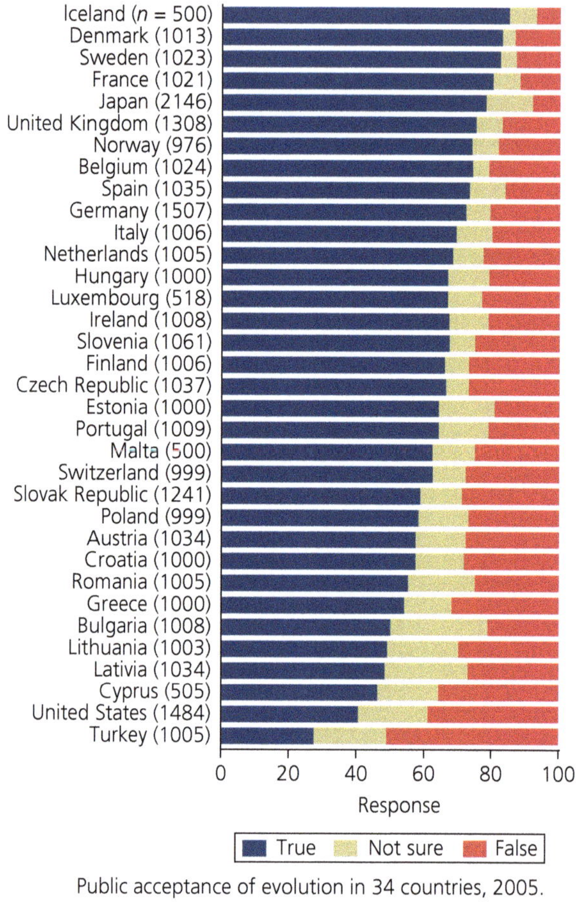

Figure 12.10 Public acceptance of evolution by the public in 29 countries in 2005
Miller, J. D., E. C. Scott, & S. Okamoto (2006). Science communication. Public acceptance of evolution. *Science* 313(5788): 765–6. doi: 10.1126/science.1126746.

48% of medical schools and in 10%, such controversy inhibited integration of evolution with its subcomponent of genetics (Hidaka et al., 2015). With a 40% response rate to the questionnaire by medical deans, these numbers may be biased in either direction. They do, however, point to areas of continued struggle with evolution and genetics—motivated and competent instructors, and acceptance of evolutionary theory and, by default, genetic control of human outcomes (Miller, Scott, & Okamoto, 2006). Equally, a challenge is the acceptance of evolution by only 40% of the general United States population. An equal number (39%) reject evolution with the remainder undecided (Figure 12.10). This remains a challenge which physicians must consider as they develop ways to implement genomic technologies and provide patients' education on all aspects of medical care, including approaches to sustain health, precision genetic diagnoses, and personalized treatments.

12.3 Conclusions

An ongoing focus of genomic education at all levels is understanding the concept of genetic variation. As shown previously, with advances in molecular techniques, progressively higher resolution of the human genome has revealed new information about genetic variation. On one hand, there is remarkable similarity between different species but yet striking variation between individuals. Genomic variation so far is particularly prevalent within the exome which is responsible for the majority of protein regulation, genetic health, and inherited disease. Variation itself is not necessarily disease producing but represents the ongoing process of opportunities for change. Evolutionary theories and Darwinian medicine would predict this high level of variation in the human genome. Classic genetic sequence variation is promulgated through increased reproductive fitness which requires numerous generations and often specific drivers (such as bottle neck), to be sustained. Alternatively, more immediate genomic change in response to the environment involves the epigenome. These heritable alterations in control of gene transcription and not base sequence changes provide immediate, first response to a changing environment. Also a source of variation between individuals, these heritable epigenomic alterations can alter phenotypes in as few as two generations.

Currently, the new genomic technologies, their applications, and the interpretation of the results are challenged by the high degrees of variation at the human genomic level. To address these concerns, human gene sequence data is being collected on two fronts—from normal, healthy populations, and from those with specific health concerns such as genetic disease, cancer, responses to pharmacologic agents, and conditions of adult onset involving almost every body system. Merging this information with the genetic variations of the healthy population is in its early stages although rapidly accelerating. A key to the future utilization of the tremendous amount of DNA sequence data available remains the ability to discern which changes represent benign variants and which will disadvantage the individual's reproductive fitness in their current environment. Bringing the concept of genetic variation as an expected, essential component of Darwinian medicine is needed at all levels.

Consideration of individual versus population benefits from DNA sequencing is an area of genomic education relevant not only to medicine but also to genomic based public health policy. Advanced genetic sequencing technology is spurring the quest for altering the genetic code of many species, including humans. In one form, the gene drive, alteration of the genomic content of an entire species with undesirable traits is possible (such as eradication of species of mosquitos carrying human infectious disease). And in humans, artificial selection for perceived unfavorable traits is rapidly growing. With preimplantation genetic testing, deleterious genes theoretically can be removed and corrected with CRISRP-cas9. The individual benefits to these interventions are currently and will likely remain the focus of this advancing work. From an overall educational perspective, however, and especially in the context of the impact of medical evolution, the potential population and species impact of these interventions should be considered.

As presented in this chapter, the rapid pace of gene discovery, initiated with human genome mapping over 15 years ago, represents an educational challenge across all people—the general population, patients, and health care providers. While increased genetics and technology education in primary and secondary schools will benefit future generations, the current population needs access to increased genetic literacy. This could be provided through public health policy efforts, community education programs, and their physicians. Genetic fundamentals and literacy are being addressed through large-scale, nationwide programs aimed across the population. However, challenges remain with a scarcity of informed educators and a population in which a belief in evolution is not held by the majority. A goal of developing an infrastructure based upon evolutionary medicine upon which to incorporate future genomic information deserves the same level of attention now being afforded to increasing baseline knowledge of basic genetic principles.

References

Abdallah, N. A., C. S. Prakash, & A. G. McHughen (2015). Genome editing for crop improvement: Challenges and opportunities. *GM Crops Food* 6(4): 183–205. doi: 10.1080/21645698.2015.1129937.

Abzhanov, A., M. Protas, B. R. Grant, P. R. Grant, & C. J. Tabin (2004). Bmp4 and morphological variation of beaks in Darwin's finches. *Science*, 305(5689): 1462–5. doi: 10.1126/science.1098095.

Amendola, L. M., G. P. Jarvik, et al. (2016). Performance of ACMG-AMP variant-interpretation guidelines among nine laboratories in the clinical sequencing exploratory research consortium. *Am J Hum Genet*, 99(1):247. doi: 10.1016/j.ajhg.2016.06.001.

Avard, D. & B. M. Knoppers. (2009). Genomic medicine: considerations for health professionals and the public. *Genome Med* 1(2): 25. doi: 10.1186/gm25.

Baars, M. J., A. J. Scherpbier, L. W. Schuwirth, L. Henneman, F. A. Beemer, J. M. Cobben, R. C. Hennekam, M. M. Verweij, M. C. Cornel, & L. P. Ten Kate (2005). Deficient knowledge of genetics relevant for daily practice among medical students nearing graduation. *Genet Med*, 7(5): 295–301. doi: 10.109701.GIM.0000162877.87333.9A.

Barker, D. J. & C. Osmond (1988). Low birth weight and hypertension. *BMJ*, 297(6641): 134–5.

Barker, D. J., C. Osmond, J. Golding, D. Kuh, & M. E. Wadsworth (1989). Growth in utero, blood pressure in childhood and adult life, and mortality from cardiovascular disease. *BMJ*, 298(6673): 564–7.

Bateson, P., P. Gluckman, & M. Hanson (2014). The biology of developmental plasticity and the Predictive Adaptive Response hypothesis. *J Physiol*, 592(11): 2357–68. doi: 10.1113/jphysiol.2014.271460.

Boeke, J. D., G. Church, et al. (2016). GENOME ENGINEERING. The Genome Project–Write. *Science*, 353(6295): 126–7. doi: 10.1126/science.aaf6850.

Charlesworth, B. & D. Charlesworth (2018). Neutral variation in the context of selection. *Mol Biol Evol*, 35(6): 1359–61. doi: 10.1093/molbev/msy062.

Christensen, K. D., T. E. Jayaratne, J. S. Roberts, S. L. Kardia, & E. M. Petty (2010). Understandings of basic genetics in the United States: Results from a national survey of black and white men and women. *Public Health Genomics*, 13(7–8): 467–76. doi: 10.1159/000293287.

Cong, L., F. A. Ran, D. Cox, S. Lin, R. Barretto, N. Habib, P. D. Hsu, X. Wu, W. Jiang, L. A. Marraffini, & F. Zhang (2013). Multiplex genome engineering using CRISPR/Cas systems. *Science*, 339(6121): 819–23. doi: 10.1126/science.1231143.

Duncan, E. J., P. D. Gluckman, & P. K. Dearden (2014). Epigenetics, plasticity, and evolution: How do we link epigenetic change to phenotype? *J Exp Zool B Mol Dev Evol*, 322(4): 208–20. doi: 10.1002/jez.b.22571.

Gibson, M. A. & R. Mace (2006). An energy–saving development initiative increases birth rate and childhood malnutrition in rural Ethiopia. *PLoS Med*, 3(4): e87. doi: 10.1371/journal.pmed.0030087.

Ginsburg, G. S. (2008). Genomic Medicine: "Grand challenges" in the translation of genomics to human health. *Eur J Hum Genet*, 16(8): 873–4. doi: 10.1038/ejhg.2008.115.

Gluckman, P. D. & C. T. Bergstrom (2011). Evolutionary biology within medicine: A perspective of growing value. *BMJ*, 343: 343. d7671. doi: 10.1136/bmj.d7671.

Gluckman, P. D. & M. A. Hanson (2004). The developmental origins of the metabolic syndrome. *Trends Endocrinol Metab*, 15(4): 183–7. doi: 10.1016/j.tem.2004.03.002.

Gluckman, P. D., M. A. Hanson, et al. (2009). Towards a new developmental synthesis: adaptive developmental plasticity and human disease. *Lancet*, 373(9675): 1654–7. doi: 10.1016/S0140-6736(09)60234-8.

Guttmacher, A. E., M. E. Porteous, & J. D. McInerney (2007). Educating health-care professionals about genetics and genomics. *Nat Rev Genet*, 8(2): 151–7. doi: 10.1038/nrg2007.

Heather, J. M. & B. Chain (2016). The sequence of sequencers: The history of sequencing DNA. *Genomics* 107 (1):1–8. doi: 10.1016/j.ygeno.2015.11.003.

Hidaka, B. H., A. Asghar, C. A. Aktipis, R. M. Nesse, T. M. Wolpaw, N. K. Skursky, K. J. Bennett, M. W. Beyrouty, & M. D. Schwartz (2015). The status of evolutionary medicine education in North American medical schools. *BMC Med Educ*, 15: 38–47. doi: 10.1186/s12909-015-0322-5.

Hoelzel, A. R., J. Halley, S. J. O'Brien, C. Campagna, T. Arnbom, B., Le Boeuf, K. Ralls, & G. A. Dover (1993). Elephant Seal genetic variation and the use of simulation models to investigate historical population bottlenecks. *J Hered*, 84(6): 443–9.

Houwink, E. J., S. J. van Luijk, L. Henneman, C. van der Vleuten, G. Jan Dinant, & M. C. Cornel (2011). Genetic educational needs and the role of genetics in primary care: A focus group study with multiple perspectives. *BMC Fam Pract*, 12:5. doi: 10.1186/1471-2296-12-5.

Ingram, V. M. (2004). Sickle–cell anemia hemoglobin: The molecular biology of the first "molecular disease"—the crucial importance of serendipity. *Genetics* 167(1): 1–7.

Korf, B. R. 2011. Genetics and genomics education: The next generation. *Genet Med*, 13(3): 201–2. doi: 10.1097/GIM.0b013e31820986cd.

Korf, B. R., M. Irons, & M. S. Watson (2011). Competencies for the physician medical geneticist in the 21st century. *Genet Med*, 13(11) :911–12. doi: 10.1097/GIM.0b013e31823331f9.

Kronauer, D. J. & J. D. Levine (2017). The ultimate and proximate underpinnings of social behavior. *J Exp Biol*, 220(Pt 1): 4–5. doi: 10.1242/jeb.152785.

Lea, D. H., K. A. Kaphingst, D. Bowen, I. Lipkus, & D. W. Hadley (2011). Communicating genetic and genomic information: Health literacy and numeracy considerations. *Public Health Genomics*, 14(4–5): 279–89. doi: 10.1159/000294191.

Lin, X., I. Y. Lim, Y. Wu, A. L. Teh, L. Chen, I. M. Aris, S. E. Soh, M. T. Tint, J. L. MacIsaac, A. M. Morin, F. Yap, K. H. Tan, S. M. Saw, M. S. Kobor, M. J. Meaney, K. M. Godfrey, Y. S. Chong, J. D. Holbrook, Y. S. Lee, P. D. Gluckman, N. Karnani, & Gusto study group (2017). Developmental pathways to adiposity begin before birth and are influenced by genotype, prenatal environment and epigenome. *BMC Med*, 15(1): 50. doi: 10.1186/s12916-017-0800-1.

Ma, H., N. Marti–Gutierrez, S. W. et al (2017). Correction of a pathogenic gene mutation in human embryos. *Nature*, 548(7668): 413–19. doi: 10.1038/nature23305.

Metcalfe, S., R. Hurworth, J. Newstead, & R. Robins (2002). Needs assessment study of genetics education for general practitioners in Australia. *Genet Med*, 4(2): 71–7. doi: 10.109700125817-200,203,000-00004.

Miller, J. D., E. C. Scott, & S. Okamoto (2006). Science communication. Public acceptance of evolution. *Science*, 313(5788): 765–6. doi: 10.1126/science.1126746.

Molster, C., T. Charles, A. Samanek, & P. O'Leary (2009). Australian study on public knowledge of human genetics and health. *Public Health Genomics*, 12(2): 84–91. doi: 10.1159/000164684.

Nelson, E. A. & A. L. McGuire (2010). The need for medical education reform: Genomics and the changing nature of health information. *Genome Med*, 2(3): 18. doi: 10.1186/gm139.

Nesse, R. M (2001). On the difficulty of defining disease: A Darwinian perspective. *Med Health Care Philos*, 4(1): 37–46.

Nesse, R. M., C. T. Bergstrom, P. T. Ellison, J. S. Flier, P. Gluckman, D. R. Govindaraju, D. Niethammer, G. S. Omenn, R. L. Perlman, M. D. Schwartz, M. G. Thomas, S. C. Stearns, & D. Valle (2010). Evolution in health and medicine Sackler colloquium: Making evolutionary biology a basic science for medicine. *Proc Natl Acad Sci USA*, 107 Suppl 1: 1800–7. doi: 10.1073/pnas.0906224106.

O'Donald, P. (1967). The evolution of selective advantage in a deleterious mutation. *Genetics*, 56(3): 399–404.

Ormond, K. E., L. Hudgins, J. M. Ladd, D. M. Magnus, H. T. Greely, & M. K. Cho (2011). Medical and graduate students' attitudes toward personal genomics. *Genet Med* 13(5): 400–8. doi: 10.1097/GIM.0b013e31820562f6.

Paabo, S. (2015). The diverse origins of the human gene pool. *Nat Rev Genet*, 16(6): 313–14. doi: 10.1038/nrg3954.

Plunkett-Rondeau, J., K. Hyland, & S. Dasgupta (2015). Training future physicians in the era of genomic medicine: Trends in undergraduate medical genetics education. *Genet Med*, 17(11): 927–34. doi: 10.1038/gim.2014.208.

Richards, S., N. Aziz, S. Bale, D. Bick, S. Das, J. Gastier-Foster, W. W. Grody, M. Hegde, E. Lyon, E. Spector, K. Voelkerding, H. L. Rehm, & Acmg Laboratory Quality Assurance Committee (2015). Standards and guidelines for the interpretation of sequence variants: A joint consensus recommendation of the American College of Medical Genetics and Genomics and the Association for Molecular Pathology. *Genet Med*, 17(5): 405–24. doi: 10.1038/gim.2015.30.

Selkirk, C. G., S. M. Weissman, A. Anderson, & P. J. Hulick (2013). Physicians' preparedness for integration of genomic and pharmacogenetic testing into practice within a major healthcare system. *Genet Test Mol Biomarkers*, 17(3): 219–25. doi: 10.1089/gtmb.2012.0165.

Slatkin, M (2004). A population-genetic test of founder effects and implications for Ashkenazi Jewish diseases. *Am J Hum Genet*, 75(2): 282–93. doi: 10.1086/423146.

Stanek, E. J., C. L. Sanders, K. A. Taber, M. Khalid, A. Patel, R. R. Verbrugge, B. C. Agatep, R. E. Aubert, R. S. Epstein, & F. W. Frueh (2012). Adoption of pharmacogenomic testing by US physicians: Results of a nationwide survey. *Clin Pharmacol Ther*, 91(3): 450–8. doi: 10.1038/clpt.2011.306.

Stern, D. L., & R. E. Dawes-Hoang (2010). Michael Akam and the rise of evolutionary developmental biology. *Int J Dev Biol*, 54(4): 561–5. doi: 10.1387/ijdb.092908ds.

Tang, L., Y. Zeng, H. Du, M. Gong, J. Peng, B. Zhang, M. Lei, F. Zhao, W. Wang, X. Li, & J. Liu (2017). CRISPR/Cas9-mediated gene editing in human zygotes using Cas9 protein. *Mol Genet Genomics* 292(3): 525–33. doi: 10.1007/s00438-017-1299-z.

Tsui, L. C. & R. Dorfman. (2013). The cystic fibrosis gene: A molecular genetic perspective. *Cold Spring Harb Perspect Med*, 3(2): a009472. doi: 10.1101/cshperspect.a009472.

Uauy, R., & J. Kain (2002). The epidemiological transition: Need to incorporate obesity prevention into nutrition programmes. *Public Health Nutr*, 5(1A): 223–9. doi: 10.1079/PHN2001297.

Vassy, J. L., K. E. O'Brien, J. L. Waxler, E. R. Park, L. M. Delahanty, J. C. Florez, J. B. Meigs, & R. W. Grant (2012). Impact of literacy and numeracy on motivation for behavior change after diabetes genetic risk testing. *Med Decis Making*, 32(4): 606–15. doi: 10.1177/0272989X11431608.

Wells, J. C. K., R. M. Nesse, R. Sear, R. A. Johnstone, & S. C. Stearns (2017). Evolutionary public health: Introducing the concept. *Lancet*, 390(10093): 500–9. doi: 10.1016/S0140-6736(17)30572-X.

Wiedenheft, B., S. H. Sternberg, & J. A. Doudna (2012). RNA-guided genetic silencing systems in bacteria and archaea. *Nature*, 482(7385): 331–8. doi: 10.1038/nature10886.

Wilkins-Haug, L., K. Erickson, L. Hill, M. Power, G. B. Holzman, & J. Schulkin (2000). Obstetrician-gynecologists' opinions and attitudes on the role of genetics in women's health. *J Womens Health Gend Based Med*, 9(8): 873–9. doi: 10.1089/152460900750020900.

Wilkins-Haug, L., L. D. Hill, M. L. Power, G. B. Holzman, & J. Schulkin (2000). Gynecologists' training, knowledge, and experiences in genetics: a survey. *Obstet Gynecol*, 95(3): 421–4.

Williams, T. N., T. W. Mwangi, S. Wambua, N. D. Alexander, M. Kortok, R. W. Snow, & K. Marsh (2005). Sickle cell trait and the risk of Plasmodium falciparum malaria and other childhood diseases. *J Infect Dis* 192(1): 178–86. doi: 10.1086/430744.

Wolyniak, M. J., L. T. Bemis, & A. J. Prunuske (2015). Improving medical students' knowledge of genetic disease: A review of current and emerging pedagogical practices. *Adv Med Educ Pract*, 6: 597–607. doi: 10.2147/AMEP.S73644.

CHAPTER 13

Ecological Rationality and Evolutionary Medicine: A Bridge to Medical Education

Shabnam Mousavi and Jay Schulkin

While traditionally medicine has talked of health and disease, normal and abnormal as dichotomous categories, a more nuanced contextual consideration is needed. **Gluckman et al. (2009)**

13.1 Introduction

Medical education involves shaping and sharpening capabilities for clinical diagnosis and ensuing treatments. In the vocabulary of decision theory, diagnosis of a medical condition falls into the category of judgment, and prescription of a treatment constitutes a decision. For theorizing the processes of judgment and decision-making, social scientists have developed different notions of rationality. A notion of rationality represents a structure that can be used to explain, evaluate, and recommend a certain course of action. *Full rationality* builds such a structure based on the rules of logic and statistical inference (von Neumann & Morgenstern, 1944; Savage, 1954). *Bounded rationality*, on the other hand, focuses on the interaction between the human mind and the task environment (Simon, 1956, 1972; Selten, 1999; Gigerenzer & Selten, 2001). Full rationality establishes benchmarks that are logically perfect, independent of the context, whereas bounded rationality explains the processes that lead to good-enough solutions for a given situation. The study of functional matches between the solutions and the ecology in which they arise constitutes the study of the ecological form of rationality (Gigerenzer et al., 1999; Smith, 2008).

Darwinian-Spencerian "survival of fittest" (a term coined by Spencer, 1864) can be formulated in the full rationality framework as long as fitness is well specified. Fitness, however, is only meaningful in correspondence to the biological inclinations of the species, which take shape in response to their environment. In other words, fitness is inherently contextual, whereas a formulation based on full rationality requires a universal decontextualized conception. Maximizing such a fluid concept of fitness, albeit mathematically elegant, will demand case-by-case elaboration and extensive justification for meaningfully connecting it to real-world phenomena. Thus, deriving perfect solutions (and recommendations) from a natural selection-based analysis can generate a coherent theoretical view but in practice amounts to hardly more than a myth. Neither the general public nor physicians satisfy

Mousavi, S. and Schulkin, J., *Ecological Rationality and Evolutionary Medicine: A Bridge to Medical Education* In: *Integrating Evolutionary Biology into Medical Education—for maternal and child healthcare students, clinicians, and scientists*.
Edited by: Jay Schulkin and Michael L. Power, Oxford University Press (2020). © Oxford University Press.
DOI: 10.1093/oso/9780198814153.003.0014

the requirements of full rationality, such as collecting and evaluating all relevant information for arriving at, benefiting from, or using these theoretically generated perfect solutions.

The bounds on human cognition, time pressure, and the diminishing value of extra considerations after a good-enough solution is found characterize both public and medical decision making in the real world. In general, a bounded notion of rationality that evaluates solutions with respect to the ecology is both realistic and practical for making sense of the ways in which humans come up with solutions, e.g., how physicians diagnose and treat diseases. Moreover, an evolutionary perspective of humans' capabilities clarifies the origins of humans' inclinations and the conditions for functional matches of these capabilities to the environment. Together, ecological rationality and evolutionary perspective can demythologize the understanding, development, and teaching of good-enough decision-making processes.

In this chapter, we outline concepts of rationality in judgment and decision making that have practical relevance to clinical practice and medical education and elaborate on the compatibility of these concepts with the evolutionary approach to medicine. After a general overview of the ecological rationality research program, we zoom in on the design of two specific tools pertaining to medical education, namely fact boxes and fast-and-frugal trees. A fact box can be designed to facilitate the communication of information between physicians and patients, and a fast-and-frugal tree can elucidate the processes that underlie medical judgments.

We suggest that evolutionary medicine can employ a framework based on ecological rationality for developing effective education tools that are accessible to and can be well retained by physicians and their patients alike. When designed and used purposefully, these tools can enhance health decisions of both groups. By the same token, a mutually beneficial exchange of concepts and findings can be extended to evolutionary public health, especially in dealing with inevitable cases where the direction of fitness does not align with that of improving health or reproduction. To explain this last element, we provide a summary of a series of articles in *The Lancet* on evolutionary public health. The last section concludes.

13.2 Ecological rationality, the adaptive toolbox, and heuristics

In line with Darwinian natural selection as an adaptive process (Darwin, 1859), the ecological notion of rationality views the mind as an adaptive toolbox. The term heuristics, in this context, refers to adaptive decision tools inside this toolbox that benefit from evolutionary capacities. Careful examination of heuristics have produced a framework with functional models for capturing the ways in which humans collect and combine information to make judgments and decisions in changing environments under time constraints and uncertainty. In this framework, a heuristic is considered ecologically rational if it constitutes a functional match with the environment in which it is used and can therefore produce a good-enough solution. Notably, this very characterization is closely compatible with the evolutionary approach to medicine, which involves identification of mismatches between naturally evolved inclinations and the ecology in which they are activated. A mismatch in this case produces a clinical condition that requires medical treatment.

Investigation and exploration of ecological conditions for successful heuristic-based problem solving shaped the focus of research at the Center for Adaptive Behavior and Cognition of the Max Planck Institute for Human Development from its inception in 1997 until conclusion in 2017.[1] The study of ecological rationality comprises a research program with three strands: descriptive accounts of heuristics, the normative status of ecological rationality, and the design of frugal heuristic tools and of environments conducive to achieving certain purposes. The first strand, description of heuristics, involves exploration of the types of rules of thumb that people, whether novices or experts, use. The normative strand aims at developing benchmarks for good practice with respect to the match between environmental (ecological) conditions and the rules (heuristics) used. The last strand has direct connec-

[1] https://www.mpib-berlin.mpg.de/en/research/concluded-areas/adaptive-behavior-and-cognition

tions to education. In this section, we introduce this research program with respect to both general mechanisms of decision making and particular connections to medical decision making. Thereafter, the next section will focus on two decision tools that have been developed in the third strand of ecological rationality studies.

In response to ecological conditions, evolutionary capacities of humans have been shaped over millennia through Darwinian mechanisms of natural selection. The mind engages in different modes of information processes to shape a judgment, and in doing so uses its adaptive, evolutionary, and learned tools. Thus, the mind can be viewed as an adaptive toolbox filled with reasoning methods, heuristic tools, and evolutionary capacities. An accurate judgment of the nature of a disease coupled with the choice of a fitting treatment constitutes an ecologically rational decision process (Marewski & Gigerenzer, 2012). Irrationality in the ecological sense characterizes a functional mismatch between the problem at hand and the attempted solution, such as a treatment that would not alleviate the ailment. Another example is the host of mismatches that arise between an evolutionarily developed inclination that serves general fitness but works against purposes of individual health or longevity (Gluckman et al., 2005, 2009, 2010; Perlman, 2013). Many examples of conditions caused by mismatches, including acne, diaper rash, hemorrhoids, and lower back pain can be found in the introductory chapter of the current book.

The study of ecological rationality has particularly focused on decision processes that use heuristics as opposed to full deliberation and exhaustive information analysis. Heuristic decision-making has three consecutive components: a *search rule* for examining cues that provide relevant information about the problem at hand, e.g., consideration of present symptoms and health history of a patient, leading to a *stopping rule* for this search, and finally producing a solution, e.g., prescription of a treatment by utilizing a *decision rule* (Gigerenzer & Gaissmaier, 2011). In almost every field, professionals develop and frequently use heuristics, especially when decisions need to be made under time pressure (Gigerenzer, 2014). A prime example of situations where decisions need to be made under extreme time constraint with frugal use of information are those made in an emergency room. Consider a patient with acute chest pain who has just arrived in the emergency room. A decision must be made whether to direct her to the coronary care unit or to a regular nursing bed. As we shall see later in this chapter, a fast-and-frugal tree that sequentially examines limited information provides an effective protocol for making this call (Green & Mehr, 1997). The use of heuristics is not limited to professionals. Patients, too, use rules of thumb to make health decisions. Trusting one's physician is a popular heuristic for the choice between available treatments for a medical condition (Wegwarth & Gigerenzer, 2013; Mousavi & Gigerenzer, 2011). Such heuristics that are used by physicians and/or patients in decision making about health issues must be identified to develop effective policies, rules, and regulations for public health (Gigerenzer & Gray, 2011). Improving health decisions thus requires specification not only of desirable goals but also of practical benchmarks. Practical benchmarks are those that relate to the actual ways in which physicians and patients process information and make decisions, which often involves a variety of heuristic rules. This is a case in point for why the study of heuristics and their ecological rationality is conducive to medical training for clinical practice.

13.2.1 Shared concepts of ecological rationality and evolutionary medicine

Evolutionary medicine defines normality, abnormality, and disease not in absolute terms but in relation to the ecological context and the individual variation in phenotypes (Gluckman, Low, & Buklijas (2010); Hanson & Beedle, 2009). Similarly, a heuristic is not good or bad per se; its rationality or goodness is assessed with respect to its match with the circumstances (ecology) in which it has been (or will be) used. This retrospective assessment of rationality is only the starting step in the study of heuristics and investigation of their ecological rationality. The systematic study of when and why a heuristic is ecologically rational has produced a flourishing area for developing decision aids in many fields (Gigerenzer & Gaissmaier 2010: Mousavi, Gigerenzer, & Kheirandish, 2017).

Structural similarities between the principles of ecological rationality and evolutionary medicine, such as contextuality, allow for the former to provide a conceptual framework for the latter. Consider, for example, the ways in which evolutionary insights benefit clinical judgment. Gluckman et al., (2009) stress that evolution sides with fitness mechanisms above all others, and often at the expense of health or reproduction. They provide several examples of the ways in which an evolutionary perspective can positively revise medical judgment and choice of treatment:

> The symptoms associated with infection provide an illustrative example. Coughing, mucus secretion, and diarrhea may be seen as evolved mechanisms for expelling the infectious microbe, and while it seems counter-intuitive to leave these symptoms untreated, there is some suggestion that blocking these normal defenses may extend the illness duration. However, if the severity of the symptoms exceeds that which is adaptive, then medical intervention would become necessary. **(Ibid., p. 252)**

In other words, intervening medicinal treatment is an ecologically rational decision only if the body is considered unable to overcome the condition through natural mechanisms. As previously mentioned, there are three main lines of inquiry pursued in ecological rationality research. The first of these explores cognition and behavior in order to locate heuristic tools, and then constructs categorizations of these heuristics to produce a descriptive account of heuristic-based decision making. The following examples illustrate how an evolutionary perspective informs the study of ecological rationality of heuristics.

13.2.2 Heuristics that use evolved capacities

Heuristics are effective decision tools that successfully exploit learned or evolved capacities. These capacities encapsulate the essence of adaptive behaviors in interaction with the environment throughout the development of humankind and during the life cycle of every individual. Along the continuum of evolution, exposure to certain environmental situations triggers these capacities in a variety of combinations. A prime example that demonstrates exploitation of an evolved capacity is the gaze heuristic (see Hamlin, 2017 for an extensive review of the use of this heuristic across domains). The capacity of gazing at an object that is moving against a noisy background is an evolved capacity in human infants (and many animals). The brain has developed this ability for a prey to avoid or a predator to intercept another moving body. A machine with the same task, lacking evolved capacities but replete with calculative abilities, would opt to solve this interception problem by calculating the trajectory of the moving target and strives for accuracy by accounting for many external variables that can affect this path. In contrast, the brain is hardwired to only keep the angle of gaze constant for achieving interception, as illustrated in Figure 13.1. The angle of gaze is the angle between the horizon and the line that connects the observer's eye to the moving object. This simple configuration of the interception problem has been extremely consequential when its adoption by British fighter pilots turned the wheels in the Second World War against German bombers (Hamlin, 2017).

Another example of humans' extraordinary built-in abilities is language acquisition, a task so complex and yet so readily and regularly performed and perfected by human children. The process of language acquisition, which is an evolutionary capacity, involves learning from error at an efficiency level comparable to best-engineered feedback loops (Gigerenzer, 2005; Mousavi & Gigerenzer, 2011).

Figure 13.1 Baseball catcher uses gaze heuristic to catch a flyball

More importantly, this learning process demonstrates an ability to imitate with high levels of sophistication by both human children and adults. Imitation is a decision tool that can be understood and modeled as a (social) heuristic, with the three components corresponding to the general configuration in terms of three elements as discussed in the previous section and as spelled out next. Imitation involves the following processes: 1) Be cognizant of the practiced modes and methods in the environment, 2) Stop when a reliable source (an adult for language, the majority for social conduct, the successful agent for personal pursuit) is found, 3) Adopt the reliable source's method.

As a human behavior, language acquisition manifests itself in manifold ways and is rooted in several aspects of human existence. Besides imitation, there exists a vast social dimension of cooperation associated with the learning, use, and spread of language. For example, from the viewpoint of archeology, humans are distinguished from other primates through several properties, including advanced linguistic consciousness (ALC, Davies, 2016). Such distinctions constitute the founding concepts of the discipline of anthropology. In this regard, psychology would focus on a cognitive order associated with language acquisition as elaborated previously, whereas the discipline of linguistics would focus on the origins of meaning:

> In the search for what is distinctive in human language, and so in ALC, the theme of sequencing, as a "recursion," has been prominent. Recursive functions take their last output as their next input, generating what the nineteenth century linguist von Humboldt referred to as 'the infinite use of finite means.' **(Fitch 2010, p. 90)**

> ...[T]he practical and mimetic acquisition of language and its constant semantic re-honing seems to have a strong relation to locality....In genus Homo, mimetic language acquisition and its continuing heuristic application is a species-wide training in co-locating with others, through our particularity, and so, by implication, in developing consciousness of a shared world. **(Ibid, p. 96)**

Much can be said and has been researched in relation to language acquisition. This vast area deserves its own many articles beyond the scope of the current chapter. (Interested readers can search for instance "Humboldt + language"). Next, we return to the ecological nature of decision-making processes, which can be well understood with respect to an ecological notion of rationality.

13.2.3 Medical decision making is contextual

Recall that full rationality generates benchmarks for optimal judgment and decision-making based on logic and statistical inference that are independent of the context. Medical decision-making, in contrast, is heavily contextual. The second strand of ecological rationality research program, as mentioned before, strives to establish a normative status for an ecological notion of rationality as a primary benchmark for success at par with the existing statistics- and logic-based norms of judgment. A heuristic tool that matches the environmental structure is ecologically rational, but not unique. There could exist more than one tool that fits a given situation. This corresponds directly to the reality of medical judgment, where more than one treatment can usually be envisioned for a given diagnosed condition. A particular sign can be symptomatic of different health conditions. This multiplicity of possible mappings calls for an active exchange of information between physicians and patients (see e.g., Hertwig et al., 2011, Table 19.2).

In addition, patients benefit more from certain forms of representations of information (Rao & Kanter, 2010). Finally, the amount of imparted benefits has a positive correlation with patients' initial level of numeracy (Peters et al., 2007; Peters, 2012; Dallacker et al., 2016; Peters, 2017).

As we have seen several heuristic tools can provide good-enough solutions for a problem in a given situation. Conversely, one heuristic tool can be used in a variety of environments, and successfully so (see, for example, Mousavi & Gigerenzer, 2017). There is no unique correct answer to any given problem, rather what is meaningful is that one of two potential tools can provide a better match to a particular environment and will thus possess a higher level of ecological rationality. This scaled notion of rationality speaks directly to physicians, who in their daily practice recurrently face situations where multiple treatments come to mind for a given medical condition, and where each treatment can be considered preferable from various perspectives in terms of evidence or importance. For example, treatment A may appear to be superior age-wise but treatment B appears so gender-wise, and so on. Where to stop and commit to a treatment is thus a judgment call.

So far, we looked at the first and second strands of the study of ecological rationality—descriptive and normative studies—in relation to medical judgment and practice. The third line of work on the ecological rationality of heuristic-based decision making applies the findings from these two strands to enhance problem-solving ability by generating easy-to-use decision aids and designing environmental structures conducive to achieving certain ends. We elaborate on this in the next section.

13.3 Tools for effective communication of risk

The ecological notion of rationality is neither absolute nor moves across situations and contexts. In contrast to the full notion of rationality, it is always and strongly context-dependent. This reflects the profound fact that effective communication needs to take account of the context. Based on this view, numeracy enhancing and other decision aids for patients and doctors must provide information in formats that the human mind can particularly easily access (Wegwarth et al., 2012; Wegwarth & Gigerenzer, 2018). Developing these types of communication tools has been the focus of focus of the Harding Center for Risk Literacy,[2] founded in 2006 at the Max Planck Institute for Human Development, Berlin. In 2016, The Winton Center for Risk and Evidence Communication, a sister organization of the Harding Center, was founded at the University of Cambridge, England,[3] with a similar mission. In what follows, we introduce some tools for effective communication of risk developed or enhanced at and promoted by these centers.

Two main sets of tools come in either the form of fact boxes or fast-and-frugal decision trees. A fact box provides an accessible representation of information on available treatments, thereby empowering patients to make informed decisions, as well as enhancing the quality of communication between doctors (and other health providers, such as insurance companies), and patients. The transparency of information representation in fact boxes benefits from close attention to the ways in which our adaptive mind best absorbs information from the environment. As such, these tools are prime candidates for inclusion in an evolutionary medicine module in medical education that targets enhancement of clinical practice through highlighting adaptive abilities of humans. Next, we provide specifics about fact boxes and fast-and-frugal trees.

13.3.1 Fact boxes

Originally introduced by Eddy (1990) as a balance sheet of benefits and harms, a fact box summarizes relevant risks, pros and cons, and other information about a treatment or drug in an easily accessible way, as can be seen in Figure 13.2 for mammography. Here, after the title, the population is first specified and then the most important aspect of any treatment is addressed, namely whether mammography saves lives. The benefits and harms are compared for treatment versus no-treatment subsets of the population. All numerical information is in the following format: how many out of 1000. Such representation of occurrence of a certain outcome, known as a natural frequency format, is considered an easily accessible representation format. For example, the number of women over 50 years old who died from breast cancer and did not participate in screening is 5 out of 1000, as opposed to 4 out of 1000 with screening.

Table 13.1 provides an overview of methods for creating a fact box from reported facts. The first column presents the original format in which data have been reported in four categories: frequency data, relative risk or odds ratios, number needed to treat (NNT), and continuous outcomes. The second column outlines step-wise instructions for summarizing the data in a given category to fit a fact-box format. For instance, in the case of frequency data, 1. Use percentage or frequency formats, 2. Keep the denominator consistent, 3. Report whole numbers. For each step, examples and explanations are provided to clarify the meaning of instructions. The

[2] https://www.harding-center.mpg.de/en

[3] https://wintoncentre.maths.cam.ac.uk

Breast Cancer Early Detection
by mammography screening

HARDING CENTER FOR **RISK LITERACY**

Numbers for women aged 50 years or older* who either did or did not participate in mammography screenings for 10 years or more.

Mammography screening reduced the number of women who died from breast cancer. However, it had no effect on overall cancer deaths. Among all women taking part in screening, some were overdiagnosed with non-progressive cancer and unnecessarily treated.

Benefits	**1,000 women without screening**	**1,000 women with screening**
How many women died from breast cancer?	5	4
How many women died from all types of cancer?	21	21
Harms		
How many women experienced false alarms and received additional treatments or had biopsies?	-	about 100
How many women without non-progressive cancer had unnecessary partial or complete breast removal?	-	5

*Where no data for women above 50 years of age are available, numbers refer to women above 40 years of age.

Source: Gøtzsche & Jørgensen. Cochrane Database Syst Rev 2013(6):CD001877.

Last update: March 2014

Figure 13.2 Mammography explained in easy terms in a fact box

Table 13.1 Transforming risk outcomes into accessible formats.

Reporting Format in Original Article	How to Summarize Outcomes in a Fact Box	Example Presentation
Frequency data	Use percentage or frequency formats: for example, "x out of 100." Use smallest denominator (of base 10) to best represent the evidence while maintaining features of the data (e.g., risk reduction ratios in whole numbers). Keep denominator consistent (e.g., do not use "x in 100" for one outcome and "x out of 1,000" for another), or if multiple Fact Boxes are to be compared (e.g., different treatments for knee osteoarthritis), the denominator across Fact Boxes should also be consistent. Report whole numbers, round if necessary. If a frequency is smaller than 1 but not 0, consider reporting as "less than 1" or "<1" or increasing the denominator for all outcomes (e.g., from <1 in 100 to 7 in 1,000).	Figure 1: Mammography screening, all outcomes
Relative risks or odds ratios	If absolute numbers are also available, proceed as for frequency data with the following exceptions: a) *Relative risk*: If data are weighted according to sample size or study quality, calculate absolute numbers for control group and apply relative risk ratio to determine absolute number for intervention group.	Online Appendix Figure 2 calculation for: "How many patients had pain 4 to 7 days after diagnosis?"

(*Continued*)

Table 13.1 Continued

Reporting Format in Original Article	How to Summarize Outcomes in a Fact Box	Example Presentation
	b) *Odds ratio*: Calculate the risk difference by first converting the control group risk to odds. Multiply by control group odds by the odds ratio, and then convert back to risks. Add/subtract the difference from the baseline risk.	Treatment effect: Relative Risk ratio (RR) = 0.79
	If absolute numbers are not reported:	*Placebo risk*: 11 out of 100
	1. Attempt to obtain absolute numbers from authors. 2. Obtain base rate data for control group from another reputable source (e.g., representative baseline risk) and apply a) or b) calculation to the estimate as above. 3. Look for alternative sources that can be used to illustrate effects (e.g., select a good-quality randomized controlled trial from a meta-analysis and extract estimates from this study if magnitude of effect is in the same ballpark).	*Treatment risk*: 11 × RR (0.79) = 9 out of 100
Number needed to treat (NNT)	NNT is not as easily understood as absolute risks, although it is mathematically equivalent.	To date, NNT has not been used to communicate risks in Fact Boxes
	a) For lay audiences, use alternative formats:	
	If absolute numbers are available, proceed as for frequency data.	
	If absolute numbers are not available, proceed as for relative risks	
	b) For professional audiences, NNT may be appropriate for reporting some outcomes.	
Continuous outcomes	*Means or medians*	
	Report aggregated data and reference scale anchors. If numbers are weighted according to sample size or study quality, incorporate weights into a calculation of the difference using the control group, and apply difference to intervention group. Medians may be more appropriate for non-symmetric distributions.	Online Appendix Figure 3: "How did patients rate their knee pain?"
	Mean differences or change scores	For example, "On a scale assessing pain (0 = worse to 100 = better), how many points did patients improve?"
	Attempt to obtain mean scores to summarize as above. If these cannot be obtained, report the outcome as "points improvement" or similar in reference to the type of scaled used.	
	If rating scales are reported only with reference to verbal anchors rather than numbers, report the average scale anchor as the outcome.	Online Appendix Figure 3: "How did patients rate their knee function?"

(Source: McDowell et al. (2016), Table 1)

third and last column provides examples for each category.

A wide range of fact boxes are generated and used by health care and insurance providers to help patients make informed decisions. The fact box illustrated in Figure 13.3 provides information on vaccination to help senior citizens decide whether and how often they want to vaccinate themselves against flu. This fact box is a result of a collaboration between the Harding Center and the major German insurance provider AOK.

Similar fact boxes have been developed for a wide range of health conditions, all providing easily accessible information that aids citizens in making

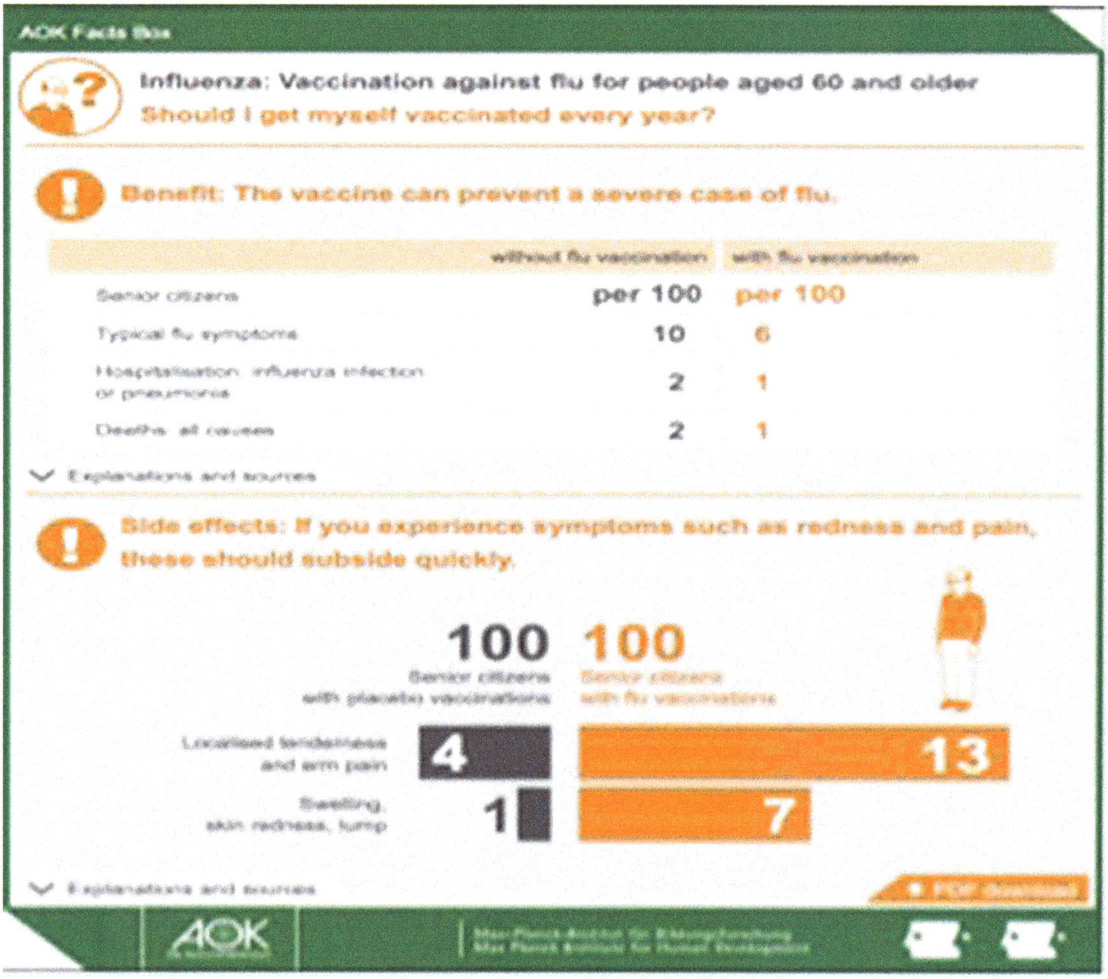

Figure 13.3 Communicating information about flu vaccination to senior citizens

their treatment and other health-related decisions. The conditions covered by AOK include facts about available treatments for ovarian cancer, dietary supplements, and shock waves. Many healthcare providers use similar tools to facilitate transfer of information to patients. (For extensive discussions on health literacy and its many aspects, see an edited volume by Gigerenzer & Gray (2011).

Fact boxes are geared towards patient education and can be used by health institutions or physicians to provide clarity on treatment options. Next, we turn to techniques that are particularly useful for the education of physicians by elucidating the underlying processes of clinical judgment.

13.3.2 Fast-and-frugal trees

A decision tree, in general, depicts a tree-like path of examining cues and associated outcomes with each of its "branches." A fast-and-frugal tree captures the lexicographic nature of decisions that are made under time pressure and in the absence of complete information (Martignon et al., 2003). A wide range of medical conditions lend themselves

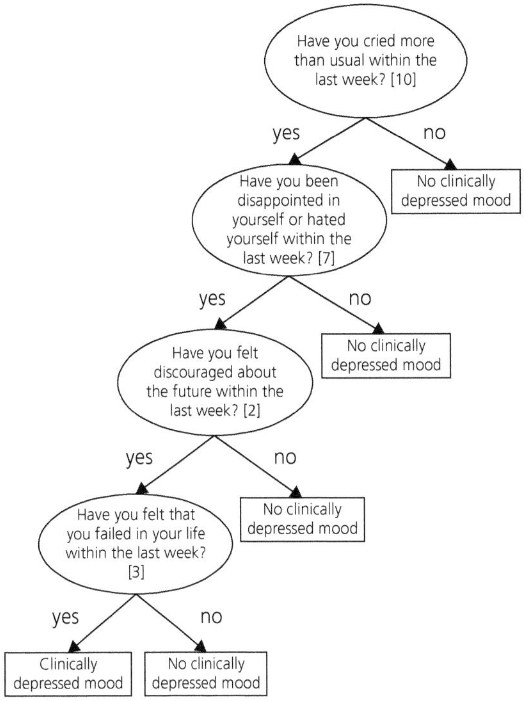

Figure 13.4 Fast-and-frugal tree screening for depressed mood (according to the total BDI score). The numbers in brackets indicate the position of the respective item in the BDI. For this figure, the full wording of the DBI has been translated into binary questions

to this approach. For example, Jenny et al., (2013), used the Beck Depression Inventory (BDI) to create a diagnostic tool for clinical depression, as illustrated in Figure 13.4. Here, the numbers in brackets correspond to the Likert scale of the BDI translated into the binary questions in this fast-and-frugal diagnostic tree. For example, the first binary question, 'Have you cried more than usual within the last week?' presents the binary form of the following scores from BDI [10]: 0 = I do not cry any more than usual; 1 = I cry now more than I used to; 2 = I cry all the time now; 3 = I used to be able to cry, but now I can not cry even though I want to. The remaining wordings for corresponding BDI scores [7], [2], and [3] that are turned into binary questions in the branches of Figure 13.4 can be found in Jenny et al., (2013, Table 1).

It is particularly important to develop easy-to-follow protocols that protect patients against defensive medical decision making. Coronary Care Unit

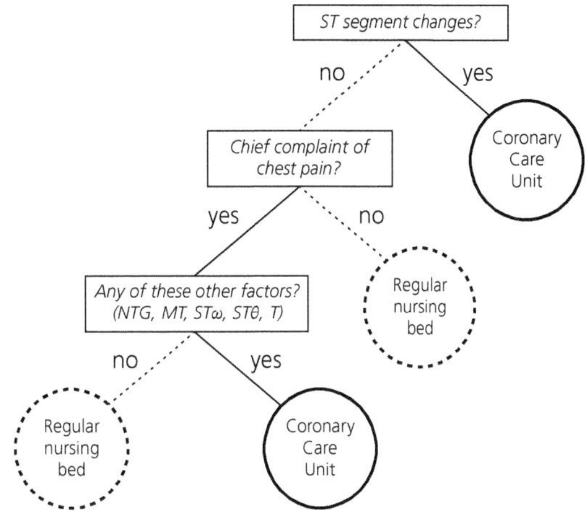

Figure 13.5 A fast-and-frugal decision tree for coronary care unit allocations

(CCU) allocation is a case in point. Fearful of malpractice suits, physicians are inclined to conservatively send to CCU about four times as many patients in the emergency room complaining of chest pain than necessary (Green & Mehr, 1997; Green & Smith, 1988). This could prove a life-threatening situation, given the considerably higher chances of acquiring secondary infections from exposure to other patients in the CCU. Nonetheless, death from infection is much less subject to a lawsuit than death resulting from a heart attack that was not treated in time because of the lack of monitoring in a regular bed compared to the CCU. Figure 13.5 presents the arguably best known example of a medical decision tree. It simplifies the decision whether a patient should be sent to the CCU or a regular bed and was developed by Green & Mehr (1997).

The heuristics approach to decision-making opens a third way of thinking about information acquisition that falls between pure intuition and sophisticated inclusive statistical models. Gigerenzer & Kurzenhäuser (2005) report that in a study with an ABAB reversal design at a research hospital, physicians were asked to decide first by intuition (condition A) before being given the Heart Disease Predictive Instrument (HDPI) with some fifty probabilities (condition B). Expectedly, HDPI reduced their false-positive responses from above

90% to less than 60%. However, their responses showed no further improvement in subsequent A and B conditions. Because it is unlikely that the physicians memorized all fifty HDPI probabilities in the first B condition, the attained and retained performance can be understood as a heuristic process wherein physicians chose a small number of probabilities from the first decision using the HDPI and simply continued to use these in the subsequent A and B conditions. The pieces of information chosen and used by the physicians were then transformed into a fast-and-frugal tree that I presented in Figure 13.5. Building this decision tree involves the following steps: ordered search, a fast stopping rule, and one-reason decision making. Notice that cues are considered one at a time, and a decision to stop or continue search is made at each stage. Because a decision is made only after search and stopping have occurred, models of heuristic decision-making capture the process of decision-making by accounting for all steps. This is in contrast with both statistical modeling that exclusively focuses on the final decision and pure intuition that simply ignores all information. As illustrated in Figure 13.5, a fast-and-frugal tree for deciding whether to send a patient with chest pain complaints to the CCU boils down to asking a small number of yes/no questions, similar to the questions about depression.

One might observe that time pressure applies only to a subset of medical conditions, primarily at the emergency room. However, medical judgments are more often than not made in the absence of complete information; collecting all possible pieces of relevant information would be prohibitive not only in terms of time but also cost-wise. A fast-and-frugal tree can do more than summarize data on observed procedures exercised by physicians to describe the common practice: It can be constructed on the basis of the best and most successful practices and then used as a protocol to promote efficiency and accuracy of diagnosis and treatment. In both forms, fast-and-frugal trees constitute powerful tools for medical research and fruitful products for clinicians.[4]

Fast-and-frugal trees are products of the ecological rationality approach to understanding the processes of judgment and decision making. Similarly, fact boxes are tools inspired by insights from the same line of studies, which always builds on the interrelationship between the environment and the mind. Fact boxes account for representations that are readily accessible to the mind because they appeal to evolutionary (or otherwise developed) capacities in the human mind. In general, numeracy, i.e., the ability to understand numbers, can be considerably improved through ecologically suitable choices of representation of information.

13.3.3 Other techniques

Considerable research has been done in lines closely related to the topics discussed in this chapter. There exists an extensive body of work on the importance of the mode of representing information (Peters, 2012, 2015, 2017); Anderson & Schulkin, (2014) offer a collection of work on numerical reasoning in the medical context. Physicians' interpretation of statistical information (e.g., Obrecht et al., 2012; Gaissmaier et al., 2014), their perceptions of patients' conditions (in the case of breast cancer see, e.g., Anderson & Schulkin, 2011), or their statistical literacy (e.g., Anderson et al., 2014) are among topics investigated in the general area of medical decision making. Finally, for diagnostic practices related to human immunodeficiency virus (HIV) patients, the role of physicians' perceptions, literacy, and inclinations are studied among others by Leddy et al. (2015) and Prinz et al. (2015).

Another line of work in medical decision-making is focused on detecting pitfalls and developing methods to alleviate these (for an extensive discussion of rational medical decision making, see Rao, 2007). For example, if the patient emotionally rejects the potentially threatening knowledge, say about a critical condition, the type of representation and level of accessibility will make no difference in communication of information. Only after the fear of knowing is overcome can information be accessed. In a series of blogs,[5] behavioral scientists

[4] A full tutorial package with numerous examples is available online for constructing fast-and-frugal trees: https://cran.r-project.org/web/packages/FFTrees/vignettes/guide.html

[5] https://www.envolvehealth.com/news/how-to-better-communicate-health-risks-and-improve-health-outcom.html

Juarez and O'Brien at the Center for Advanced Hindsight at Duke University[6] suggest that self-affirmation techniques offered before information in transparent formats increase attention and compliance among patients. Another strategy is story telling. In addition, making the matter completely personal and easing the patient into receiving the information can overcome their deliberate ignorance towards unpleasant information about their health status. It is only when patients are receptive to information that the accessibility of the representation format becomes the primary focus. This is just a small sample of a great deal of work in progress that can be viewed as congruent with the contents of this chapter and can be brought to bear in advancing medical education and clinical practice. Interested readers can visit, for instance, the following websites: envolvepeoplecare.com, advanced-hindsight.com. For educational material on how behavioral science can have an impact on medical practice, consult the edited volume by Sahler & Carr (2017).

https://www.envolvehealth.com/news/how-to-better-communicate-health-risks-and-improve-health-outcom0.html.

[6] http://advanced-hindsight.com/

13.4 Interventions can benefit from evolutionary insights

A comprehensive picture of biology includes an evolutionary-based understanding of fitness, reproduction, and endurance mechanisms in the human body. Physiology and biology are pillars of medicine. Thus, by transitivity, an evolutionary approach to medicine holds a tangible promise of enriching clinical practice by providing additional insights and solutions.

In part, human conditions develop in response to environmental stimuli. This fact is illustrated in Figure 13.6, which details the pathway from interaction between the environment and existing infections to the development of immunoregulation mechanisms and other metabolic functions.

The insights from an evolutionary perspective are not limited to inter-organism mechanisms but can be extended to inform the design of public health interventions (Redelmeier, 2013). In this area, *The Lancet* recently published a series of three

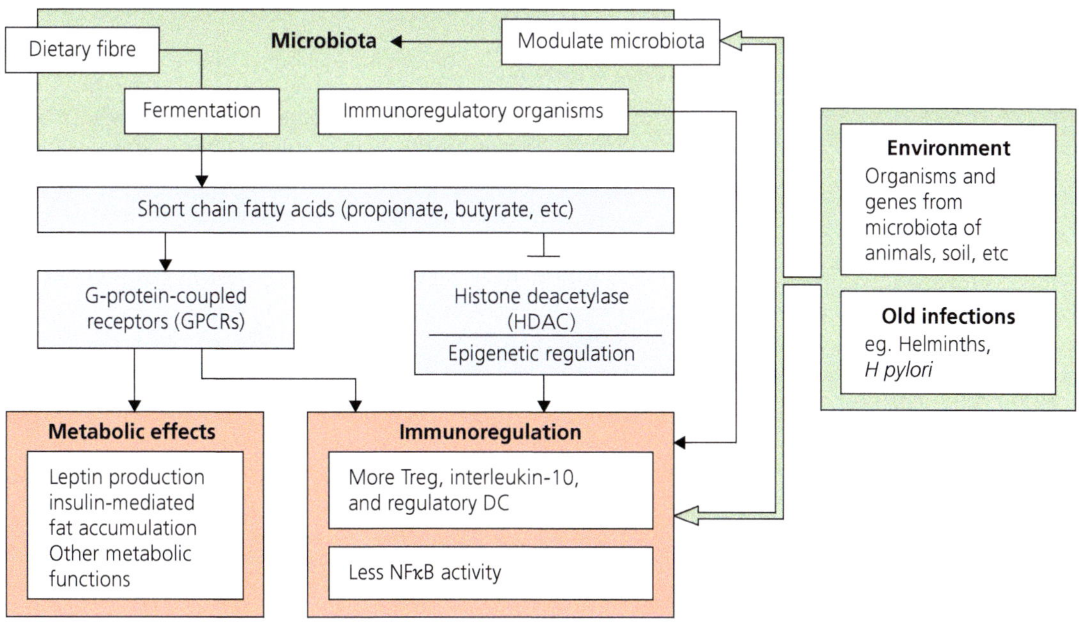

Figure 13.6 Some major pathways involved in immune and metabolic regulation by organisms with which human beings co-evolved

papers on evolutionary public health, which we summarize next.

The first paper, by Wells, Nesse, & Stearns (2017), explains that according to life history theory, the basis of evolutionary analysis, organisms allocate their limited energy between maintenance, growth, reproduction, and defense, where reproductive success is favored by evolution. More often than not, reproductive enhancement comes at the cost of other functions, and as such can reduce longevity or health. Health policy interventions are expensive to implement and extensive in impact. It is advisable therefore to precede an intervention with mathematical modeling that allows for outcomes to first emerge in a virtual controlled environment, assessed and modified for improvement before actual implementation in vivo. The authors strongly advocate that evolutionary theory can revolutionize medicine and significantly help prevent diseases:

> A key benefit of evolutionary approaches is the availability of solid overarching theory. Most natural sciences have a strong theoretical basis—e.g., quantum theory in physics and molecular theory in chemistry. Evolution is also a "basic science," and it is no exaggeration to suggest that its application in medicine could revolutionize the discipline. . . . the global burden of ill health is shifting towards non-communicable diseases (NCDs), for which individuals' constitutions and behavior are key to susceptibility and prevention . . . . By shedding more light on how physiology and behavior respond to such compound stresses, evolutionary approaches could improve societal efforts to prevent NCDs, just as they already help reduce the burden of infectious disease.
>
> **(Ibid., p. 501)**

Evolutionary public health policies benefit from the life history theory that produces testable hypothesis and predicts trade-offs related to public health, such as that between longevity and reproduction.

Health care is viewed as cultural evolution, which interacts with biological evolution and generates outcomes that cannot be predicted in isolation. A major concern is that of societal stresses without biological preparation, including processed food and sedentary environments. Whereas conventional public health models humans as deliberative agents, the life history theory simply makes predictions about behavior with no such assumption. Evolutionary principles remain relevant to understanding the associations between behavior and health outcomes even in cases where the environment and biology have progressed incongruently. For example, delay in or altogether refraining from reproduction among higher-income populations in high-income societies can be understood in light of extrinsic mortality risk. Another, more general, implication is that societies with different resource availability and extrinsic mortality risk cannot be expected to produce the same outcome for the same intervention. Thus, adjusting interventions for the level of income (or poverty) becomes a major consideration in the design of global health interventions towards increasing efficacy.

The second paper in the evolutionary public health series (Jasienska et al., 2017) focuses on human reproduction, observing that a trade-off exists between reproduction and health maintenance. The authors argue and provide evidence that evolutionary theory affords public health policy-making with perspectives that could reduce the risk of some non-communicable diseases such as hormone-sensitive cancers (also see Schulkin, 2000, 2008).

The last paper in this series (Rook et al., 2017) deals with human–microbe interaction and plasticity. For example, when antibiotics are used there are at least three penalties that need to be considered, one for losing microbes expected by our regulatory systems, another for microbes that will cost us energy when encountered, and a third one if the pathogens become stronger as a result of medical strategies. They conclude that microbiota have significant relevance for global public health programs:

> Perhaps as the understanding of the human ecosystem and its evolved requirements increases, and targeted hygiene is developed that does not exclude essential organisms, human beings will learn to modulate our microbial exposures in ways that reduce developmental, inflammatory, and infectious disorders with less reliance on anti-inflammatory treatments, antibiotics, and vaccines.
>
> **(Ibid, p. 528)**

Since the early 2000s, behavioral decision science has been increasingly applied to policymaking across the board following the popularization of behavioral insights through bestselling books such as Gladwell's *Blink*, Levitt and Dubner's *Freakonomics*, and Thaler and Sunstein's *Nudge*. Relatedly, Redelmeier (2013) suggests that the failure of the Human Genome Project to develop therapies for healthcare also played an indirect role in the emergence of a behavioraly rooted approach. After all, if behavior can be redirected, many diseases can be prevented or controlled. For example, if smokers quit smoking, their chances of lung cancer will diminish, or if people with diabetes are enticed to alter their sedentary habits, their diabetes will be better controlled.

Clinical health delivery can benefit in many ways from applying behavioral insights. This is precisely the task of behavioral design teams (BDTs) in hospitals and health organizations. In the 2017 Issue Brief of the Commonwealth Fund, a non-profit organization,[7] the main areas in which behavioral insights can improve delivery of health care in every step of a patient's journey are outlined, as illustrated in T. Robertson, M. Darling, J. Leifer et al., Behavioral Design Teams: The Next Frontier in Clinical Delivery Innovation? The Commonwealth Fund, Nov. 2017. The fund recommends integration of BDTs in every organization that provides health care at every stage of the process, and also promotes taking advantage of technology not only to develop user-friendly interfaces for patients but as a platform for promoting innovation in health care.

To better understand how an intervention would work, it is essential to know which underlying mechanisms will be triggered in response to an intervening policy. Evolutionary forces, however, do not necessarily align with typical goals of public health policies such as improvement of longevity and prevention of diseases.

Whereas evolution sets the goal as reproduction, public health is an individual pursuit of endurance. Insights from the trade-offs involved in evolution provide a valuable perspective and can be used to shape behavior for desired goals. Interventions must recognize and treat evolutionary inclinations to redirect behavior towards modern ends such as better health or "maintenance" and longer life or "endurance." In sum, an evolutionary approach to human behavior and medicine can provide many useful insights for health-related policies. These insights can in turn be understood in a framework of contextualized behavior such as that of the study of ecological rationality (see Gigerenzer, 2007, 2014).

13.5 Concluding remarks

We introduced ecological rationality, a concept of rationality in judgment and decision making with practical relevance to clinical practice and medical education. The study of ecological rationality includes the concepts of mind as an adaptive toolbox and heuristics (rules of thumb) as efficient decision tools. We elaborated on the compatibility of these concepts with the elements of an evolutionary approach to medicine. This compatibility is rooted in the common approach of both disciplines to evaluation and validation of judgment with respect to the environment.

We suggested that the current medical curricula can be improved by adding modules on risk communication through accessible representations of information and attention to behavioral aspects of judgment and decision making. The study of ecological rationality has produced effective risk communication tools for doctors and patients that can be used to build such modules. Accounting for natural and evolutionary inclinations can enhance presentation and communication of medical information. We introduced two techniques for developing such communication tools that have benefited from the study of processes of heuristic decision making: fact boxes and fast-and-frugal trees. Other related techniques include checklists and strategies to alleviate behavioral pitfalls that hinder the consideration of information, independent of the format in which it is presented.

More often than not, physicians have to treat their patients by making good-enough decisions. Good-enough decisions neither exhaust all relevant information nor aim for optimized solutions. The ecological notion of rationality provides a framework for the study of processes involved in making

[7] http://www.commonwealthfund.org/publications/issue-briefs/2017/nov/behavioral-design-teams-delivery-innovation. Also see http://www.ideas42.org/wpcontent/uploads/2017/11/Robertson_behavioral_design_teams_ib.pdf

such decisions, and specifies the criteria for success as well. Many of these processes involve the use of rules of thumb or heuristics that are closely connected to humans' evolutionary capabilities. The understanding of how evolutionary capabilities emerged and their effect on the human organism, on the other hand, is the bread and butter of evolutionary medicine (Trevathan et al., 1999; Nesse, 2001; Nesse et al., 2010; Perlman, 2013). Thus, through this common evolutionary approach, the two frameworks of ecological rationality and evolutionary medicine can come together. The compatibility between the two suggests that findings and products from one can be fruitfully used in the other. In this chapter, we suggested adopting ecological rationality as a theoretical framework for equipping evolutionary medicine with insights and tools that facilitate the practice of effective clinical medicine thanks to their common root in the evolutionary approach.

References

Anderson, B. L., Gigerenzer, G., Parker, S., & Schulkin, J. (2014). Statistical literacy in obstetricians and gynecologists. *Journal of Healthcare Quality*, 36(1): 5–17.

Anderson, B. L. & Schulkin, J. (2011). Physicians' perceptions of patients' knowledge and opinions regarding breast cancer: Associations with patient education and physician numeracy. *Breast Care*, 6(4): 285–8.

Anderson, B. L. & Schulkin, J. (2014). *Numerical Reasoning in Judgment and Decision Making About Health.* Cambridge: Cambridge University Press.

Beck, A. T., Ward, C. H., Mendelson, M., Mock, J., & Erbaugh, J. (1961). An inventory for measuring depression. *Archives of General Psychiatry*, 4: 561–71. doi: 10.1001/archpsyc.1961.01710120031004

Dallacker, M., Hertwig, R., Peters, E., & Mata, J. (2016). Lowering parental numeracy is associated with children being under- and overweight. *Social Science and Medicine*, 161: 126–33.

Darwin, C. (1859). *On the Origin of Species by Means of Natural Selection, Or, The Preservation of Favoured Races in the Struggle for Life.* London: John Murray.

Davies, O. (2016). Niche construction, social cognition, and language: Hypothesizing the human as the production of place. *Culture and Brain*, 4: 87–112. doi: 10.1007/s40167-016-0039-2

Eddy D. M. (1990). Comparing benefits and harms: The balance sheet. *JAMA*, 263(18): 2493.

Fitch, W. T. (2010). *The Evolution of Language*. Cambridge, UK: Cambridge University Press.

Gaissmaier, W., Anderson, B. L., & Schulkin, J. (2014). How do physicians provide statistical information about antidepressants to hypothetical patients? *Medical Decision Making*, 34(2): 206–15.

Gigerenzer, G. (2005). I think, therefore I err. *Social Research*, *72*: 195–218.

Gigerenzer, G. (2007). *Gut Feelings: The Intelligence of the Unconscious*. New York: Viking Press.

Gigerenzer, G. (2014). *Risk Savvy: How to Make Good Decisions*. New York: Viking Press.

Gigerenzer, G. & Gaissmaier, W. (2011). Heuristic decision making. *Ann Rev Psych*, 62: 451–82.

Gigerenzer, G. & Gray, J. A. M. (2011). (eds.) *Better Doctors, Better Patients, Better Decisions: Envisioning Health Care 2020*. Strüngemann Forum Report, Vol. 6. Cambridge, MA: MIT Press.

Gigerenzer, G. & Kurzenhäuser, S. (2005). Fast and frugal heuristics in medical decision making. In R. Bibace, J. D. Laird, K. L. Noller, & J. Valsiner (eds.) *Science and Medicine in Dialogue: Thinking Through Particulars and Universals*, pp. 3–15. Westport, CT: Praeger.

Gigerenzer, G. & Selten, R. (eds.) (2001). *Bounded Rationality: THE Adaptive Toolbox*. Cambridge, MA: MIT Press.

Gigerenzer, G., Todd, P. M., & the ABC Research Group. (1999). *Simple heuristics that make us smart*. New York: Oxford University Press.

Gladwell, M. (2005). *Blink: The power of Thinking Without Thinking*. New York: Little, Brown and Co.

Gluckman, P. D., Beedle A., & Hanson, M. (2009). *Principles of Evolutionary Medicine*. Oxford: Oxford University Press.

Gluckman, P. D., Hanson, M. A., & Spencer, H. G. (2005). Predictive adaptive responses and human evolution. *TRENDS in Ecology and Evolution*, 20(10): 527–33.

Gluckman, P. D., Low, F. M., Buklijas, T., Hanson, M. A., & Beedle, A. (2010). "How evolutionary principles improve the understanding of human health and disease." *Evolutionary Applications*, 4(2): 249–63.

Green, L. & Mehr, D. R. (1997). What alters physicians' decisions to admit to the coronary care unit? *The Journal of Family Practice*, 45(3): 219–26.

Green, L. & Smith, M. (1988). Evaluation of two acute cardiac ischemia decision-support tools in a rural family practice. *The Journal of Family Practice*, 26(6): 627–32.

Hamlin R. P. (2017). "The gaze heuristic:" Biography of an adaptively rational decision process. *Topics in Cognitive Science*, 9(2): 264–88. doi: 10.1111/tops.12253.

Hautz, S. C., Hautz, W. E., Feufel, M. A., & Spies, C. D. (2015). Comparability of outcome frameworks in medical education: Implications for framework development. *Medical Teacher*, 37(11): 1051–9. doi:10.3109/0142159X.2015.1012490

Hertwig, R., Buchan, H., Davis, D. A., Gaissmaier, W, Härter, M., Kolpatzik, K., Légaré, F., Schmacke, N., & Wormer, H. (2011). How will health care professionals and patients work together in 2020? A manifesto for change. In G. Gigerenzer & J. A. Muir Gray (eds.) *Better Doctors, Better Patients, Better Decisions: Envisioning Health Care 2020,* (pp. 317–37). Cambridge, MA: The MIT Press.

Hood, E. & Jenkins, K. P. (2008). Evolutionary medicine: A powerful tool for improving human health. *Evolution: Education and Outreach*, 1(2): 114–20.

Jasienska, G., Bribiescas, R. G., Furberg, A., Helle, S., & Núñez-de la Mora, A. (2017). Human reproduction and health: An evolutionary perspective. *The Lancet*, 390(10093): 510–20.

Jenny, M. A., Pachur, T., Williams, S. L., Becker, E., & Margraf, J. (2013). Simple rules for detecting depression. *Journal of Applied Research in Memory and Cognition*, 2(3): 149–57.

Juarez, L. & O'Brien, J. (2018). How to better communicate health risks and improve health outcomes: story-telling for more impact. https://advanced-hindsight.com/blog/better-communicate-health-risks-improve-health-outcomes-story-telling-impact/

Juarez, L. & O'Brien, J. (2018). How to better communicate health risks and improve health outcomes: Self-affirmation and feeling good. https://advanced-hindsight.com/blog/how-to-better-communicate-health-risks-and-improve-health-outcomes-self-affirmation-and-feeling-good/

Keinan, A. & Clark, A. G. (2012). Recent explosive human population growth has resulted in an excess of rare genetic variants. *Science*, 336(6082): 740–3.

Leddy, M. A., Anderson, B., & Schulkin, J. (2015). Obstetrician-gynecologists' practices, statistical literacy, and risk communication with regard to sexually transmitted infections. *Current HIV Research*, 13(5): 391–8. doi: 10.2174/1570162X13666150511141942.

Levitt, S. D. & Dubner, S. J. (2005). *Freakonomics: A Rogue Economist Explores the Hidden Side of Everything*. New York: William Morrow.

Marewski, J. N. & Gigerenzer, G. (2012). Heuristic decision making in medicine. *Dialogues Clin Neurosci*, 14(1): 77–89.

Martignon, L., Vitouch, O., Takezawa, M., & Forster, M. R. (2003). Naive and yet enlightened: From natural frequencies to fast and frugal decision trees. In D. Hardman and L. Macchi (eds.), *Thinking: Psychological Perspective on Reasoning, Judgment, and Decision Making*, (pp. 189–211). Chichester: Wiley.

McDowell, M., Rebitschek, F. G., Gigerenzer, G., & Wegwarth, O. (2016). A simple tool for communicating the benefits and harms of health interventions: A guide for creating a fact box. *MDM Policy & Practice*, 1(1): 2,381,468,316,665,365.

Mousavi, S. & Gigerenzer, G. (2011), Revisiting the "error" in studies of cognitive errors, in D. A. Hofmann & M. Frese (eds.), *Error in Organizations* (pp. 97–112). New York, USA: Taylor & Francis.

Mousavi, S. & G. Gigerenzer (2017). Heuristics are tools for uncertainty. *Journal of Homo Oeconomicus* 34: 361–79.

Mousavi, S., Gigerenzer, G., & Kheirandish, R. (2017), Rethinking behavioral economics through fast-and-frugal heuristics. In R. Frantz, S. Chen, K. Dopfer, & S. Mousavi (eds.) *Routledge Handbook of Behavioral Economics*, (pp. 80–96). London: Routledge.

Nesse, R. M. (2001). How is Darwinian medicine useful? *Western Journal of Medicine*, 174(5): 358–60.

Neese, R. M. and Stearns, S. C. (2008), The great opportunity: Evolutionary applications to medicine and public health. *Evolutionary Applications*, 1: 28–48.

Nesse, R. M. & Williams, G. C. (1995). *Why we Get Sick: The New Science of Darwinian Medicine*. New York Times Books: A division of Random House, Inc.

Neese, R. M., Bergstrom, C. T., Ellison, P. T., Flier, J. S., Gluckman, P., Govindaraju, D. R., Niethammer, D., Omenn, G. S., Perlman, R. L., Schwartz, M. D., Thomas, M. G., Stearns, S. C., & Valle, D. (2010). Making evolutionary biology a basic science for medicine. *Proceedings of the National Academy of Sciences*, 107: 1800–7. www.pnas.org□/cgi/doi/10.1073/pnas.0906224106

Obrecht, N. A., Anderson, B. Schulkin, J. & Chapman, G. B. (2012). Retrospective frequency formats promote consistent experience-based Bayesian judgments. *Applied Cognitive Psychology*, 26(3): 436-40.

Perlman, R. L. (2013, 2014) *Evolution and Medicine*. Oxford: Oxford University Press.

Peters, E. (2008). Numeracy and the perception and communication of risk. In W. T. Tucker, S. Ferson, A. M. Finkel, and D. Slavin (eds.). *Strategies for Risk Communication: Evolution, Evidence, Experience. Annals of the New York Academy of Science*, 1128: 1–12.

Peters, E. (2012). Beyond comprehension: The role of numeracy in judgment and decisions. *Current Directions in Psychological Science*, 21(1): 31–5. doi: 10.1177/0963721411429960.

Peters, E. (2017). Educating good decisions. *Behavioural Public Policy*, 1(2): 162–76.

Peters, E. & Bjalkebring, P. (2015). Multiple numeric competencies: When a number is not just a number. *Journal of Personality and Social Psychology*, 108(5): 802–22.

Peters, E., Hibbard, J., Slovic, P., & Dieckmann, N. (2007). Numeracy skills and the communication, comprehension, and use of risk-benefit information. *Health Affairs*, 26 (3): 741–7. doi 10.1377/hlthaff.26.3.741

Peters, E., Shoots-Reinhard, B., Tompkins, M.K., Schley, D., Meilleur, L., Sinayev, A., Tusler, M., Wagner, L., & Crocker, J. (2017). Improving numeracy through values affirmation enhances decision and STEM outcomes. *PLoS ONE*, 12(7): e0180674.

Prinz, R., Feufel, M. A., Gigerenzer, G., & Wegwarth, O. (2015). What counselors tell low-risk clients about HIV test performance. *Current HIV Research 13*: 369–80. doi:10.2174/1570162X13666150511125200.

Rao, G. & Kanter, S. L. (2010). Physician numeracy as the basis for an evidence-based medicine curriculum. *Academic Medicine* 85(11): 1794–9. doi: 10.1097/ACM.0b013e3181e7218c

Rook, G., Bäckherd, F., Levin, B. R., McFall-Ngai, M., & McLean, A. R. (2017). Evolution, human-microbe interactions, and life history plasticity. *The Lancet*, 390(10093): 521–30.

Rao, G. (2007). *Rational Medical Decision Making: A Case-Based Approach*. New York, NY: McGraw-Hill.

Rao, G. & Kanter, S. L. (2010). Physician numeracy: Essential skills for practicing evidence-based medicine. *Family Medicine*, 40(5): 354–8.

Redelmeier, D. A. (2013). Behavioral decision science applied to health-care policy. In E. Shafir (ed.) *The Behavioral Foundations of Public Policy*. Princeton: Princeton University Press.

Sahler, O. J. Z. & Carr, J. E. (2017). *The behavioral sciences and health care*. Cambridge, MA: Hogrefe & Huber Publisher.

Savage, L. J. (1954). *The foundations of statistics*. New York, USA: Wiley.

Schulkin, J. (2000). Decision sciences and evidence-based medicine – two intellectual movements to support clinical decision making. *Academic Medicine*, 75(8): 816–18.

Schulkin, J. (2008). *Medical Decisions, Estrogen, and Aging*. New York: Springer.

Selten, R. (1999). What is bounded rationality? *Dahlem Conference 1999* (SFB Discussion Paper B-454). https://pdfs.semanticscholar.org/8237/d1a44256daddab131078bd09d622216c90d8.pdf

Simon, H. A. (1956). Rational choice and the structure of the environment. *Psychological Review*, 63 (2): 129–38.

Simon, H. (1972). Theories of bounded rationality. In C. B. McGuire, & R. Radner (eds.), *Decision and Organization*, (pp. 161–76). Amsterdam: North-Holland.

Smith, V. L. (2008). *Rationality in Economics: Constructivist and Ecological Forms*. Cambridge: Cambridge University Press.

Spencer, H. (1864). *Principles of biology*. Edinburgh: William and Norgate.

Thaler, R. H. & Sunstein, C. R. (2009). *Nudge: improving decisions about health, wealth, and happiness*. Rev. and expanded ed. New York: Penguin Books.

Trevathan, W. R., Smith, E. O., & McKenna, J. J. (eds.) (1999). *Evolutionary Medicine*. Oxford: Oxford University Press.

von Neumann, J. & O. Morgenstern (1944). *Theory of Games and Economic Behavior*, Princeton, NJ, USA: Princeton University Press.

Wegwarth, O. & Gigerenzer, G. (2013). Trust-your-doctor: A simple heuristic in need of a proper social environment. In R. Hertwig, U. Hoffrage, & the ABC Research Group, *Simple Heuristics in a Social World*, (pp. 67–102). New York, USA: Oxford University Press.

Wegwarth, O., Schwartz, L., Woloshin, S., Gaissmaier, W., & Gigerenzer, G. (2012). Do physicians understand cancer screening statistics? A national survey of primary care physicians in the United States. *Annals of Internal Medicine*, 156: 340–9.

Wegwarth, O. & Gigerenzer, G. (2018). The barrier to informed choice in cancer screening: Statistical illiteracy in physicians and patients. In U. Goerling & A. Mehnert (eds.), *Psycho-Oncology (Recent Results in Cancer Research No. 210)* 2nd ed., (pp. 207–21). Cham: Springer.

Wells, J. C. K., Nesse, R. M., Sear, R., Johnstone, R. A., & Stearns, S. C. (2017). Evolutionary public health: introducing the concept. *The Lancet*, 390(10093): 500–9.

Index

Tables, figures, and boxes are indicated by an italic *t*, *f*, and *b* following the page number.

C

D

E